Sustainable Approaches and Strategies for E-Waste Management and Utilization

A. M. Rawani
National Institute of Technology, Raipur, India

Mithilesh Kumar Sahu
O.P. Jindal University, India

Siddharth S. Chakarabarti
O.P. Jindal University, India

Ajit Kumar Singh
GLA University, India

A volume in the Practice, Progress, and Proficiency in Sustainability (PPPS) Book Series

IGI Global
PUBLISHER of TIMELY KNOWLEDGE

Published in the United States of America by
IGI Global
Engineering Science Reference (an imprint of IGI Global)
701 E. Chocolate Avenue
Hershey PA, USA 17033
Tel: 717-533-8845
Fax: 717-533-8661
E-mail: cust@igi-global.com
Web site: http://www.igi-global.com

Library of Congress Cataloging-in-Publication Data

Names: Rawani, A. M., editor. | Sahu, Mithilesh Kumar, editor. | Chakrabarti, Siddharth S., editor. | Singh, Ajit Kumar, editor.
Title: Sustainable approaches and strategies for E-waste management and utilization / edited by A. M. Rawani, Mithilesh Kumar Sahu, Siddharth Sudhansu Chakrabarti, Ajit Kumar Singh.
Description: Hershey, PA : Engineering Science Reference, [2023] | Includes bibliographical references and index. | Summary: "The objectives of this book are listed below: 1. To assess the activities involved in e-waste generation, its type, nature, and estimated volumes. 2. To identify the potential impacts of E-waste on society, economy, and environment. 3. To review and recommend appropriate E-waste handling and disposal measures following the rules of regulatory bodies. 4. To formulate the strategies for effective utilization of E-waste for sustainable development"-- Provided by publisher.
Identifiers: LCCN 2022050829 (print) | LCCN 2022050830 (ebook) | ISBN 9781668475737 (hardcover) | ISBN 9781668475744 (paperback) | ISBN 9781668475751 (ebook)
Subjects: LCSH: Electronic waste.
Classification: LCC TD799.85 .S87 2023 (print) | LCC TD799.85 (ebook) | DDC 621.381028/6--dc23/eng/20221220
LC record available at https://lccn.loc.gov/2022050829
LC ebook record available at https://lccn.loc.gov/2022050830

This book is published in the IGI Global book series Practice, Progress, and Proficiency in Sustainability (PPPS) (ISSN: 2330-3271; eISSN: 2330-328X)

British Cataloguing in Publication Data
A Cataloguing in Publication record for this book is available from the British Library.

For electronic access to this publication, please contact: eresources@igi-global.com.

Practice, Progress, and Proficiency in Sustainability (PPPS) Book Series

ISSN:2330-3271
EISSN:2330-328X

Editor-in-Chief: Ayman Batisha International Sustainability Institute, Egypt

MISSION

In a world where traditional business practices are reconsidered and economic activity is performed in a global context, new areas of economic developments are recognized as the key enablers of wealth and income production. This knowledge of information technologies provides infrastructures, systems, and services towards sustainable development.

The **Practices, Progress, and Proficiency in Sustainability (PPPS) Book Series** focuses on the local and global challenges, business opportunities, and societal needs surrounding international collaboration and sustainable development of technology. This series brings together academics, researchers, entrepreneurs, policy makers and government officers aiming to contribute to the progress and proficiency in sustainability.

COVERAGE

- ICT and knowledge for development
- Eco-Innovation
- Global Content and Knowledge Repositories
- Strategic Management of IT
- Socio-Economic
- Knowledge clusters
- Green Technology
- E-Development
- Sustainable Development
- Intellectual Capital

IGI Global is currently accepting manuscripts for publication within this series. To submit a proposal for a volume in this series, please contact our Acquisition Editors at Acquisitions@igi-global.com or visit: http://www.igi-global.com/publish/.

The Practice, Progress, and Proficiency in Sustainability (PPPS) Book Series (ISSN 2330-3271) is published by IGI Global, 701 E. Chocolate Avenue, Hershey, PA 17033-1240, USA, www.igi-global.com. This series is composed of titles available for purchase individually; each title is edited to be contextually exclusive from any other title within the series. For pricing and ordering information please visit http://www.igi-global.com/book-series/practice-progress-proficiency-sustainability/73810. Postmaster: Send all address changes to above address.

Titles in this Series

For a list of additional titles in this series, please visit:
http://www.igi-global.com/book-series/practice-progress-proficiency-sustainability/73810

Handbook of Research on Solving Societal Challenges Through Sustainability-Oriented Innvation
Luísa Cagica Carvalho (Instituto Politécnico de Setúbal, Portugal & CEFAGE, University of Évora, Portugal) Paulo Bogas (Instituto Politécnico de Setúbal, Portugal) Jordana Kneipp (Federal University of Santa Maria, Brazil) Lucas Avila (Federal University of Santa Maria, Brazil) and Elis Ossmane (Universidade Aberta, Portugal)
Engineering Science Reference • © 2023 • 400pp • H/C (ISBN: 9781668461235) • US $295.00

Handbook of Research on Socio-Economic Sustainability in the Post-Pandemic Era
Jozef Oleński (High School Technology and Economics, Jaroslaw, Poland) Jeffrey Sachs (Columbia University, USA) Masayuki Susai (Nagasaki University, Japan) Yannis Tsekouras (University of Macedonia, Greece) and Arjan Gjonça (London School of Economics and Political Science, UK)
Information Science Reference • © 2023 • 400pp • H/C (ISBN: 9781799897606) • US $270.00

Handbook of Research on Promoting Sustainable Public Transportation Strategies in Urban Environments
Zafer Yilmaz (TED University, Turkey) Silvia Golem (University of Split, Croatia) and Dorinela Costescu (Polytechnic University of Bucharest, Romania)
Engineering Science Reference • © 2023 • 401pp • H/C (ISBN: 9781668459966) • US $295.00

Energy Transition in the African Economy Post 2050
Olayinka Ohunakin (Covenant University, Nigeria)
Engineering Science Reference • © 2023 • 300pp • H/C (ISBN: 9781799886389) • US $215.00

Implications of Industry 5.0 on Environmental Sustainability
Muhammad Jawad Sajid (Xuzhou University of Technology, China) Syed Abdul Rehman Khan (Xuzhou University of Technology, China) and Zhang Yu (ILMA University, Pakistan)
Business Science Reference • © 2023 • 328pp • H/C (ISBN: 9781668461136) • US $250.00

IGI Global
PUBLISHER of TIMELY KNOWLEDGE

701 East Chocolate Avenue, Hershey, PA 17033, USA
Tel: 717-533-8845 x100 • Fax: 717-533-8661
E-Mail: cust@igi-global.com • www.igi-global.com

Table of Contents

Detailed Table of Contents

Dheeraj Kumar, National Institute of Technology, Durgapur, India
Amit Kumar, Sri Eshwar College of Engineering, Coimbatore, India
Nagendra Kumar, National Institute of Technology, Durgapur, India
Abhimannyu Sharma, National Institute of Technology, Durgapur, India
Ramit Choudhury, National Institute of Technology, Durgapur, India
Nadeem Faisal, Central Institute of Petrochemicals Engineering and Technology, Balasore, India
Ravi Kant Singh, National Institute of Technology, Durgapur, India
Bhagyashri Krishnat Mane, Dattajirao Kadam Technical Education Society's Textile and Engineering Institute, India

The safe handling of electronic and electrical waste (E-waste/WEE) is a growing concern in many nations' world. Particularly, developing nations have a variety of challenges regarding the creation, transboundary movement, and control of greenhouse gases. Examining the types of technology used to handle e-waste, evaluating the difficulties associated with implementing sustainable methods, and identifying the gaps across formal and informal e-waste employees were among the goals of the research. The following research inquiries were framed: What standard of state of the art is currently used to handle e-waste? What obstacles are preventing the implementation of sustainable innovations? The components of e-waste, its classification, worldwide and Indian e-waste circumstances, potential recoverable, recyclable, and hazardous materials discovered in e-waste, adverse effect on both the environment and human health best possible techniques, recycling, and regeneration used, as well as the risks to the environment and the workforce that they pose are all covered in this paper.

Management of electronic garbage (e-waste) is presently a major problem, particularly in emerging economies where poor waste disposal results in financial loss as well as detrimental effects on the environment and public health. Improper handling and processing practices of e-waste are related with numerous environmental and healthiness issues. The purpose of this study is to elaborate the different programmes that can bring awareness to the people for household e-waste, environmental related issues, inappropriate disposal, and eagerness to contribute to e-waste management. The study revealed useful e-waste programme and recommendation run by government for educational interaction and increasing public awareness of e-waste. A practical strategy to managing e-waste is also presented in the article. This chapter can assist legislators and government organizations of India and other developing countries to decide awareness programme and citizen responsibility for e-waste management.

This chapter elaborates on the various e-waste production courses from multiple human activities. The impacts of various e-wastes and human diseases are clearly illustrated. The different methods of e-wastes management methods, such as Reduce, Reuse, Recycle, and e-waste recycling practices in India are also explained. The characteristics of e-waste, the quantity of e-waste, the nature of e-waste, the composition of e-waste, non-hazardous wastes, and hazardous wastes have been discussed. As per the deep survey of e-waste management systems, the scope and proposed e-waste management methods have been driven.

In the last few decades, there has been massive growth in the electrical and electronic industries. The demand for electronic products is growing enormously, resulting

(extended producer responsibility) schemes. Although developed nations have implemented efficient legislative frameworks and regulations, emerging nations suffer due to their plans. E-waste management systems differ in developed and developing countries; thus, this study evaluates the differences between the management systems and outlines the areas where the developing nations lack effective e-waste management and the advantages developed countries enjoy. Therefore, the current study results are crucial for comprehending the severe hazard posed by improper management of e-waste and the viability of future research into creating strategies to address these problems of developing nations.

The usage of electronic devices increased rapidly. The over-usage of electronic devices always led to a huge amount of waste in the form of electronic products like monitors, wires, discs, etc. e-waste is hazardous, involving things containing toxic components that are dangerous to human health, such as mercury, lead, cadmium, polybrominated flame retardants, barium, and lithium. The negative health effects of these toxins on humans include brain, heart, liver, kidney, and skeletal system damage. Every organization follows a throwback policy of dumping all these electronic wastes into the environment. This chapter proposes some tips and tricks on how to recycle hazardous e-wastes into beautiful reusable products, and methods for the distribution of these objects by creating an e-waste management system.

Electronic waste (e-waste) represents a fast-growing disposal scenario worldwide. The ever-increase in e-waste volume from electronic industries and households has been creating health and environmental concerns since this waste contains some toxic substances. Additionally, the shipment of toxic wastes has generated a huge public outcry. E-waste is the discarded electrical or electronic devices or equipment. Used electronics which are destined for refurbishment, reuse, resale, salvage recycling through material recovery, or disposal are also considered e-waste. Electronic scrap components, such as CPUs, contain potentially harmful materials such as lead, cadmium, beryllium, or brominates flame retardants. Recycling and disposal of e-waste involve significant risk to health and environment. E-waste is electronic products nearing the end of their useful life. Computers, televisions, VCRs, stereos,

Decision Making Trial and Evaluation Laboratory (DEMATEL) technique has been used to integrate the most influencing factors of sustainable supply chain flexibility and circular economy.

In this chapter, the authors discuss the utilization of e-waste in the concrete for civil construction activities. Various tests have been used to investigate the effects of e-waste mixed with concrete. The various percentages of e-waste have been mixed with concrete to improve the strength of buildings. An e-waste concrete beam has a maximum tensile strength of 6.23 MPa under sulfuric curing conditions, and the highest flexural strength at 10% e-waste replacement during the hydrochloride curing process. The compressive strength is at its highest value when e-waste replaces 10% of it. After 28 days of curing, the concrete cylinder's maximum split tensile strength was 15%. Thus, the e-waste could be effectively utilized for civil construction purposes to reduce its environmental impacts.

E-waste is a general term for damaged, surplus, or excess electrical and electronic equipment. Due to quick technological advancement and inexpensive beginning costs, there is a rapidly expanding global surplus of e-waste. If not managed appropriately, the many sorts of compounds and chemicals included in e-waste can seriously harm both human health and the environment. Large volumes of e-waste may be recycled in the concrete sector, solving both the disposal and reuse issues. The most practical application of e-waste is thus thought to be recycling and reuse in the geopolymer concrete (GPC) industry. The experimental study is carried out to

Preface

The world has become increasingly reliant on technology in recent decades, leading to a proliferation of electronic devices. These devices have become ubiquitous in our daily lives, making communication, work, and leisure activities more efficient and enjoyable. However, this dependence on technology has also created a significant problem in term of its disposal commonly known as the electronic waste or e-waste. E-waste is a rapidly growing issue worldwide due to the hazardous materials it contains, such as lead, cadmium, and mercury. When these materials are improperly disposed of, they can contaminate the environment and harm human health. Moreover, the unutilised e waste leads to significant loss of valuable resources as E-waste also contains precious metals and rare earth elements, that could be recycled and reused. The issue of e-waste management is particularly pressing due to the large and growing population of the world, as well as the increasing use of electronic devices among the livelihoods. As technology advances and consumers replace their devices at an ever-increasing pace, the amount of e-waste generated continues to rise. In this regard, this book, "Sustainable Approaches and Strategies for e-waste Management and Utilization," aims to provide a comprehensive overview of the current trends, regulations, challenges, and management strategies of e-waste, along with innovative solutions for e-waste management and utilization.

The book is divided into 13 chapters that cover a range of topics related to e-waste management. The first chapter of the book covers an introduction to the current trends and regulations governing e-waste management in India. This chapter provides an overview of the e-waste management landscape in India, including the key stakeholders involved, the regulations and policies governing e-waste management, and the challenges faced by the industry. It explains that the increasing amount of electronic and electrical waste (E-waste/WEE) is a global concern, especially in developing nations, where the creation, transboundary movement, and greenhouse gas emissions are difficult to control. The main objectives of this research are to examine the current state-of-the-art technologies used to handle e-waste, evaluate the challenges faced in implementing sustainable methods, and identify gaps in the formal and informal e-waste management sector. The study aims to address two research questions: What are the most effective and sustainable methods for handling e-waste in developing nations? What are the obstacles hindering the implementation of sustainable innovations? The paper will cover various aspects of e-waste, including its components, classification, global and Indian e-waste scenarios, recoverable, recyclable, and hazardous materials present in e-waste, their adverse effects on the environment and human health, best practices in recycling and regeneration, and the associated risks to the workforce and the environment.

The second chapter delves into the awareness and handling practices of electronic waste in Indian households and society. These chapters discuss the various sources of e-waste, the awareness levels of Indian households regarding e-waste, and the different practices used to handle e-waste. It explains that the improper management of electronic waste (e-waste) is a significant issue, especially in emerging economies where inadequate waste disposal practices result in financial losses and adverse impacts on the environment and public health. The aim of this study is to explore various programs that can educate people about household e-waste, environmental concerns related to inappropriate disposal, and encourage them to participate in e-waste management. The research highlights effective e-waste programs and recommendations implemented by the government to promote public awareness and education about e-waste. Furthermore, the paper offers a practical strategy for managing e-waste. The findings of this study can assist lawmakers and government organizations in India and other developing countries in designing awareness programs and encouraging citizen responsibility towards e-waste management.

The third chapter of the book covers the source of e-waste and its disposal in detail. This chapter provides an in-depth analysis of the production of e-waste resulting from various human activities and their associated impacts on human health. It further delves into the different approaches to managing e-waste, including the 3R principle (Reduce, Reuse, and Recycle) and the specific e-waste recycling practices implemented in India. The chapter also covers important aspects related to e-waste, such as its characteristics, quantity, nature, and composition, including both non-hazardous and hazardous wastes. Drawing on extensive research into existing e-waste management systems, the chapter concludes with a proposal for effective e-waste management methods and their potential scope for implementation. To reduce the harmful effects of e-waste, it is crucial to adopt sustainable waste management practices, such as the 3R principle (Reduce, Reuse, and Recycle), the implementation of strict regulations for e-waste management, and the promotion of eco-friendly products. Increasing awareness and educating people about the proper disposal of e-waste can also help to mitigate its adverse impacts on the environment. By taking these steps, we can effectively manage e-waste and protect our environment for future generations.

The fourth chapter covers the "External Barriers Affecting E-Waste Remanufacturing in Indian Context". This chapter aims to identify the critical barriers to e-waste remanufacturing in India and to establish logical relationships between them using the ISM and MICMAC techniques. The results reveal that the lack of collection centres, customer reluctance to return products, and uncertainty in demand are major driving barriers, while incomplete recycling systems, sorting policies, consumption attitudes, and uncertainty in quality are significant dependent barriers. To overcome these barriers, proper awareness programs should be organised to educate people about the harmful effects of e-waste, which can increase customer eagerness to return used products and boost demand for remanufactured products. This can help to improve the recycling system and sorting policies, ultimately contributing to sustainable e-waste management in India.

The fifth chapter of the book is related with "Impact of e-waste on the Environment." This chapter explains about that the E-waste is a major environmental hazard, with an annual production of up to 50 million tons globally. This amount is expected to increase even further as electronics continue to play a crucial role in our daily lives, from small gadgets to large-scale machines. E-waste contains a variety of hazardous chemicals, including lead, mercury, brominated flame retardants, beryllium, and cadmium, which can have severe impacts on both human health and the environment. Improper disposal of electronic waste can lead to the release of these chemicals into the air, water, and soil, causing contamination and long-term damage to ecosystems. This chapter highlights the significant impacts of e-waste on the environment and presents possible solutions for its management and disposal.

The sixth chapter of the book covers "A Comparative Study on E-Waste Management Systems in Developed and Developing Countries: Legislative Compliances and Initiatives." Which explains that the issue of e-waste remains a persistent problem, particularly in developing nations, where poor recycling practices, transboundary movement, improper management, the lack of environmentally sound management (ESM) programs, and ineffective extended producer responsibility (EPR) schemes have taken a significant toll on the environment and human health. While developed nations have implemented efficient legislative frameworks and regulations for managing e-waste, emerging nations struggle with ineffective plans. This study evaluates the differences between e-waste management systems in developed and developing countries, highlighting the areas where developing nations lack effective e-waste management, and the advantages enjoyed by developed countries. The results of this study are crucial for understanding the severity of the hazards posed by improper e-waste management and the need for future research to develop strategies to address these problems in developing nations. To improve e-waste management, developing countries need to implement more effective ESM programs and EPR schemes, invest in recycling infrastructure and technology, and improve awareness and education among the public about the dangers of improper e-waste disposal. By taking these steps, we can address the growing e-waste problem and protect our environment and communities from the harmful effects of this hazardous waste stream.

The seventh chapter of the book covers the "E-Waste Management System (EMS): An Online Platform to Sell Crafts Made From Tech Waste." This chapter explains that in today's world, the usage of electronic devices has skyrocketed, leading to an increase in the production of electronic waste (e-waste) in the form of discarded monitors, wires, discs, and other products. E-waste contains toxic components such as mercury, lead, cadmium, polybrominated flame retardants, barium, and lithium that pose a significant threat to human health, causing damage to the brain, heart, liver, kidney, and skeletal system. However, organizations often follow a throwback policy of dumping e-waste into the environment, which can have devastating consequences for both humans and the environment. This chapter proposes tips and tricks to recycle hazardous e-waste into beautiful reusable products and methods for creating an e-waste management system to distribute these products. By adopting a sustainable approach to e-waste management, we can reduce the environmental impact of these toxic materials and prevent harm to human health. The chapter suggests using a three-step approach: reducing, reusing, and recycling e-waste. This approach involves reducing the production of e-waste by using electronic devices responsibly, reusing electronics by repairing or refurbishing them, and recycling e-waste by extracting valuable materials and turning them into new products. The chapter also proposes creating an e-waste management system that involves collecting, segregating, and transporting e-waste to authorized recycling centres, which can help prevent e-waste from ending up in landfills or being illegally exported to developing nations. Overall, the chapter emphasizes the importance of responsible e-waste management and encourages individuals and organizations to take a sustainable approach to reduce their environmental impact and protect human health.

The eighth chapter explains about strategies for waste recycle, reuse, and reduce. This chapter explains about the increasing amount of e-waste generated from electronic industries and households is a growing concern due to the presence of toxic substances. This has led to health and environmental risks, as well as public outcry about the shipment of toxic waste. E-waste refers to discarded electrical or electronic devices that are no longer in use. This can include products intended for refurbishment, reuse, resale, salvage recycling, or disposal. Some electronic components such as CPUs contain hazardous materials like lead, cadmium, beryllium, or flame retardants. Proper recycling and disposal of e-waste is critical to reducing the risks to health and the environment. Common electronic products that contribute to e-waste include computers, televisions, VCRs, stereos, copiers, printers, cell phones, and fax machines. Many of these products can be reused, refurbished, or recycled.

The ninth chapter of the book explains "A Study on Printed Circuit Board (PCB) E-Waste Recycling Process." Which explains that the Recycling e-waste is essential for resource preservation and environmental sustainability, as electronic devices contain a significant amount of valuable metals that can be recovered through recycling. The process of recycling e-waste involves the separation of different components, including metals, plastics, and glass, which can then be reused or recycled to make new products. The recovery of precious metals, such as gold, silver, and platinum, from electronic devices is particularly important, as these metals are scarce and valuable resources that can be used in a variety of industrial applications. One of the most effective methods for e-waste recycling is high-temperature pyrolysis, which involves heating the waste materials to a high temperature in an inert atmosphere to recover valuable metals and other components. This process not only recovers valuable metals but also reduces the amount of waste that needs to be further processed or managed. In addition, the use of pyrolysis in e-waste recycling can help to reduce greenhouse gas emissions and other harmful pollutants that are generated during the disposal of electronic waste.

The tenth chapter of the book covers the "Integration of Sustainable Supply Chain Flexibility (SSCF) and Circular Economy (CE): Waste Minimization Technique." The chapter explains that the circular economy (CE) is an essential element of sustainable economic growth and development. However, achieving a fully functional CE requires closed-loop material and energy fluxes, which can pose operational difficulties for companies. To address these difficulties, it is necessary to incorporate Sustainable Supply Chain Flexibility (SSCF) initiatives into supply networks. This operational competence must not compromise performance, increase expenses, or lower product quality. Designers of supply chains understand that SSCF initiatives are critical to achieving CE-targeted performance. Material and energy recycling, a closed-loop supply chain, and minimal environmental impact all contribute to a product's performance aimed at achieving CE status. To integrate the most influential factors of SSCF and CE, the Decision-Making Trial and Evaluation Laboratory (DEMATEL) technique has been used in this study. DEMATEL is a method that helps to identify the causal relationships between different factors and prioritize them based on their impact on a particular issue. In this case, the technique has been used to identify and prioritize the most influential factors in SSCF and CE. The results of the study show that the most critical factors for achieving SSCF and CE are supplier relationships, environmental regulations, product design, and process innovation. To achieve SSCF and CE, companies must focus on developing sustainable supplier relationships that prioritize closed-loop material and energy fluxes. Additionally, they must comply with environmental regulations and prioritize product design and process innovation that minimizes environmental impact. These factors are interdependent and addressing them is necessary to achieve SSCF and CE.

In conclusion, the integration of SSCF and CE is critical to achieving sustainable economic growth and development. The DEMATEL technique is an effective tool for identifying and prioritizing the most influential factors in SSCF and CE, which can help companies develop sustainable supply chains that prioritize closed-loop material and energy fluxes and minimal environmental impact.

The eleventh chapter of the book explains about the "Utilization Process for Electronic Wastes in Eco-Friendly Concrete: Experimental Study." This chapter focus on the utilization of e-waste in concrete for civil construction activities is a potential solution to the growing e-waste problem. This chapter discusses the various tests conducted to investigate the effects of e-waste mixed with concrete and its impact on the strength of buildings. Different percentages of e-waste have been mixed with concrete to improve the strength of the resulting material. The tests show that the maximum tensile strength of an e-waste concrete beam is 6.23 MPa under sulfuric curing conditions, while the highest flexural strength is observed at 10% e-waste replacement during the hydrochloride curing process. Additionally, the compressive strength of the concrete is highest when 10% of the material is replaced by e-waste. After 28 days of curing, the maximum split tensile strength of the concrete cylinder was observed to be 15%. These findings suggest that e-waste can be effectively utilized for civil construction purposes while reducing the environmental impacts of waste disposal. The use of e-waste in concrete is a sustainable solution that helps to conserve natural resources and reduce landfill waste. It also has the potential to reduce the carbon footprint of the construction industry by reducing the energy required to produce traditional building materials. In conclusion, the utilization of e-waste in concrete for civil construction activities is a promising approach to tackling the e-waste problem. The results of various tests conducted in this study show that e-waste can improve the strength of concrete when added in appropriate proportions. This approach can contribute to sustainable development by reducing environmental impacts, conserving natural resources, and reducing the carbon footprint of the construction industry.

The twelfth chapter of the book explains about the "Utilization of E-Waste as Coarse Aggregates in Geopolymer Concrete," which explains that e-waste is a growing global issue, and if not handled properly, it can pose serious threats to both human health and the environment. The concrete industry has the potential to contribute to the solution by recycling and reusing large volumes of e-waste. Geopolymer concrete (GPC) has been identified as a promising application for e-waste recycling and reuse. In this chapter, an experimental study is conducted to improve the mechanical properties of M30 grade GPC by replacing the coarse aggregates with varying percentages of e-waste. Cubes, cylinders, and beams are cast to evaluate the effect of e-waste on the strength of the concrete. Additionally, an empirical regression-based model is formulated to evaluate the strength of GPC at varying percentages of e-waste coarse aggregate (EWCA). The results of the study show that the replacement of coarse aggregates with e-waste improves the mechanical properties of GPC. The compressive strength of the concrete is found to increase with an increase in the percentage of e-waste replacement up to a certain limit. Beyond this limit, the compressive strength decreases. The flexural and split tensile strengths also improve with an increase in e-waste replacement. The regression-based model developed in this study provides a convenient and accurate tool for predicting the strength of GPC at different percentages of e-waste coarse aggregate. This model can be used in the design and development of sustainable concrete mixtures using e-waste as a raw material. In conclusion, the use of e-waste in concrete has the potential to address both the disposal and reuse issues associated with e-waste. The experimental study and regression-based model presented in this chapter demonstrate that e-waste can improve the mechanical properties of GPC when used as a replacement for coarse aggregates. This approach can contribute to sustainable development by reducing waste and conserving natural resources.

The last chapter of the book explains the "E-Waste Management Toward Environment Sustainability in Malaysia," which covers that the Effective management of e-waste is a critical global challenge, and Malaysia is no exception. The country faces significant challenges in managing e-waste, including inadequate infrastructure, insufficient regulations and policies, and a lack of public awareness. To address these challenges, there is a need for comprehensive, structured policy measures to be developed and implemented both nationally and globally. The effects of e-waste on the environment are significant and far-reaching. Toxic substances in e-waste can contaminate soil, air, and water, leading to serious health problems for humans and animals. In Malaysia, the government and other organizations have taken various initiatives to manage e-waste, including establishing e-waste management centres, conducting awareness campaigns, and developing policies and regulations. However, there are still many obstacles to overcome in achieving sustainable e-waste management in Malaysia. These include inadequate funding, a lack of technical expertise, and the need for stronger regulations and policies. Additionally, effective collaboration between various stakeholders, including the government, private sector, and civil society organizations, is essential to achieving sustainable e-waste management. In conclusion, the management of e-waste is a critical issue that requires urgent attention in Malaysia and globally. By implementing comprehensive policies, developing infrastructure and technical expertise, raising public awareness, and fostering collaboration among stakeholders, Malaysia can achieve sustainable e-waste management and mitigate the environmental and health risks associated with e-waste.

This book is intended to serve as a valuable resource for researchers, academicians, professionals, and policymakers in the field of e-waste management and sustainability.

The book highlights the importance of addressing e-waste in a sustainable and responsible manner to ensure a cleaner, healthier, and more sustainable future for all. We hope that the insights and solutions presented in this book will inspire further advance the research and innovative approaches to e-waste management, contributing to a more sustainable future.

A. M. Rawani
National Institute of Technology, Raipur, India

Mithilesh Kumar Sahu
O.P. Jindal University, India

Siddharth S. Chakarabarti
O.P. Jindal University, India

Ajit Kumar Singh
GLA University, India

Acknowledgement

We would like to express our deepest gratitude to all the contributors and reviewers who have made this book project possible. Your invaluable contributions and insightful feedback have been instrumental in shaping the direction and quality of this book. Special thanks to the authors for their dedication and hard work in conducting the research and writing the manuscript and considering this book project as the best suitable source for the publication. Your expertise and knowledge in this field have greatly enhanced the book and made it a valuable contribution to the literature. We would also like to acknowledge the reviewers who provided critical feedback and suggestions for improvement off all the book chapters. Your thoughtful comments and constructive criticisms have helped us to refine our ideas and strengthen the each book chapters and come to this level of quality.

Finally, we would like to express our gratitude to all the individuals and organizations who have supported and provided resources for this book project. Your contributions have been crucial in enabling us to undertake this important book project on e-waste management.

Thank you all for your valuable contributions and support.

Chapter 1
Current Trends, Regulations, Challenges, and Management Strategies of E-Waste in India

Dheeraj Kumar
https://orcid.org/0000-0002-3998-6079
National Institute of Technology, Durgapur, India

Amit Kumar
Sri Eshwar College of Engineering, Coimbatore, India

Nagendra Kumar
National Institute of Technology, Durgapur, India

Abhimannyu Sharma
National Institute of Technology, Durgapur, India

Ramit Choudhury
National Institute of Technology, Durgapur, India

Nadeem Faisal
Central Institute of Petrochemicals Engineering and Technology, Balasore, India

Ravi Kant Singh
National Institute of Technology, Durgapur, India

Bhagyashri Krishnat Mane
Dattajirao Kadam Technical Education Society's Textile and Engineering Institute, India

ABSTRACT

The safe handling of electronic and electrical waste (E-waste/WEE) is a growing concern in many nations' world. Particularly, developing nations have a variety of challenges regarding the creation, transboundary movement, and control of greenhouse gases. Examining the types of technology used to handle e-waste, evaluating the difficulties associated with implementing sustainable methods, and identifying the gaps across formal and informal e-waste employees were among the goals of the research. The following research inquiries were framed: What standard of state of the art is currently used to handle e-waste? What obstacles

DOI: 10.4018/978-1-6684-7573-7.ch001

are preventing the implementation of sustainable innovations? The components of e-waste, its classification, worldwide and Indian e-waste circumstances, potential recoverable, recyclable, and hazardous materials discovered in e-waste, adverse effect on both the environment and human health best possible techniques, recycling, and regeneration used, as well as the risks to the environment and the workforce that they pose are all covered in this paper.

INTRODUCTION

E-waste is now the waste stream with the highest growth rate in the world. Rapid socioeconomic growth and technical progress are the primary forces behind this trend. The harmful chemical constituents of e-waste have the potential to cause harm. Effects on ecosystems and human health if not adequately managed. This is an immediate obstacle to accomplishing sustainable development objectives. Although technologically advanced nations are the primary producers of e-products and generators of e-waste, due to travel and transfer from e-waste source nations, the amount of garbage created in emerging and transitioning nations has also increased. Consequently, developing nations are in a precarious position owing to a lack of inventory data and waste management. Environmentally sound management practices and modern technologies (Alam et al., 2022)

This research intends to illustrate that the current global e-waste scenarios and associated health risks may delay the fulfilment of sustainable development goals. This research demonstrates situations from several vantage points and highlights issues about e-waste. Identifies information gaps and offers a foundation for knowledge and awareness building and technical advancement to promote long-term sustainable development on a global scale (Arya et al., 2020).

What is E-Waste?

E-waste presents a serious threat to the environment, wildlife, and people. Plastics, metals, cathode ray tubes (CRTs), printed wires, circuit boards, and other materials are often found in e-waste. A schematic is shown in Figure 1. for the sources of E waste generation. Once the e-waste is properly handled, precious metals including Cu, Hg, Au, and Pt may be recycled. It is extremely dangerous if e-waste is disassembled and synthesized in an imprecise manner using rudimentary techniques because it contains toxic materials like liquid crystal, Li, Hg, Ni, Se, polychlorinated biphenyls (PCBs), toxic chemicals, Ba, Br fire retardants, heavy metals, Cr, Cu and Zn, brass, and Pb. Fractions of metals presents in various sources are shown in Figure 2. Then they are no longer suitable for use, electronic devices such as computer, mainframes, Printing and scanning machines, compact discs (CDs), copiers, calculators, batteries, cellular phones, medical equipment, iPods, household appliances become e-waste. Even in trace amounts, heavy metals and extremely poisonous chemicals like Hg, Pb, Be, and Cd represent a serious hazard to the environment (Askarzai et al., 2011).

Figure 1. Sources of E waste generation through daily needful things

Figure 2. Percentage of metal fractions present in electrical and electronic equipment

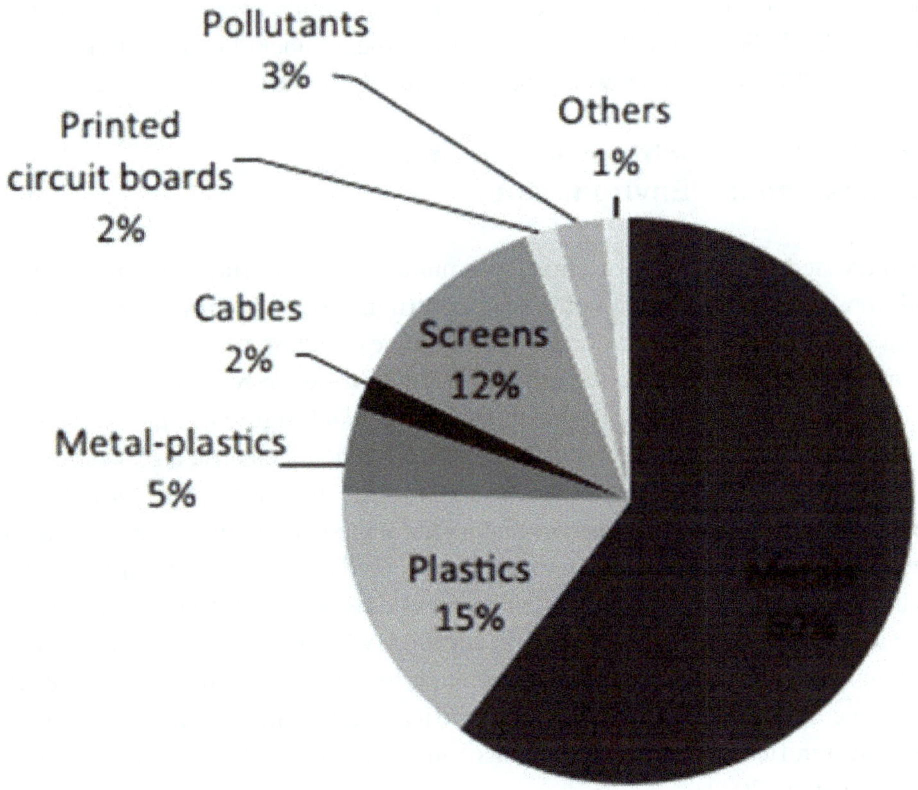

E-Waste Categories

The effective management of electronic trash requires both an accurate definition of "e-waste" and an accurate identification of its many kinds.

FORECASTS FOR THE DEVELOPMENT OF E-WASTE

Employing a model of economics that is circular for e-waste may result in a rising number of economic advantages. The disposal of electronic trash is estimated to total more than 50 million tons annually, according to current forecasts. Only around 20% of this garbage gets recycled at any one time (Babel et al., 2020).

The quantity of electronic waste that is thrown away increases by 2.5 million tons on an annual basis. The generation of electronic garbage is forecast to see a meteoric rise in the years to come. It is anticipated that by the year 2030, the electronic trash sector would increase to approximately 75 million tons, which will more than double the quantity of garbage that is presently generated yearly. This expansion may most likely be attributable to an increase in the number of people buying goods with shorter life cycles and fewer repair alternatives for older technological devices.

The Effects of Electronic Waste on Both Humans and the Environment

The increased access to information, communication, and educational possibilities made possible by electronic items may contribute to an overall improvement in living standards. On the other hand, the rapidly expanding worldwide demand for electronics is soon outpacing the industry's ability to securely dispose of or recycle obsolete items. A schematic is shown in Figure 3 and Figure 4. Showing various diseases due to present metals particles in electronic and electricals goods.

Most electronic gadgets are not biodegradable; this means that they will not break down naturally or in a manner that is safe if they are thrown away in landfills. Electronic garbage that has been thrown away tends to accumulate in the natural environment, where it may contaminate ecosystems and release chemicals that are toxic to humans, animals, and plants. In addition, when electronic trash is heated to very high temperatures, hazardous compounds are discharged into the atmosphere. Electronic trash presents a number of significant environmental problems, one of the most significant of which is the emission of pollutants and harmful substances (Debnath et al., 2022).

Figure 3. Presence of metals particles in electronic and electricals goods harmful for human body parts

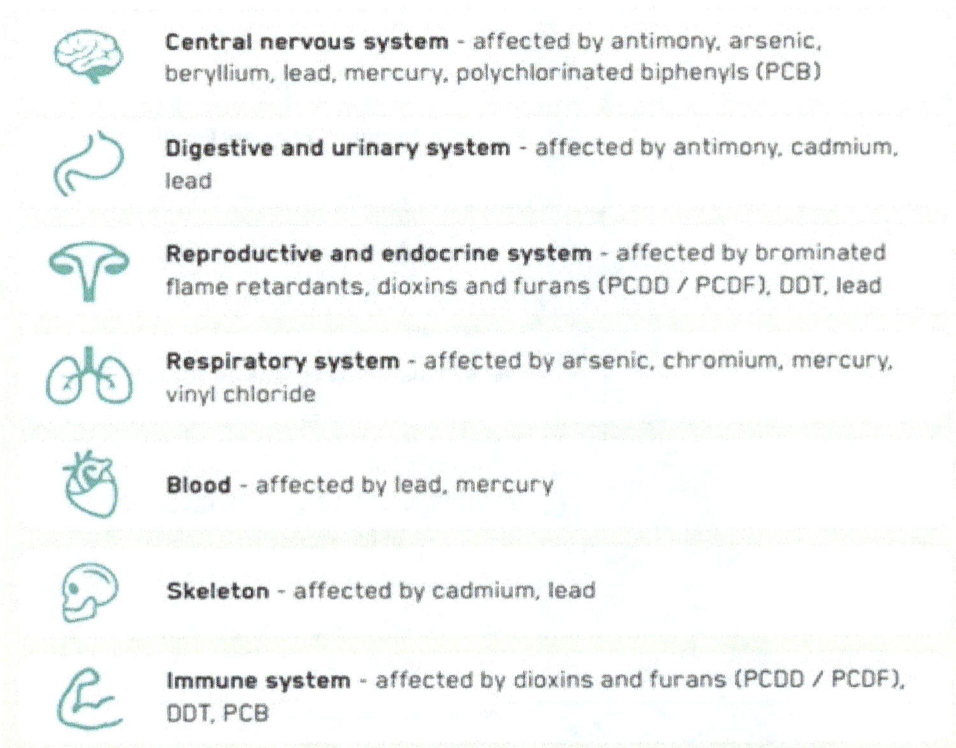

	E-waste-related inorganic toxicants	Reported health effects
Environmental matrices	Fe, Al, Zn, Cu, Pb, Co, Cd, Cr, Mn, Ba, Sn, V, As, Sb, Br, Cl, Hg,	-Abnormal thyroid and reproductive function, Reduction of gonadal hormone, Injuries and respiratory symptoms, Cough and chest pain, Headache, stomach discomfort, miscarriage, and cancer
Human matrices	As, Cd, Cr, Hg, Fe, Ag, Al, Be, Cd, Co, Hg, Mn, Pb, Sb, V, Ga, Mn, Mo, Sr, Zn, Cs, Bi, Ba, Ag,	-Reduction of gonadal hormone - Depletion of antioxidant reserves - Injuries and respiratory symptoms - Cough and chest pain

Figure 4. Various metals affecting human and environment matrices

How Their Presence Degrading the Quality of Environment as Well Human Health?

The effects of inappropriate e-waste disposal in landfills or other non-dumping locations pose substantial risks to present public health and have the potential to contaminate ecosystems for future generations. Toxic chemicals are generated when electronics are incorrectly disposed of and wind up in landfills, which has an effect on the environment's air, soil, and water—and eventually, human health.

Negative Effects on Air Quality

When electronic trash is disposed of informally by disassembling, shredding, or melting the components, dust particles or chemicals, including dioxins, are released into the atmosphere, causing air pollution and harming respiratory health. Burning is a common practice for e-waste that has little value, but it may also be used to extract precious metals like copper from devices. Burning e-waste increases the chance of developing chronic illnesses and cancer because it produces tiny particles that may travel hundreds of kilometres and pose a number of dangers to both human and animal health. Highly integrated electronics often have higher value components like gold and silver removed using acids, desoldering, and other chemicals, which can create fumes in locations where recycling is not well controlled. The people who handle this material are most at risk from the harmful air consequences of informal e-waste recycling, although the pollution may travel thousands of kilometres from recycling centers.

Particular animal species are more negatively impacted by e-air waste's pollution than others, which might put these species and the biodiversity of some heavily contaminated areas in peril. Air pollution over time may harm plant species, soil quality, and water availability, causing irreparable harm to ecosystems. By extracting precious metals from e-waste, for instance, parties developed an informal recycling centre in India. As a result, the area has exceptionally high lead levels in the air, which are breathed and then consumed when returned to water and soil. Larger animals, wildlife, and nearby people may suffer disproportionate brain damage as a result.

The Harmful Effects on the Soil

Both heavy metals and flame retardants may seep straight into the soil during inappropriate e-waste disposal in ordinary landfills or in locations where it is illegally deposited. This can contaminate nearby or nearby crops, as well as the groundwater underneath them. Heavy metal contamination of the soil makes crops more susceptible to absorbing these poisons, which may lead to a number of ailments and reduce the productivity of the field.

Due to their size and weight, big particles that are liberated during the burning, shredding, or disassembly of e-waste rapidly re-deposit to the ground and pollute the soil. Temperature, soil type, soil pH levels, and soil composition are just a few of the variables that affect how much soil is polluted. These contaminants may linger in the soil for a very long time and affect both plants and soil microbes. In the end, animals and wildlife that depend on nature for survival will eat impacted plants, leading to interior health issues.

Negative Effects on Water

Following soil contamination, e-waste heavy metals including mercury, lithium, lead, and barium subsequently flow even farther through the earth to reach groundwater. These heavy metals ultimately find their way into ponds, streams, rivers, and lakes once they enter groundwater. Even if a community is kilometres distant from a recycling facility, these channels cause acidification and toxification in the water, making it unhealthy for animals, plants, and communities. Finding clean water to drink becomes difficult.

Freshwater and marine creatures may die from acidification, which can also disrupt biodiversity and damage ecosystems. Acidification may harm ecosystems to the point that recovery is difficult, if not impossible, if it affects water sources.

The Harmful Effects on People

As previously indicated, harmful materials found in electronic trash include lithium, barium, lead, cadmium, mercury, and polybrominated flame retardants. Humans exposed to these chemicals may suffer harm to their brain, heart, liver, kidney, and skeletal systems. Additionally, it has a significant impact on the neurological and reproductive systems of humans, which may result in illness and birth problems. It is crucial to raise awareness about the hazard posed by improper e-waste disposal since it poses a very serious risk to the ecosystem worldwide. It is essential to correctly e-cycle electronic trash so that products may be recycled, repaired, resold, or repurposed in order to prevent these hazardous impacts of e-waste. If people are not informed about the proper methods of disposal, the expanding stream of e-waste will only become worse.

E-WASTE MANAGEMENT CHALLENGES IN INDIA

Electronic waste (sometimes referred to as e-waste) is produced when any electrical or electronic equipment is no longer suitable for its intended purpose or has passed its expiration date. The creation of newer electronic equipment and the quick evolution of technology make it simple to swap out outdated versions with more modern ones. Particularly in India, it has caused e-waste to rise exponentially. People have a tendency to migrate to newer models and popular technology, and with time, product lifespans grow shorter. But e-waste management in India and its difficulties remain the problem. In India, the recycling of e-waste mostly occurs in the unorganized sector. Numerous low-income families survive by salvaging supplies from trash sites. For trash paper, plastic, clothes, or metal, middle-class urban homes often recycle by selling it to small-scale, unregulated purchasers known as "kabadiwalas," who then sort it and utilize it as a raw material for artisanal or industrial processors (Dervash et al., 2020)

An analogous tendency may be seen in India's treatment of e-waste. Numerous metropolitan families are employed by an unregulated e-waste recycling industry to gather, sift, repair, refurbish, and disassemble outdated electrical and electronic devices. The situation is different in developed nations, and there is no idea of people willingly giving outdated electrical and electronic equipment at official e-waste recycling facilities in India (Dervash et al., 2020). Additionally, the idea of customers paying for the disposal of the electronic garbage they produce does not exist. The major causes for the E waste generation are listed in Figure 5.

These major issues are brought on by the e-waste recycling industry's excessive dependence on the unorganized sector, as listed below:

- First off, it is useless to try to penalize people financially for breaking the laws for how to handle and manage e-waste.
- Second, since fewer well-paid employees who do this task lack sufficient training, the general public is less informed about market pricing and health and safety expenses associated with e-waste recycling.
- Third, despite the enormous growth in the amount of e-waste produced each year, there is a dearth of large-scale industrial infrastructure investment in recovery and recycling.

Few Recent Updates on Indian Scenario on E-Waste Management

Out of the 10,14,961.21 tons of e-waste produced in India in 2019–20, only 22.7% was gathered, disassembled, recycled, or properly disposed. This e-waste is made up of 21 different categories of electrical and electronic equipment (EEE) that were reported in accordance with the 2016 E-Waste (Management) Rules. Approximately 21.35 percent of the entire amount of created e-waste was handled, collected, deconstructed, and recycled in India in 2018–19, compared to only 9.79 percent in 2017–18.

By obtaining Extended Producer Responsibility (EPR) authorization, manufacturers are now responsible for managing a system of e-waste collection, storage, transportation, ecologically responsible disassembly, and recycling. The regulations also support and facilitate the creation of a successful e-waste collecting system.

Figure 5. Issues/Causes related to the generation E-waste

However, the findings indicate that just 25% of the created e-waste is being appropriately handled. E-waste must be thrown in open areas and aquatic bodies, according to experts, much like untreated solid garbage.

Inadequate Infrastructure for Recycling E-Waste

India has minimal infrastructure that can manage massive amounts of e-waste. Only roughly 1/5th of the nation's total e-waste produced annually is recycled at the relatively few government-approved facilities. The co-funded grant program, which the Indian government provides, covers 25% to 50% of the project expenditures for e-waste disposal facilities and developing the capacity of e-waste enterprises. However, virtually little of this plan has been adopted. Additionally, there is a dearth of legally certified e-waste recycling plants, since the bulk of irregular e-waste collectors in India have poorly established supply networks that cause them to operate considerably below their capabilities (Frazzoli et al., 2022).

However, in India, the official recycling industry can only manage e-waste via mechanical disassembly and human sorting. The large-scale recovery of valuable and alloying elements now requires industrial e-waste management with appropriate environmental controls, which are lacking. Several young Indian enterprises recover metals from e-waste, but their processing power is limited. The majority of the e-waste that is handled on a formal basis is shipped from other nations that have the massive infrastructure required for metal extraction. In contrast, the informal industry collects metals through risky processes, including acid leaching and open-air incinerator that worsen environmental contamination and health concerns (Ghosh et al., 2020).

The formal and informal sectors have primarily focused on metal recovery and less on the bottle, plastic materials, and ceramic materials that encompass the substantial portion of e-waste, even though the E-waste guidelines provide a variety of techniques for reprocessing and manufacturing different types and electronics of e-waste. Including fire retardants and other chronic organic contaminants makes recycling plastics e-waste considerably more difficult (Hassan et al., 2022).

Financial Incentives and A Lack of Awareness

In India, there is a dearth of public understanding of the risks associated with e-waste, and as a result, recycling is relatively low. The majority of customers are either unaware of or have a limited understanding of the dangers associated with e-waste components and the consequences of incorrect disposal. They are unaware that state or local governments in India manage e-waste (Godbole et al., 2011).

Few specialized collection depots or official recycling facilities exist in many cities where customers may freely drop off their e-waste. When buying new electrical or electronic devices from small-scale retail stores, most individuals and urban home customers used to sell their old electronics or get a discount. Consumers have little financial incentives to dispose of their e-waste safely because they lack market knowledge on pricing for e-waste and different e-waste constituents.

Less Evidence on The Rates of E-Waste Generation

It is recognized that there aren't any e-waste inventories, and the different State Pollution Control Boards are solely responsible for state-by-state e-waste inventories (SPCBs). An essential component in estimating the amount of e-waste is the sales information for electronic items. It is often aggregated at the national level, which makes it challenging to generate inventories at the state level. In addition to being produced domestically, e-waste is often illegally purchased from industrialized nations (Gollakota et al., 2020).

The kind and volume of electronic trash brought into the nation are less well known. The systems are necessary for efficient waste creation and composition knowledge as well as collecting, conveyance, and preparation.

Marketing Poor Management for End-of-Life Goods

The capacity of private actors to establish e-waste management systems in a formal industry is constrained by the difficulty of consistently obtaining sufficient amounts of e-waste to generate economies of scale. In India, for example, adopting efficient recycling methods for e-waste management may need substantial up-front capital costs, which private firms cannot justify in the lack of confidence about where to get sufficient amounts of e-waste. Information obstacles also hamper these marketplaces.

First off, because the recycling of electronic trash is a relatively young industry, there may be a market barrier due to a lack of knowledge about efficient recycling methods. Second, the operation of markets is impacted by poor knowledge, which is partially caused by customers' access to credible information on e-waste management. However, improved markets for e-waste will be made possible by public policy, which will have a bigger impact (Ikhlayel et al., 2018).

Informal Sector Practices That Are Unsustainable for the Environment

Even while the official recycling and dismantling industry is expanding, relatively little garbage is being handled there. Due to a lack of adequate waste sources, most of these formal facilities work below their permitted limits. The unwillingness of household and institutional customers to return their garbage to the official sector is being reduced by the lack of understanding about e-waste and the fees associated with returning end-of-life equipment to formal collection centers. Most crucially, compared to the official sector, which has not yet invested in reliable collection and processing equipment, the informal sector makes it appealing for consumers to return their garbage via the ease of home pickup and financial incentives (Jangre et al., 2022).

Millions of individuals, many of them are members of the most disadvantaged groups, have found work in the unregulated e-waste industry. Contrarily, the industry's waste management methods seriously endanger the environment and the health of the general public and the employees. The continued success of any e-waste management in India would depend on finding a moral solution to this possible moral impasse (Jayapradha et al., 2015).

Poorly Designed and Enforced Regulations

The required take-back system for manufacturers had minimal impact on e-waste management procedures because it lacked related collection objectives, which served as incentives for producers to accept responsibility. Several adjustments were suggested that increased regulatory clarity by defining progressively tougher collection objectives. The regulatory structure imposes a heavy load on the regulatory bodies, which are already under-resourced. The regulators are responsible for reviewing the producers' EPR plan, approving it, and enforcing its rules (Khatun et al., 2022).

As well as requiring the authorities to strictly comply with specified criteria, the laws must also include extensive requirements and procedures for other businesses such as dismantlers, collectors, recyclers, and bulk consumers. The government regulators must profit from lax enforcement, a lack of openness, and a reluctance to make information on regulation and compliance activities available to the public. E-waste restrictions are not an exception to this long-standing problem with environmental regulatory implementation in India. It represents the biggest challenge to public policy facing India's potential e-waste management (Khetriwal et al., 2019).

ENHANCEMENT OF INDIA'S CURRENT PROCEDURES FOR THE REDUCTION OF ELECTRONIC WASTE

E-waste management may be improved in India in a number of ways. Nevertheless, in a nutshell, there are five essential elements that may be connected to enhance India's e-waste governance. Underneath is a discussion on it:

Providing Price Information on The Market for E-Waste

It is a well-established market for the trading of e-waste both within the informal sector and between operators in the official sector. However, urban consumers are not well-informed on or made aware of the pricing for e-waste & its parts. Weekly price updates are necessary because they would serve as a strong market signal for clients who sell their e-waste to nearby merchants (Koli et al., 2019).

The pricing list must include all e-waste materials, including different kinds of glass, metals, plastics, ceramics, batteries, and bulk e-waste. Local newspapers and urban towns must publish the statistics on specific websites, much as they do with commodities pricing lists or foreign currency rates. The pricing list must take current market conditions for e-waste component demand into account and must allow informal sector collectors to purchase and sell e-waste to corporate processing units or govt recycling and disassembly facilities at fair market rates (Nikabadi et al., 2018).

Promoting the Recycling of Formal E-Waste

To encourage formal enterprises to route their e-waste via government-approved recycling facilities, the Indian government has implemented a point-based compensation system known as E-waste Recycling Credits (ERCs). E-waste items including laptops, PCs, and cell phones are already classified and coded by the E-Waste Rules. Various ERC reward levels must be associated for these groups. Organizations are required to get the necessary ERCs, which may be utilized to reduce energy utility costs, relying on the category of e-waste delivered. Such a program will also provide a significant incentive for unlicensed e-waste enterprises to formalize their operations and create supply chain connections with authorized recycling facilities.

The ERCs might be piloted for a period of 3 to 5 years to evaluate effectiveness and make adjustments in preparation for future adoption. More than 70% of the e-waste produced in large cities comes from the industrial and governmental sectors. In Mumbai, Delhi, or Bangalore, a few sizable businesses and governmental agencies may test the ERCs (Mehta et al., 2019).

By jointly sponsoring infrastructure improvements and handling systems at current administration approved recycling facilities, the Indian government may also increase formal e-waste recycling potential. It is able to offer governments with co-funding opportunities for the establishment of new recycling facilities via public-private partnerships with significant enterprises that produce electronic trash (Mohd et al., 2018). State governments might also create incentive programs to encourage small-scale, improvised e-waste recycling facilities to modernize their infrastructure and conform to environmental and occupational health and safety requirements. States are eligible to apply for national financial programs for urban development, which may be utilized to connect the extensive informal sector networks of decentralized collecting and micro recycling units with commercial recycling facilities on a big scale. Figure 6. Shows the market management strategies for E waste (Murthy et al., 2022).

Companies in the Informal Sector Are Trained and Upgraded

Especially for handling and disassembling hazardous products, the bulk of an informal e-waste recycling labor need training and support. It must assure worker and ecosystem health and safety, as well as a connection between the supply and formal sector manufacturers. It is promoted by the National Skill Development Mission of the Indian government. The cumulative knowledge of the Electronics Sectoral Committee, the Green Energy Sector Skill Council, and regulatory bodies like the Central and State Pollution Control Boards can be used to create creative short

Figure 6. Schematic representation of the E waste management market having 3R's compilation

lessons and teaching programs that are specifically tailored for e-waste collectibles, operators, and designated disposal (Panda et al., 2022).

both training to raise public understanding of the risks connected with e-waste, the regional and state authorities should launch a coordinated, countrywide program to upskill employees in the consultative sector. The significance of the unofficial sector in the disposal of e-waste and the locations of official e-waste collection facilities that have been authorized by the ministry (Pariatamby et al., 2020).

Utilizing Ready-to-Use and Advanced Recycling Technologies

To increase the recycling effectiveness of the vast quantities of e-waste management in India, it is necessary to implement mature recycling technology alongside the country's current manual approaches. The plastics processing industry in India is quite vast and developed, and it can recover plastic from e-waste (Patil et al., 2020).

The establishment of substantial industrial e-waste recycling facilities requires the Indian government to encourage cooperative partnerships between local and foreign businesses. These endeavours may be financed by a mix of public and private funding.

Creating New Technologies & Approaches for Handling New Types of E-Waste

Due to the introduction of new electronic gadgets into the market, e-waste is continuously changing in composition. In order to prepare India's e-waste policy and management for the future, major investment is needed in research and development for creative recycling techniques and technologies. For instance, during the last five years, cell phone use has increased significantly in India, but the li-ion batteries that drive the devices are still not covered by any e-waste recycling regulations (Rai et al., 2020).

The future generations of electronic products will be made using several new batteries and chemical technologies. The Indian administration must thus encourage and support research that creates cutting-edge, futuristic technology for recycling and converting fresh streams of e-waste into high-value goods.

Six Strategies to Reduce Your E-Waste

A significant amount of energy inputs is used in the creation of these gadgets and the usage of rare materials. Reducing our use of resources and energy from the world by minimizing e-waste is beneficial.

- Re-evaluate. Honestly, do you require that additional device? Look for a single gadget that can do many tasks.
- Make your gadgets last longer.
- Purchase a case, maintain your gadget spotless, and steer clear of overcharging the battery.
- Buy devices that are ecologically friendly. Look for goods with the Energy Star or Electronic Products Environmental Review Tool certifications (EPEAT).
- Make a donation of old gadgets to social programs to support issues like protecting kids from harm, helping victims of home abuse, the ecosystem, and more. For a priority mail for your mobile phone or ink cartridge, ask

your students REP. The Global Wildlife Foundation will earn $1 for each item sent.

- Utilize the e-waste recycling containers dotted throughout campus to recycle batteries and devices. Large devices may be placed in the bigger bins that are available in your building.

HOW CAN INDIA'S E-WASTE BE MANAGED EFFECTIVELY?

We are always assessing the efficacy of e-waste legislation and making the required adjustments. The government is crucial in bringing together different system players. We have a few options that might be taken into consideration to advance. A schematic is shown in Figure 7. Showing various steps to strengthen the controlling power of E waste generation.

Boost the Informal Sector

Explicitly recognizing the informal sector as a commercial entity in any future e-waste system might be the first step. A deeper knowledge of the industry's motivations and difficulties is necessary to address the issue of informal sector e-waste practices. A starting approach may be to engage with organizations and workers in the informal sector in a way that respects their right to their livelihoods, fosters trust, and increases knowledge of the issues and possible solutions. The government must set up a forum to encourage talks between diverse stakeholders, including employees in the unorganized sector, NGOs that support unorganized workers, outsiders, private companies, registered recyclers, and manufacturers (Roy et al., 2022). At a particular level within the State Administration, the forums might be established under the Ministry of Environment, Forestry, and Global Warming.

Instruments of Policy Under EPR

Extended Producer Responsibility

Extended Produced Responsibility, often known as EPR, is a new initiative for the management of electronic trash that was launched by the Ministry of Environment, Forest, and Climate Change. It is an essential component in the management of electronic waste. It was established so that electronic trash may be collected, redirected, recycled, and managed more effectively. It adheres to the principles of circularity, which require producers to take more responsibility for the total management of each product they sell in the marketplace. There is a possibility that EPR may accelerate

Figure 7. Steps to strengthen the controlling power of E waste generation

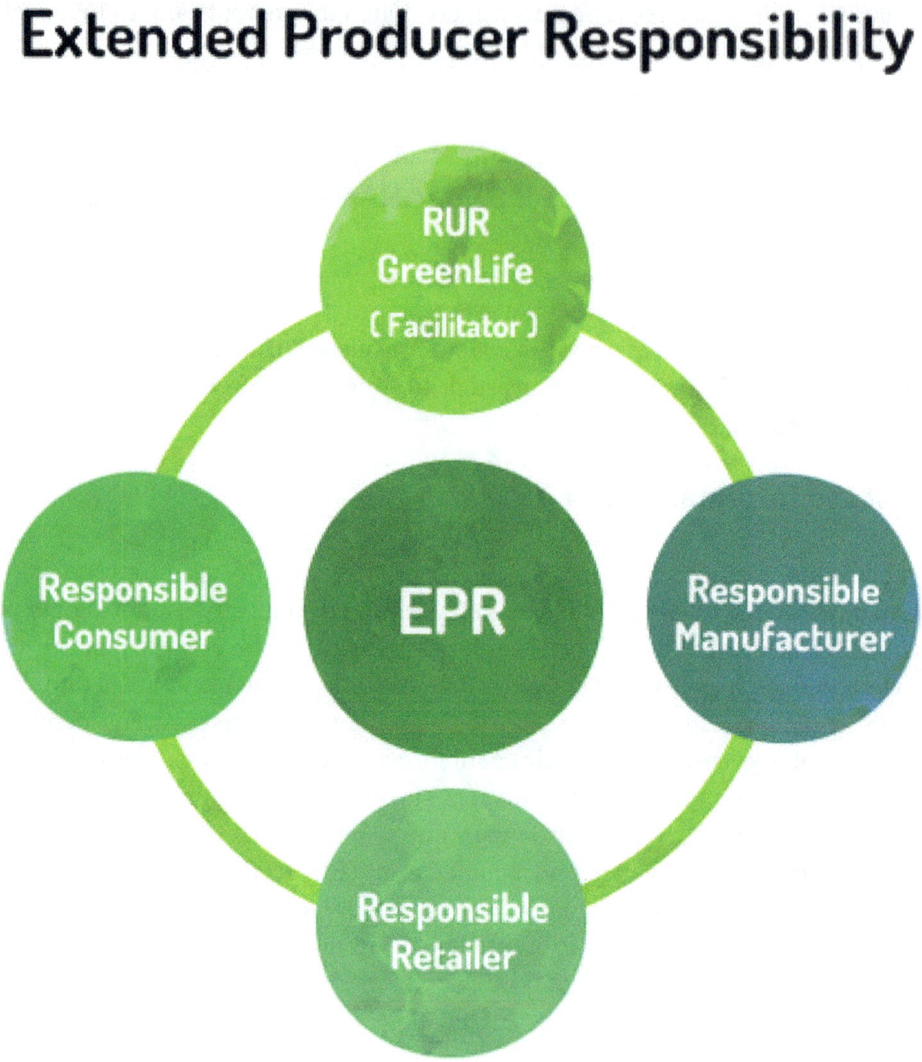

beneficial transformation. It has also been found to have a favourable influence on the management of electronic trash (Sakhuja et al., 2022). The development of a computerized system for efficient EPR monitoring should be prioritized. Bringing transparency to the EPR implementation process requires tracking and digitizing every step of the process. Duties of EPR is shown in Figure. 8.

Figure 8. Schematic representation of extended produced responsibility (EPR)'s duties

The policy tools used by the government in the EPR methodology need to be reconsidered. It needs capabilities in collection logistics given the existence of the informal sector. The best tool is not an obligatory take back with collecting goals.

The concept of producer responsibility encompasses more than just take-back requirements.

Producers will no longer be physically responsible for collection thanks to economic tools like efficient and advanced recycling fees or refined disposal fees on per unit of goods delivered in the marketplace. The money made from these fees may be utilized to create markets for obsolete or end-of-life items. There are several uses for the income that go into a savings account (Sengupta et al., 2022).

A design for environmental improvements in product design, which has been one of the primary objectives of the EPR method internationally, may also be encouraged by the appropriate cost. In the long term, the charge may be dependent on variables like the ease of recycling, disassembly, and environmental effect of components utilized in a piece of machinery in order to further encourage reforms. The legislative framework should also place an emphasis on the establishment of homegrown technologies and/or the transferring existing systems in order to stimulate the wider deployment of environmentally friendly e-waste recycling process (Surekha et al., 2022).

Regulatory Compliance

By eliminating the need to control producers, switching to economic tools like an ADF may also reduce the burden of regulation. The ability to redirect the ADF on electronic devices to a different fund should be simple given the lengthy history of tax collecting.

It will still be necessary for the Regional and Central Pollution Control Board to keep an eye on and ensure adherence to the requirements set out for pickup facilities, dismantlers, recycling plants, and PROs. The MoEFCC must make all e-waste regulations public. The public should have access to regulatory activities such approvals and their circumstances, information on assessments of recognized facilities, and the compliant status of examined facilities.

A few SPCBs now make some of these records available to the public on their webpages, but these procedures need to be codified in the laws of the nation. The creation of an inventory of district-level e-waste production volumes broken down by e-waste type (such as mobiles, laptops, and appliances), waste proportions, and circulation will also be crucial for compliance.

Imports of E-Waste

E-waste may be transported for reuse and recycling but not for ultimate disposal, according to the current legislation. Because our nation lacks the necessary recycling networks, we should strongly explore prohibiting all imported goods. Data on imported goods must be combined with an e-waste inventory in order to provide precise predictions of e-waste.

Public Education on the Management of E-Waste

The manufacturers are required by the present e-waste legislation to give real info on the effects of e-waste, acceptable disposal methods, and other concerns on the websites. There is also a necessity for periodic awareness efforts. Despite the fact that many manufacturers have already posted information on the websites, research indicates that bulk customers' general levels of knowledge are still low. Stricter instructions for the manufacturers regarding the frequency and style of these awareness programs may help the problem. As a substitute, the makers must be required to implement these programs via local e-waste groups. The government must, for its part, think about fusing awareness campaigns on e-waste with those about other waste streams, such batteries and municipal solid garbage (Widmer et al., 2005).

The government may also have a role in evaluating awareness programmes and doing research on the best message strategies. These awareness campaigns must be focused on lowering long-term usage of electronic items and ensuring safe treatment of e-waste. Initiatives to raise public awareness should, in general, be built on cooperation and partnerships amongst diverse stakeholders.

System for Managing E-Waste

Environmentally friendly e-waste management approaches must be promoted via increased informational initiatives, capacity development, and understanding. To lessen any unlawful e-waste trade, further efforts are urgently needed to strengthen present procedures including collection plans and management standards. Since it will aid in preventive, reducing the quantity of dangerous compounds in e-products will also benefit the particular e-waste sources.

The majority of e-waste is recycled in India through unorganized facilities, requiring a sizable workforce. The riskiest action is the recovery of metals from PCBs using traditional methods. To provide individuals who rely on this for a living better tool, proper training, attention, and, most crucially, alternatives cost-effective technologies, must be made available (Yong et al., 2019). To solve India's e-waste management problems, a comprehensive strategy is required. To integrate big units in the formal sector and smaller squads in the informal economy into a discrete value chain, the appropriate mechanism must be developed. In our strategy, the organized sector might handle metal extraction, recycling, and disposal, while informal sector businesses would focus on gathering, disassembling, and sorting.

CONCLUSION

The proliferation of information technology has resulted in an increased need for machinery that utilizes electrical and electronic components. The rise in the usage of electronic items has also led to an increase in the amount of garbage that is classified as "e-waste." A significant increase in the amount of e-waste contributes to the myriad of challenges associated with the management of electronic trash. At the international level, electronic waste disposal is among India's top five priorities. In addition to their own garbage, a significant amount of electronic waste is dumped here, and it is also the destination for electronic waste imported from other wealthy countries. The handling of electronic trash may be accomplished more cheaply in India because to the country's skilled labor force.

In addition to all of the challenges and negative effects, there are also some opportunities available for the central government regarding the management of electronic waste. These opportunities include extended producer responsibility (EPR), e-waste policy and regulations, e-waste recycling, capacity building, training, and awareness program. The handling of electronic trash in India is a significant obstacle for the government entities of most developing economies. It is growing tremendously every day and is turning into a major public health problem. E-waste has to be retrieved separately, handled carefully, and dumped of. Additionally, it avoids using open burning and traditional landfills. Integrating the informal sector with the official sector is crucial. In developing nations like India, the appropriate organizations must set up procedures for managing and processing e-waste in a sustainable approach.

There is an immediate need for a comprehensive analysis of the actual situation as well as the potential outcomes, with consideration given to quantification, characteristics, existing disposal techniques, environmental implications, and so on. Encouragement should be given to private businesses and manufacturers to form partnerships for the purpose of establishing centers for the collection, exchange, and recycling of electronic waste. It is possible that a successful take-back program that provides incentives for manufacturers to develop goods that generate less waste, include fewer harmful components, and are simpler to dismantle, reuse, and recycle might contribute to the reduction of waste. When developing new electronic equipment, manufacturers should give top emphasis to planning for end-of-life management.

ACKNOWLEDGMENT

The authors sincerely acknowledge the suggestions and comments of the authors are heavily indebted to them in providing their valuable suggestions and insights.

REFERENCES

Alam, T., Golmohammadzadeh, R., Faraji, F., & Shahabuddin, M. (2022). E-Waste Recycling Technologies: An Overview, Challenges and Future Perspectives. *Paradigm shift in E-waste management*, 143-176.

Arya, S., & Kumar, S. (2020). E-waste in India at a glance: Current trends, regulations, challenges and management strategies. *Journal of Cleaner Production*, *271*, 122707. doi:10.1016/j.jclepro.2020.122707

Askarzai, W. (2011). The negative impact of ICT waste on environment and health. In *Handbook of research on green ICT: technology, business, and social perspectives* (pp. 242–255). IGI Global. doi:10.4018/978-1-61692-834-6.ch016

Babel, S., Ta, A. T., & Liyanage, T. U. H. (2020). Current situation and challenges of waste management in Thailand. In *Sustainable Waste Management Challenges in Developing Countries* (pp. 409–440). IGI Global. doi:10.4018/978-1-7998-0198-6.ch017

Debnath, B., Das, A., & Das, A. (2022). Towards circular economy in e-waste management in India: Issues, challenges, and solutions. In *Circular Economy and Sustainability* (pp. 523–543). Elsevier. doi:10.1016/B978-0-12-821664-4.00003-0

Dervash, M. A., Geelani, S. M., Bhat, R. A., Singh, D. V., & Wani, A. A. (2020). Electronic Waste: Implications on Environs and Management Strategies. In Innovative Waste Management Technologies for Sustainable Development (pp. 82-97). IGI Global.

Frazzoli, C., Ruggieri, F., Battistini, B., Orisakwe, O. E., Igbo, J. K., & Bocca, B. (2022). E-WASTE threatens health: The scientific solution adopts the one health strategy. *Environmental Research*, *212*, 113227. doi:10.1016/j.envres.2022.113227 PMID:35378120

Ghosh, S. K. (2020). Waste Management Under the Legal Framework in India. In *Sustainable Waste Management Challenges in Developing Countries* (pp. 144–165). IGI Global. doi:10.4018/978-1-7998-0198-6.ch006

Godbole, N. (2011). E-waste management: Challenges and issues. Handbook of Research on Green ICT: Technology, Business and Social Perspectives, 480-505.

Gollakota, A. R., Gautam, S., & Shu, C. M. (2020). Inconsistencies of e-waste management in developing nations–Facts and plausible solutions. *Journal of Environmental Management*, *261*, 110234. doi:10.1016/j.jenvman.2020.110234 PMID:32148304

Hassan, Z., & Dhusia, D. K. (2022). Hazardous E-Waste Recycling Practices Affecting Informal Recycler Health in India: A Case Study. [IJCEWM]. *International Journal of Circular Economy and Waste Management*, *2*(1), 1–25. doi:10.4018/IJCEWM.302205

Ikhlayel, M. (2018). An integrated approach to establish e-waste management systems for developing countries. *Journal of Cleaner Production*, *170*, 119–130. doi:10.1016/j.jclepro.2017.09.137

Jangre, J., Prasad, K., & Patel, D. (2022). Analysis of barriers in e-waste management in developing economy: An integrated multiple-criteria decision-making approach. *Environmental Science and Pollution Research International*, *29*(48), 1–15. doi:10.100711356-022-21363-y PMID:35696062

Jayapradha, A. (2015). Scenario of E-waste in India and application of new recycling approaches for E-waste management. *Journal of Chemical and Pharmaceutical Research*, *7*(3), 232–238.

Khatun, A., & Dhara, N. (2022). E-Waste Management: A New Concern for Environmental Sustainability. In Smart Cities, Citizen Welfare, and the Implementation of Sustainable Development Goals (pp. 222-238). IGI Global. doi:10.4018/978-1-7998-7785-1.ch012

Khetriwal, D. S. (2019). E-waste management in India. In Waste Electrical and Electronic Equipment (WEEE) Handbook (pp. 541-571). Woodhead Publishing. doi:10.1016/B978-0-08-102158-3.00020-3

Koli, S. K., & Hussain, A. (2019). Status of Electronic Waste Management in India: A Review. *Advanced Treatment Techniques for Industrial Wastewater*, 238-250.

Mehta, P. (2019). E-waste, chemical toxicity, and legislation in India. In *Advanced Methodologies and Technologies in Engineering and Environmental Science* (pp. 144–156). IGI Global. doi:10.4018/978-1-5225-7359-3.ch011

Mohd, S., & Kaushal, V. K. (2018). E-waste management in India: current practices and challenges. *ResearchGate A Comprehensive Study on Electrical, 355.*

Murthy, V., & Ramakrishna, S. (2022). A review on global E-waste management: Urban mining towards a sustainable future and circular economy. *Sustainability*, *14*(2), 647. doi:10.3390u14020647

Nikabadi, M. S., & Hajihoseinali, A. (2018). E-Waste Recycling System in Closed Loop Supply Chain. [IJSDA]. *International Journal of System Dynamics Applications*, *7*(2), 55–80. doi:10.4018/IJSDA.2018040104

Panda, B., Panda, G. K., & Pradhan, A. (2022). E-Waste Management. In *IoT-Based Smart Waste Management for Environmental Sustainability* (pp. 1–18). CRC Press. doi:10.1201/9781003184096-1

Pariatamby, A., Bhatti, M. S., & Hamid, F. S. (2020). Waste Management in Developing Countries: An Overview. *Sustainable Waste Management Challenges in Developing Countries*, 494-519.

Patil, R. A., & Ramakrishna, S. (2020). A comprehensive analysis of e-waste legislation worldwide. *Environmental Science and Pollution Research International*, *27*(13), 14412–14431. doi:10.100711356-020-07992-1 PMID:32162230

Rai, A., Kothari, R., & Singh, D. P. (2020). Assessment of available technologies for hospital waste management: A need for society. In Waste Management: Concepts, Methodologies, Tools, and Applications (pp. 860-876). IGI Global.

Roy, H., Rahman, T. U., Suhan, M. B. K., Al-Mamun, M. R., Haque, S., & Islam, M. S. (2022). A comprehensive review on hazardous aspects and management strategies of electronic waste: Bangladesh perspectives. *Heliyon*, *8*(7), 09802. doi:10.1016/j.heliyon.2022.e09802 PMID:35815143

Sakhuja, D., Ghai, H., Bhatia, R. K., & Bhatt, A. K. (2022). Management of e-Waste: Technological Challenges and Opportunities. Handbook of Solid Waste Management, 1523-1557.

Sengupta, D., Ilankoon, I. M. S. K., Kang, K. D., & Chong, M. N. (2022). Circular economy and household e-waste management in India: Integration of formal and informal sectors. *Minerals Engineering*, *184*, 107661. doi:10.1016/j.mineng.2022.107661

Surekha, A., & Suguna, A. (2022). E-waste management practices and perception about its health hazards. *International Journal of Preventive Medicine*, *13*(1), 123. PMID:36276896

Widmer, R., Oswald-Krapf, H., Sinha-Khetriwal, D., Schnellmann, M., & Böni, H. (2005). Global perspectives on e-waste. *Environmental Impact Assessment Review*, *25*(5), 436–458. doi:10.1016/j.eiar.2005.04.001

Yong, Y. S., Lim, Y. A., & Ilankoon, I. M. S. K. (2019). An analysis of electronic waste management strategies and recycling operations in Malaysia: Challenges and future prospects. *Journal of Cleaner Production*, *224*, 151–166. doi:10.1016/j.jclepro.2019.03.205

Chapter 2
Awareness and Handling Practices of Electronic Waste in Indian Households and Society

Dharam Singh
Maulana Azad National Institute of Technology, India

Vikash Kumar
Maulana Azad National Institute of Technology, India

ABSTRACT

Management of electronic garbage (e-waste) is presently a major problem, particularly in emerging economies where poor waste disposal results in financial loss as well as detrimental effects on the environment and public health. Improper handling and processing practices of e-waste are related with numerous environmental and healthiness issues. The purpose of this study is to elaborate the different programmes that can bring awareness to the people for household e-waste, environmental related issues, inappropriate disposal, and eagerness to contribute to e-waste management. The study revealed useful e-waste programme and recommendation run by government for educational interaction and increasing public awareness of e-waste. A practical strategy to managing e-waste is also presented in the article. This chapter can assist legislators and government organizations of India and other developing countries to decide awareness programme and citizen responsibility for e-waste management.

INTRODUCTION

Electronic garbage, or waste generated by outdated electronic devices like computers

DOI: 10.4018/978-1-6684-7573-7.ch002

and cell phones, is one of the fastest-growing waste sources in the world today. By 2020, it is anticipated that the annual global generation of e-waste would exceed 50 million tonnes. With an estimated yearly production of 2 million tonnes, India is one of the top five producers of e-waste worldwide. According to estimates, more than 90% of the garbage is handled in the informal sector, which dominates e-waste treatment in India and some other developing nations. E-waste (EW) included metals, wood, glasses, and plastic. Unscientific methods used in the e-waste processing are linked to a number of environmental and health problems. Figure 1 show the electronics devices which convert into e-waste after passing their decided lifespan.

Figure 1. Raw material for E-waste

Different researchers proposed different model and theory to explore problem causes by E-waste. Joseph. K (2007) raise critical issue such as handling of toxic material, recycling of economically valuable e-waste and materials, inadequate legislation, and lack of awareness for e-waste. Borthakur et al. (2017) explored two important issues, Disposal of e-waste and awareness for e-waste. For this they have performed extensive review of disposal and handling behaviour adopted for e-waste at global level. Significant differences were noted between the developed and developing countries for e-waste disposal behaviour. India's E-waste disposal management system was complex due to its multifaceted socio-economic & cultural differences. Almulhim I. Abdulaziz (2022) evaluate level of household awareness of e-waste by snowball sampling method. An online survey was conducted based on pre decided questionaries to 523 people. Question is based on disposal of e-waste, handling and effect of e-waste on environment, answer to theses question recorded for conclusion drawing. One-way ANOVA and t-tests was used to analyse the recorded data. Based on statistical analysis it can be concluded that 70.1% people had no idea how serious problem was created was e-waste on environment & 88.35% were interested to take part in e-waste management. All the respondent discarded e-waste with household waste due to lack of recycling option. Sharma et al. (2021) examined the issues and challenges of electronic waste urban mining (EWUM) in India. Multi-Criteria-Decision making methods (MCDM) was adopted to perform this study. For deciding issue priority & possible solution Stepwise Weight Assessment Ratio Analysis (SWARA) & Weighted Assessment Sum Product Assessment (WASPAS) methods was selected. A deeper comprehension of the problems would aid decision-makers in putting best practises into action to improve India's urban mining process. Islam et al. (2021) conducted a questionary based studied on young consumer for awareness & disposal of electrical and electronic equipment. statistical correlation between different variables like reasons for disposing of equipment, storage & disposal behaviour. They were shown that the majority of respondents had a propensity towards ethical e-waste disposal, highlighting the necessity of awareness campaigns to prevent the improper disposal of electrical & electronic e-waste. The frequency of new product purchases and the reasons for product disposal were both positively correlated with family income. The average number of years that our respondents owned their televisions was 6.21, their laptops were 4.31, and their desktop PCs were 5.47. Sahoo et al. (2021) experimentally investigate life cycle of e-waste. Blockchain approach has been implemented to address e -waste life cycle, incentive mechanism, access control, payment channel associated with disposal of e-waste. Janowski et al. (2018) studied paradigms of citizen & administration relationships, to implement e-waste rule from bureaucrats to citizens. This studied focussed on creating conceptual framework between citizen-administration to implement e-waste law. Pandey et al. (2018) examined e-waste generation patterns & green practices adopted at the household level. They conducted detail survey and discussion on nine waste material paper, plastic, glass, metal, textile, kitchen waste, garden waste, e-waste, and appliances. Reuse and recycle practices had been documented for Bhopal city. They conclude that there was a difference general practices used by various income group for reuse and recycle of e -waste. Ravindra et al. (2019) explore e-waste management practices used in Chandigarh. For this purpose, physical survey and questionnaire has been conducted to 300 household. The result of the questionnaire suggested that only 30% of the questionnaire participant familiar with E-waste & its hazards effect on environment. Only 10% of respondents know e-waste collection location in the city. The average household e-waste in Chandigarh generates was 17 kg yearly resulted an of 4100 t annual generation of E-waste. Informal handling of e-waste in Chandigarh exposes worker to face hazardous chemical and causes harmful health effect. Widmer et al. (2005) give an outlook on e-waste as a global problem. E-waste is responsible for toxic material. They discuss Basel Convention and the hurdle in its implementation. Designated e-waste as hazardous and provided a framework for limitations on the transboundary movement of such trash due to the high toxicity of these pollutants, especially when burnt or recycled in uncontrolled circumstances. The Basel Ban would go a step further by outlawing the export of e-waste from developed to industrializing nations. It is an amendment to the Basel Convention that has not yet entered into force. Danish et al. (2023) explore the influence of e-waste aggregates on the properties of concrete, including workability, fresh and dry density, mechanical properties, and thermal properties. This study deals with percentages of e-waste aggregates that can be incorporated as natural coarse aggregates to prepare sustainable concrete for structural and non-structural purposes. They concluded that e-waste with modified concrete offer a glow of hope for the safe and sound disposal of a growing quantity of e-waste. Ashokan et al. performed the model analysis of impacts of electric vehicle (EV) deployment in India and its policy to promote EVs. They projected two-wheeler EVs achieve its peak in 2030 and lead-acid battery-based waste achieve its peak in 2027.To handle e-waste they suggest extended producer responsibility (EPR) system.

The objective of present study:

1. To explore the e-waste handling norms for common people.
2. Identify policy issue and its practical implementation problem.
3. To suggest public education and awareness programme for handling and management of E-waste.
4. To suggest steps to promote recycling and reuse of e-waste.]

GLOBAL E-WASTE STATISTICS

Electric and electronic products, which include a wide range of items used in everyday life, are a crucial component that contribute to world growth. Across the world, they may be found in homes and businesses. However, ownership rates differ according to household income. E-waste production in 2019 was estimated at 53.6 million metric tonnes (Mt) (excluding PV panels), or 7.3 kg per person. In 2030, more than 74Mt of e-waste is anticipated to be produced. As a result, the amount of e-waste generated globally is rising at a startling pace of about 2 Mt years. Compared to the amount of e-waste created in 2019, the formal documented collection and recycling in 2019 total 9.3 Mt, or 17.4%. Since 2014, it has increased by 1.8 Mt, or approximately 0.4 Mt year. However, the output of electronic trash as a whole grew by 9.2 Mt, growing by over 2 Mt annually. This demonstrates that recycling efforts are not keeping up with the rise of e-waste globally. Based on information submitted by nations, statistics on e-waste collection and recycling are compiled. The most recent data on the amount of electronic trash that has been formally collected and recycled globally generally dates to 2016. The vast majority of the e-waste produced in 2019 (82.6%) was probably not properly collected and not handled in an environmentally responsible way. Most of the time, the flows are not systematically or consistently documented. The absence of information on e-waste that is properly collected and recycled suggests that the majority of the 44.3 Mt of e-waste produced in 2019 is managed outside of the official collection system, and in some cases is exported to underdeveloped nations. Small-size electronics can be disposed of with municipal solid waste in residences in better income nations by ending up in regular trash bins. As a result, it is not properly recycled, which results in a loss of materials. According to estimates, 0.6 Mt of e-waste from EU nations ends up in trash cans (2016). Small equipment (17.4 Mt), large equipment (13.1 Mt), and temperature exchange equipment make up the bulk of the world's e-waste in 2019. (10.8 Mt). Less e-waste was produced in 2019 from screens and monitors, small IT and communications equipment, and lamps (6.7 Mt, 4.7 Mt, and 0.9 Mt, respectively).Since 2014, the largest increases in e-waste weight have been seen in the Temperature exchange equipment (with an average yearly increase of 7%), Large equipment, and other categories. (+5%), and small equipment and lamps (+4%).The rising use of these items in lower-income nations, where they raise living standards, is what is driving this trend. Smaller pieces of IT and telecom equipment have been expanding more slowly, while screens and monitors have actually decreased (-1%). This drop can be attributed to the recent replacement of bulky CRT monitors and screens with lighter flat panel displays, which has reduced overall weight even as the number of parts has increased. In 2019, Asia generated the majority of e-waste (24.9 Mt), whereas Europe produced the most kg of e-waste per person (16.2 kg per capita). Additionally, Europe has the greatest known formal e-waste collection and recycling rate (42.5%) of any continent. In all other continents, the e-waste documented as formally collected and recycled is much lower than the estimated e-waste created. According to recent data, Asia came in second place globally in 2019 with 11.7%, followed by the Americas and Oceania with 9.4% and 8.8%, respectively, and Africa with 0.9%. However, because consumption and disposal habits rely on a variety of factors (such as income levels, policies in existence, the design of the waste management system, etc.), data can range significantly across various geographic areas. A national e-waste policy, law, or regulation applied to

Figure 2. statistics of different E-waste by different equipments

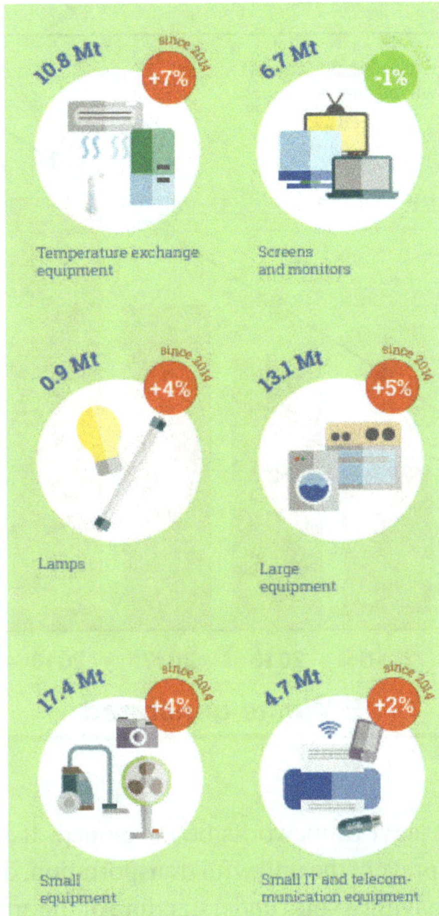

71% of the world's population as of October 2019. Since 2014, when just 44% of the population was covered, advancements have been achieved. The fact that the most populous nations, including China and India, have national legislative frameworks in place has an impact on the high coverage rate. This population coverage, however, only covers 78 of the 193 nations. As a result, less than half of all nations in the world at the moment are governed by a policy, law, or regulation (2020).

MEASURING E-WASTE STATISTICS

For defining and evaluating goals, tracking the amounts and flows of e-waste is crucial for monitoring changes over time. Only with greater e-waste statistics can

Figure 3. E-waste generated by year in kg per capita and metric ton

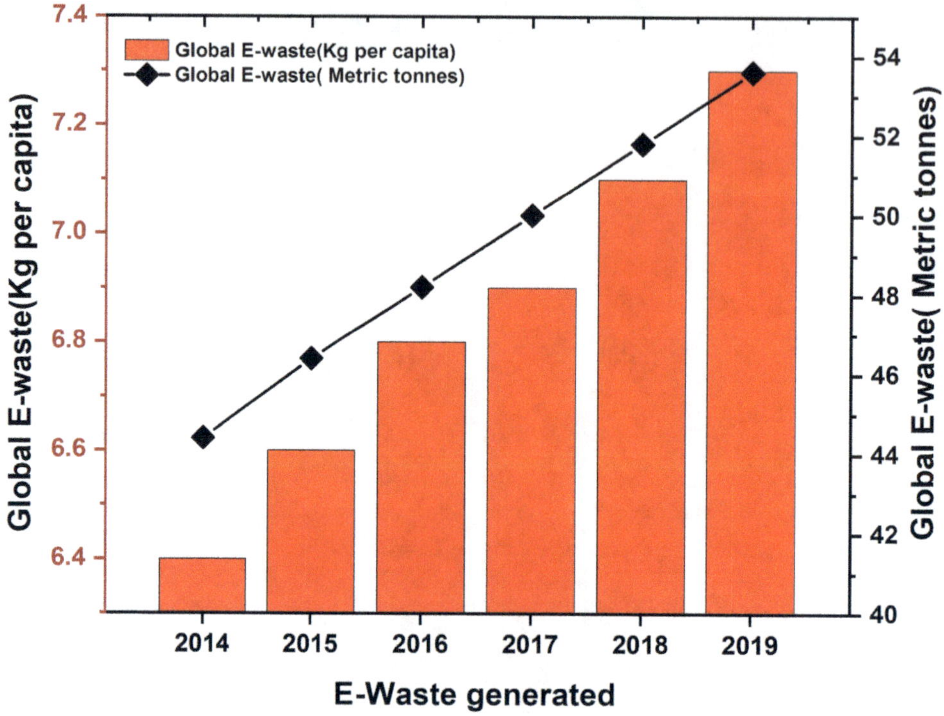

reasonable policies and legal frameworks be developed. It is possible to monitor, manage, and eventually prevent the unlawful transportation, dumping, and incorrect treatment of e-waste by having a solid understanding of the amounts and movements of e-waste. Policymakers at the national, regional, and international levels won't be able to solve these problems without any measurement of informal e-waste activities or transboundary flows.

INFLUENCE OF E-WASTE ON HUMAN

In improper e-waste recycling facilities, kids live there, work there, and play there. Children and adults may be exposed by ingestion of tainted food and drink, inhalation of poisonous gases and particle matter, skin contact with corrosive substances, and other means. Children are also at danger from other exposure pathways. During pregnancy and breast-feeding, several dangerous substances can be transferred from mothers to their offspring. Children who are playing outside or in nature run the danger of exposure since they routinely put their hands, items, and soil in their mouths. Due to their physiology, behaviour, and extra exposure pathways, foetuses, babies, children, and teenagers are more susceptible to harm from exposure to toxicants in e-waste. Studies have shown links between exposure to informal e-waste recycling and unfavourable foetal outcomes, including stillbirth, preterm birth, low gestational age, low birth weight and length, and low APGAR scores. They have also shown links between exposure to informal e-waste recycling and growth patterns, neurodevelopment, unfavourable learning and behavioural outcomes, immune system function, and lung function. Numerous research has looked at how exposure to e-waste affects a child's thyroid function, but the results have been erratic. A few studies have also shown that exposure to informal e-waste management may be linked to DNA damage, changes in gene expression, cardiovascular regulation alterations, premature start of blood coagulation, hearing loss, and olfactory memory. Additionally, tension, migraines, shortness of breath, chest discomfort, weakness, and dizziness have been observed by e-waste employees. DNA damage has been linked to exposure to chemicals in e-waste among adults working in informal e-waste management or residing in e-waste settlements. A small number of studies have also noted negative consequences from exposure to unregulated e-waste recycling on sperm quality, male reproductive and genital diseases, fasting blood glucose levels, and liver function. Over the past ten years, there has been a significant surge in study on the effects of e-waste recycling on human health.

Health Hazard of E -Waste

E-waste is a mixture of polymers and chemicals, thus poor treatment of it harms both the environment and people. India primarily relies on the unorganised sector for e-waste recycling because there are so few established e-waste recycling facilities. In the bulk of the country's urban slums, where untrained personnel perform risky processes without personal protective equipment, endangering not only their health but also the environment, over 95% of the country's e-waste is handled and processed. E-waste is manually disassembled in underdeveloped nations, releasing persistent chemicals and nondegradable polymers into the environment, lowering the quality of the air, water, and land.

Here are some instances of such primitive tactics worth mentioning:

1. physical dismantling to separate various materials with hammers, chisels, screwdrivers, and bare hands.
2. Components from printed circuit boards are removed using heat from coal-fired grills.
3. Open-pit acid baths are used to strip metals to recover gold and other metals.
4. Without adequate airflow, plastics might chip and melt.

5. Burning undesirable things outside and igniting wires to recover copper.
6. Dumping salvageable goods on riverbanks and fields.

Health Risks of Lead Exposure

One of the most widely utilised heavy metals is lead; it is used in computer and television screens as well as in the solder that holds different circuit board components in place. High amounts of lead exposure for a brief period of time might result in nausea, vomiting, diarrhoea, convulsions, coma, or even death. Lead mostly damages the neurological system, kidneys, and brain in the human body. Once you've been exposed to lead, it may linger in your body for years, either in your bones or in the blood. Due to greater absorption, children are more vulnerable to lead contamination at even lower levels of exposure. Children's hearing, stature, and conduct are just a few of the negative effects that have been seen. Lead can also pass through the placenta during pregnancy and harm the developing foetus. According to studies, female workers in the unorganised sector who are exposed to high levels of lead experience higher miscarriages and stillbirths.

E-WASTE MANAGEMENT: ISSUES & POLICY

Insufficient Data on EW Creation Rates

It can be difficult to create inventories at the state levels since the sales information for electronic devices, a crucial contribution in estimating the amount of EW, is frequently accessible at the countrywide level of accumulation. E-waste is not only produced domestically but is also frequently illegally imported from rich economies. The kind and volume of electronic garbage coming into the nation are poorly understood. Knowledge about trash formation, composition, and flows must be relatively precise in order to design systems for efficient collection, transportation, and processing.

Unsustainable Informal Sector Activities in Terms of The Environment

Despite the formal sector's expansion (EW facilities expansion), the amount of garbage that is handled there is still very low. Anecdotal data suggests that the majority of these official facilities are running well below their allowed capabilities due to a lack of trash supply. The unwillingness of household and institutional customers to return their garbage to the formal sector is being reduced by their lack of knowledge about e-waste and the expenses associated with bringing end-of-life equipment to official collection facilities. In contrast to the formal sector, which has not yet made significant investments in reliable systems of collection and processing. The informal sector was more attracting for household consumer to collect e waste from them. Millions of individuals, many of them come from the most disadvantaged groups, find work in the informal e-waste industry. On the other hand, the industry's waste management procedures pose major environmental and health risks to both the workers and the public. Any system for managing e-waste will depend on our capacity to address this moral conundrum in order to succeed over the long term.

Market Conflict for Items at The End of Their Useful Lives

The capacity of private companies, like PROs, to establish e-waste managing structures in the formal area is constrained by the difficulty to consistently procure e-waste amounts that produce economies of scale. In the lack of assurance over the availability of sufficient quantities of e-waste, implementing active recycling methods for e-waste may necessitate large upfront capital investments that may not be acceptable for private businesses. These marketplaces are also hampered by information obstacles. First, because e-waste recycling is a young industry, there is a chance that there will be a lack of knowledge about efficient recycling methods. Second, the operation of markets is impacted by poor knowledge, which is partially caused by customers' lack of access to trustworthy information on e-waste management.

Improve Regulatory Design and Its Execution

The compulsory take-back scheme in the 2012 rules. Without corresponding collection objectives, producers were not motivated to accept responsibility, which led to minimal progress in e-waste management methods. The 2016 revisions addressed this by defining progressively tighter collection objectives, which increased regulatory certainty. However, the regulatory structure adds significantly to the already inadequate workload of the regulatory authorities.

Making a Robust Enterprise Resource Management System

The government may need to play a facilitative role in bringing together diverse system stakeholders by continuously assessing the efficacy of e-waste legislation and enacting appropriate regulatory reforms. We list a few actions that ought to be taken into account in order to proceed.

Informal sector

This sector plays a crucial role in e-waste regime. The acknowledgement of informal sector can be also done by the Municipal Solid Waste Rules in 2016). A deeper knowledge of the industry's motivations and difficulties is necessary to address the issue of informal sector e-waste practises. A crucial first step is to engage with the groups and workers in the informal sector in a way that respects their right to a living, fosters trust, and creates a shared knowledge of the issues and potential solutions. The government should establish a platform that allows for dialogues between different parties, including employees in the informal sector, NGOs that engage with the sector, and outside private parties.

Policy Instruments Under EPR (Extended Producer Responsibility)

To adopt EPR strategy it is required to reconsider government policy. A forced take return with collection objectives may not be the best tool if there is an unorganised sector with expertise in logistics for collection. Other than required takeback, there are several more ways that producers may be held accountable. Producers would no longer be physically responsible for collection thanks to economic tools like advanced recycling fees (ARF) or advanced disposal fees (ADF) on every unit of the product sold in the market. The money raised could then be utilised to create markets for end-of-life items. The receipts, which are placed in a separate fund, may be applied in a number of ways.

E-Waste Imports

Imports of e-waste are permitted for reuse and recycling but not for final disposal, according to the current legislation. We should seriously explore an import restriction of all types, similar to what China just did, given the lack of suitable recycling facilities in the nation. Integrating import data with the e-waste inventory is necessary to provide precise estimates of e-waste.

Public Education

In accordance with the existing e-waste legislation, manufacturers must post information about the effects of e-waste, acceptable disposal methods, and other relevant topics on its website. Additionally, they must periodically conduct awareness programmes. Despite the fact that many companies have already posted information on their websites, statistics reveal that even among large numbers of customers, awareness levels are still low. The issue may be improved by giving manufacturers stricter instructions or rules on how frequently and how they should conduct these awareness initiatives. As an alternative, the makers should be required to implement these programmes through local e-waste groups. Combining municipal solid waste and e-waste handling and disposal was one of the possible solutions for awareness of the people. The government might also have a role in evaluating information campaigns and doing research on message strategies. These awareness campaigns have to focus on both minimising long-term usage of electronic items and ensuring safe treatment of e-waste. In general, partnerships and collaboration amongst many stakeholders should be the foundation of public awareness creation projects.

CONCLUSION

Developing nations like India faces a serious environmental dilemma as a result of the growth in electronic devices over the past ten years and the associated rapid increase in e-waste. India's seven-year-old legislation haven't made much of an impact, which highlights the difficulties the nation confronts in managing its e-waste. According to this article, some of India's problems include informal sector e-waste activities, subpar regulation design and implementation, and inadequate awareness. To create a strong e-waste management system for the future, meaningful participation of all stakeholders should be prioritised.

REFERENCES

Almulhim, A. I. (2022). Household's awareness and participation in sustainable electronic waste management practices in Saudi Arabia. *Ain Shams Engineering Journal, 13*(4), 101729. doi:10.1016/j.asej.2022.101729

Asokan, V. A., Teah, H. Y., Kawazu, E., & Hotta, Y. (2023). Ambitious EV policy expedites the e-waste and socio-environmental impacts in India. *Resources, Conservation and Recycling, 190*, 106829. doi:10.1016/j.resconrec.2022.106829

Borthakur, A., & Govind, M. (2017). Emerging trends in consumers' E-waste disposal behaviour and awareness: A worldwide overview with special focus on India. *Resources, Conservation and Recycling, 117*, 102–113. doi:10.1016/j.resconrec.2016.11.011

Danish, A., Mosaberpanah, M. A., Ozbakkaloglu, T., Salim, M. U., Khurshid, K., Bayram, M., Amran, M., Fediuk, R., & Qader, D. N. (2023). A compendious review on the influence of e-waste aggregates on the properties of concrete. *Case Studies in Construction Materials, 18*, e01740. doi:10.1016/j.cscm.2022.e01740

Forti, V., Balde, C. P., Kuehr, R., & Bel, G. (2020). The Global E-waste Monitor 2020: Quantities, flows and the circular economy potential.

Islam, M. T., Dias, P., & Huda, N. (2021). Young consumers' e-waste awareness, consumption, disposal, and recycling behavior: A case study of university students in Sydney, Australia. *Journal of Cleaner Production, 282*, 124490. doi:10.1016/j.jclepro.2020.124490

Janowski, T., Estevez, E., & Baguma, R. (2018). Platform governance for sustainable development: Reshaping citizen-administration relationships in the digital age. *Government Information Quarterly, 35*(4), S1–S16. doi:10.1016/j.giq.2018.09.002

Joseph, K. (2007). Electronic waste management in India–issues and strategies. *Eleventh international waste management and landfill symposium,* Sardinia.

Pandey, R. U., Surjan, A., & Kapshe, M. (2018). Exploring linkages between sustainable consumption and prevailing green practices in reuse and recycling of household waste: Case of Bhopal city in India. *Journal of Cleaner Production, 173*, 49–59. doi:10.1016/j.jclepro.2017.03.227

Ravindra, K., & Mor, S. (2019). E-waste generation and management practices in Chandigarh, India and economic evaluation for sustainable recycling. *Journal of Cleaner Production, 221*, 286–294. doi:10.1016/j.jclepro.2019.02.158

Rotter VS, Maehlitz P, Korf N, Chancerel P, Huisman J, Habib H, Herreras L, Ljunggren SM, and Hallberg A. (2016). ProSUM Deliverable 4.1. *Waste Flow Studies*. 1–100.

Sahoo, S., Mukherjee, A., & Halder, R. (2021). A unified blockchain-based platform for global e-waste management. *International Journal of Web Information Systems*.

Sharma, M., Joshi, S., & Govindan, K. (2021). Issues and solutions of electronic waste urban mining for circular economy transition: An Indian context. *Journal of Environmental Management, 290*, 112373. doi:10.1016/j.jenvman.2021.112373 PMID:33932756

Widmer, R., Oswald-Krapf, H., Sinha-Khetriwal, D., Schnellmann, M., & Böni, H. (2005). Global perspectives on e-waste. *Environmental Impact Assessment Review, 25*(5), 436–458. doi:10.1016/j.eiar.2005.04.001

Chapter 3
Study on the Source of E-Waste Management and Disposal Methods

Harikaran M.
https://orcid.org/0000-0003-2229-7127
Muthayammal Engineering College, India

Sampath Boopathi
https://orcid.org/0000-0002-2065-6539
Muthayammal Engineering College, India

S. Gokulakannan
Muthayammal Engineering College, India

M. Poonguzhali
Narasu's Sarathy Institute of Technology, India

ABSTRACT

This chapter elaborates on the various e-waste production courses from multiple human activities. The impacts of various e-wastes and human diseases are clearly illustrated. The different methods of e-wastes management methods, such as Reduce, Reuse, Recycle, and e-waste recycling practices in India are also explained. The characteristics of e-waste, the quantity of e-waste, the nature of e-waste, the composition of e-waste, non-hazardous wastes, and hazardous wastes have been discussed. As per the deep survey of e-waste management systems, the scope and proposed e-waste management methods have been driven.

DOI: 10.4018/978-1-6684-7573-7.ch003

INTRODUCTION

Rubbish management is the process of disposing of waste. It includes garbage collection, transportation, processing, and disposal. Waste avoidance, recycling, reuse, and recovery are critical waste management practices that reduce landfill pressure, protect natural resources, and save energy. This promotes more efficient and sustainable resource utilization. The words "e-waste" or "e-waste" refer to unwanted electrical or electronic devices. E-waste is used electronics that are intended for recovery, resale, salvage, recycling, or destruction. Electronic trash is a major problem that endangers the environment and the public. Electronic garbage disposal is a routine task everywhere. Recycling e-waste is a component of the solution to ecological and environmental problems (Ahirwar & Tripathi 2021). Because concrete is utilized in construction projects so frequently, the supply of raw materials is being called into question. It is vital to look into alternative substitute materials as a result. E-waste is one such option for concrete's aggregate replacement. Based on the findings of several studies, this study presents an overview of the potential use of electronic waste in concrete and emphasizes the likelihood that this waste will be employed as extra aggregate and have a negative impact on the environment. The number of natural aggregates used in concrete will be decreased if various byproducts are used as a substitute material (Kirthika et al., 2020). Waste recycling is also more important.

Technology has developed since the beginning of time. Many gadgets that play a key role in our day-to-day lives have been created by humans, who have achieved remarkable progress in this area. Stones were carved into knives and spears from the rock. The absence of certain tools makes life appear depressing. This type of technological advancement has considerably increased in the 20th and 21st centuries. The standard of living has increased as a result of these technological innovations. Today, a major concern is how to handle the enormous volumes of garbage that are produced because of the globalization of machines.

Anatomical waste consisting of human and animal body parts, leftover medication, sharp trash, microbiological and biotechnological waste, and garbage from bioprocessing are all produced and disposed of by stockholders. Wastes can be created in a variety of ways, including liquid, solid, organic, toxic, and recyclable materials. This research focuses on potentially hazardous electronic wastes that could seriously impact the environment and public health if they were disposed of outside. To reduce environmental pollution and protect public health, environmental sustainability will be put into effect (Elleuch et al., 2018). Numerous industrialized countries contributed greatly to the severe environmental degradation that also harmed developing countries as their economies expanded swiftly. All of these countries are currently dealing with problems like energy and resource scarcity, environmental deterioration, a lack of infrastructure development, unemployment, and health hazards for their citizens. The volume of waste has increased as a result of economic and social development, especially technological waste. Hazardous waste materials are combustible, corrosive, reactive, toxic, toxic, volatile, poisonous, and infectious, which cause long-term harm to the environment. There is still pressure on India to use open landfills and open dumps to dispose of its municipal solid waste. Hazardous waste materials are combustible, corrosive, reactive, toxic, toxic, volatile, poisonous, and infectious, which cause long-term harm to the environment (Needhidasan et al., 2014). Pressure to dispose of municipal solid trash in open dumps and landfills still exists in India. All sorts of waste electrical and electronic equipment (WEEE) are referred to as "e-waste" since their components were abandoned as waste by the producers with no thought given to possible reuse. WEEE is the waste stream with the fastest global growth. Total production of e-waste was over 41.5 million tons in 2011 and 93.5 million tons in 2016. One of the primary industries is the electronic sector, and e-waste is presently the waste stream that is evolving the fastest. Older household appliances such TVs, refrigerators, air conditioning and heating units, VCRs, Audio players, and other electrical and electronic items are included in this waste stream (PDAs).

Despite the dangers associated with e-waste, some or all of its components may be used to replace common supplies and components needed in the booming hardware industry and other industrial sectors. Because the countries that create the most e-waste have low levels of wealth, e-waste reuse and transfer can aid in the efficient use of resources. Only the most valuable and carefully separated components, such gold plating, aluminum, and copper wire, may be recovered through the controlled reuse and transfer of e-waste. Platinum, cadmium, and other excess materials must be carefully disposed of to avoid significant environmental damage (Ankit et al. 2021). Furthermore, developed countries trade e-waste with developing countries under the guise of recycling, reusing, or gifting. These exports are making e-waste problems in developing nations worse. E-waste management has become a major problem in developing nations. The e-waste administration structure is creating problems as opposed to the intricate collection, transportation, organization, destruction, and reuse processes. Japan is one such example of a developing country. Some governments have devised and drafted a few laws and procedures to regulate e-waste in order to solve the issue of e-waste and to follow the principles of industrialized nations, but their approval is still insufficient.

One of the key drivers of better E-waste management in the country is public awareness of pollution and health hazards. India has a bad knowledge of electronics waste, thus both the domestic and foreign markets must be contested. It is anticipated that a large amount of time would be needed for the envisaged laws for e-waste regulation and management to be properly implemented (Turaga et al., 2019). The industrial, commercial, and home sectors produce a staggering quantity of waste nowadays. The escalating cost of living and population expansion make this problem worse. Products become waste if they are not recycled or used again after they have completed their intended function. E-product trash consists of packaging materials and waste-producing products such as TVs, phones, refrigerators, and auto parts. Over the past two decades, India has created almost 8 lakh tons of e-waste yearly. In 2010, there were created 400 million tons of electronic garbage on average. According to the Central Pollution Control Board (CPCB), India produces 1.47 lakh tons, or 0.573 MT, of electronic waste daily. The handling of e-waste is a major issue in developing nations, particularly India. India would have created 8 lakh tons of e-waste by the end of 2012, up from 2 lakh tons in 2007. Consequently, India's ecology is under a lot of stress. The ecology and all living things may be negatively impacted by the burning and open dumping of this trash.

By the year 2025, the roughly 1.3 billion tons of waste produced annually on a global scale are expected to increase to 4.3 billion tons. Electronic waste, sometimes known as "E-waste," is produced on a global scale at a rate that is three times faster than other types of waste. It is essential to create effective recycling strategies for its decrease because it is one of the main causes of environmental contamination. According to the United Nations Environment Program (UNEP) 2010, the generation of e-waste is predicted to rise by 500% from outmoded PCs and by approximately 18 times from abandoned mobile phones by 2020 compared to the year 2007. The electronic industry is one of the biggest and fastest-growing manufacturing sectors in the world. E-waste now poses an alarming amount of environmental risk due to the nation's rising use of electronics and IT items. Electronic waste, also known as E-waste, e-scrap, or waste electrical electronic equipment (WEEE), refers to all used computers, entertainment systems, electronics, mobile phones, and other products like televisions, refrigerators, etc. (whether sold, donated, or discarded by their original owners). For the purposes of reuse, resale, salvage, recycling, or disposal, this definition includes surplus electronics that are regularly combined with loads of used electronics (good, recyclable, and non- recyclable).

Globally, e-waste has grown to be a significant problem. Electronics are subject to quick technological change, and due to their low initial cost and propensity for obsolescence, they pose a growing environmental risk. Due to environmental concerns related to WEEE management, several nations and organizations have adopted national legislation to promote reuse, recycling, and other types of recovery. WEEE recycling is a critical issue from the standpoint of waste management and the recovery of valuable commodities (Mohabuth & Miles 2005). WEEE is intricate and varied. To developing a recycling system that is both affordable and environmentally responsible, the waste stream's categorization is essential. Older PCs and other electronic gadgets produce millions of tons of electronic garbage each year.

Since 2008, the Ministry of Environment and Forest and Climate Change (MOEF and CC) handbook has addressed the numerous sorts (more than 1000 variants) of chemicals and compounds found in e-waste that represent a substantial risk to human health and the environment. Several dangerous materials, including heavy metals including lead, cadmium, mercury, arsenic, selenium, and hexavalent chromium, are present in e-waste. Mercury and cadmium are both present in electronic trash, and they combined make up around 70% of the heavy metals found in landfills (USEPA, 2000). E-waste has been linked to cancer, allergies, and brain impairment. E-waste is also found to include sizeable amounts of valuable elements, including gold, copper, and other common metals (Cherukuri et al., 2018; Turaga et al., 2019).

Arsenic, barium, copper, chrome, cobalt, lead, zinc, nickel, mercury, and silver are the primary contaminants in electrical waste. These contaminants have the capacity to significantly harm the ecosystem. Significant levels of lead, cadmium, and various poisonous gases are produced when waste is burned or dumped in an open area, and the same is true when waste is dumped in an open area. These gases can injure plants, pollute the environment, and cause lung cancer as well as eye and skin disorders. During the burning process, melted granular E-waste and ash deposition are created. This pollutant is the principal one that affects environmental ecology. The amount of waste disposed and its impact will be decreased by proper recycling and reuse of the leftovers.

Recycled e-waste could be used in the construction industry as a solution to manage e-waste while simultaneously improving building quality and protecting the environment.

It is essential to have a thorough grasp of human well-being and widespread civil society participation for sustainable development to offer practical answers (OECD, 2001). In both rich and emerging nations, the issue of how to balance socioeconomic growth with environmental conservation has become critical. It is possible to meet current demands by integrating the economy, society, and environment without jeopardizing the wellbeing of coming generations. For long-term growth, the environment's sustainability is essential. Environmental sustainability is put into practice to reduce pollution and protect public health. Many industrialized countries across the world significantly affected the environment in their early phases of growth, which also occurred in emerging nations. All countries today face difficulties such resource and energy restrictions, environmental degradation, a lack of infrastructure development, unemployment, and health hazards because of their rapid economic progress. The continual introduction of additional trash categories and unregulated E-waste recycling have made the challenges associated with solid waste management even more onerous.

E-waste contains a range of precious elements that may be recycled, giving them a higher economic worth. As a result, there may be significant financial benefits to recycling and disposing of E-waste in an ecologically responsible way. The governments of developed countries, including our own, have recognized the gravity and urgency of the e-waste problem and have been working to find a solution. Many nations have enacted national e-waste regulations to control producers' and importers' obligations to gather and discard outdated electrical and electronic equipment (Clean Japan Centre, 2002). The illegal recycling and disposal of e-waste has had a substantial negative impact on both the environment and the health of employees and residents in developing countries. Careless e-waste exports from wealthy nations and an increase in local e-waste have made managing e-waste in poor countries increasingly challenging. Due to inadequate management in poorer nations, controlling e-waste has proven difficult.

The recycled electronic debris is substituted for coarse aggregate in a range of percentages (from 0 to 20%) and sizes (10 to 12.5 mm) (crushed). It was determined what concrete's mechanical properties and chemical behavior were. Through an E-waste reaction in the concrete under different types of chemical curing, the functionality and durability of concrete are also discovered. After e-waste was substituted in high performance reinforced concrete, the behavior, performance, and durability of RC beams under cyclic loading conditions were evaluated (Ullah et al., 2022). The last part of this experiment examines the structural properties of fresh concrete, hard concrete, and high-performance RC beams created from E-waste. Additionally, it considers whether non-melted crushed E-waste can replace coarse aggregate in concrete.

GENERATION OF E-WASTE IN INDIA

The amount of electrical and electronic garbage produced worldwide each year—primarily from computers and televisions—has reached worrisome levels. WEEE, or "e-waste," will be produced by 3 billion electronic and electrical items, according to a 2006 prediction by the International Association of Electronics Recyclers (IAER). By 2010, the annual production of electronic waste will have reached 400 million units. Around 20 to 50 MT (million tons) of electronic garbage are disposed of annually globally, which is 5% of all municipal solid waste(Turaga et al., 2019).

There are estimates based on independent research conducted by NGOs or government organizations on how much garbage is produced and disposed of in India, even though there are no exact official numbers available. According to the report of the Comptroller and Auditor General, the country generates approximately 7.2 metric tons of industrial hazardous waste per year, as well as 4 lakh tons of electronic garbage, 1.5 metric tons of plastic trash, 1.7 metric tons of medical waste, and 48 metric tons of municipal waste. According to the Central Pollution Control

Board (CPCB), India generated 0.573 metric tons of electronic waste per day, or 1.47 lakh metric tons, in 2005. 4.34 lakh metric tons of e-waste had been created in India altogether by the end of 2009, per research conducted by the Electronics Industry Association of India and presented at the electronics industry exhibition "Componex Nepcon 2009." (ELCINA). By 2012, the CPCB predicts it will be more than 8 lakh tons, or 0.8 MT (Ullah et al., 2022).

Ten states contribute 70% of the nation's total e-waste, while 65 locations are responsible for more than 60% of the nation's e-waste. Delhi, Karnataka, Gujarat, Madhya Pradesh, and Punjab are the top 10 states for manufacturing e-waste, with Maharashtra coming in at #1. Tamil Nadu, Andhra Pradesh, Uttar Pradesh, West Bengal, and West Bengal round out the top 10 states. Mumbai, Delhi, Bengaluru, Chennai, Kolkata, Ahmedabad, Hyderabad, Pune, Surat, and Nagpur are the top 10 cities in terms of e-waste production.

The main producers of electronic trash in India are the public, commercial, and industrial sectors, which together account for roughly 70% of the country's total rubbish production. Individual residences contribute around 15% of the total, with manufacturing making up the remaining 85%. Although individual residences don't contribute much to the trash created by computers, they do consume a lot of consumer durables and might be viewed as potential waste generators. Televisions and computers with servers accounted for 68% and 27%, respectively, of India's total volume of e-waste, according to a 2009 survey by the Indian Market Research Bureau (IMRB) on "e-waste production at source." 1% of the total was made up of mobile phones, while 2% was made up of imports.

In January 2010, the Attero Recycling Plant in Roorkee started operating as a highly structured e-waste recycling facility. The majority of the recycling process still happens in the unorganized sector, despite the fact that 23 units are currently registered with the Government of India's Ministry of Environment and Forests/Central Pollution Control Board as e-waste recyclers or processors and have environmentally sound management facilities. The Cobalt-60 radiation event in Delhi, in Mayapuri, where one person died and six others were hospitalized, served as a wake-up call by highlighting systemic problems with waste management and drawing attention to the nation's growing volume of hazardous garbage, notably e-waste. The Hazardous Wastes (management, handling, and transboundary movement) Rules, 2008, have been released by the Ministry of Environment and Forests (MoEF) for efficient management of hazardous wastes, including e-waste, across the country. Radioactive wastes like cobalt-60 are excluded from these laws under the Atomic Energy Act of 1962(Priya Jeyaraj 2021).

Table 1. Waste generations

S.No	Types of Fields	% Waste composition	Waste generated (MT/day)
1	Domestic household	36.18	1860
2	Business establishments	6.85	345
3	Hotels and dining establishments	13.14	645
4	Organizations	2.28	130
5	Gardens and parks	1.22	70
6	Sweeping the streets	6.27	330
7	Drainage waste	3.37	170
8	Markets	9.22	477
9	Temples	0.40	32
10	Stalls selling chicken, mutton, and seafood	3.46	170
11	Movie theaters	0.30	17
12	Meeting rooms	1.44	90
13	Hospitals	0.45	32
14	Building and demolition	12.42	642

SOURCE AND TYPES OF E-WASTES

Table 1 shows the overall quantity of garbage generated by various enterprises. Table 2 lists the top 10 Indian cities by e-waste creation. Table 3 displays the environmental effects of electronic trash. Table 4 illustrates the origins of e-waste and its consequences on health. Figure 1 displays the many types of e-waste.

Polymer E-Waste

The issue with solid trash that is expanding the quickest globally is electronic junk, or E-waste. It is a catch-all phrase for old electronics that are thrown away in the waste stream and come from a range of sources. The list of commodities includes, among other things, air conditioners, refrigerators, outdated and other electronic devices, parts, and accessories. It has become evident how seriously detrimental e-waste is to the environment, animals, and humans due to the rapid rise of electronics waste and the rapid depreciation of goods (Priya Jeyaraj 2021). There is no denying that less developed nations produce a sizable amount of electronic trash. E-waste production in China and the United States was 6.3 million metric tons and 7.2 million metric tons, respectively, in 2016. (Richer, 2017). The harm they do to the environment and people's health made the landfill, open dumping, and casual recycling the least

Table 2. E waste generations in top 10 cities of India

S.No	City Electronics and Electrical waste	Tonnes per year
1	Mumbai	11017.1
2	Delhi	9790.3
3	Bangalore	4648.8
4	Chennai	4132.2
5	Kolkata	4025.3
6	Ahmadabad	3287.5
7	Hyderabad	2833.5
8	Pune	2584.2
9	Surat	1836.5
10	Nagpur	1768.9

desirable alternatives. Leaching occurs when rainwater runs through garbage in a landfill, making it a dangerous option for getting rid of waste (Figure 2).

Table 3. Environmental effects of electronic waste

Root Electronic waste	Mechanisms employed	Ecology implications
Gadgets screens or monitors	Manual removal of the copper yoke and disposal	Toxic metals from displays leach into underground water sources and discharge phosphor.
microchips and other gold-plated substance	Chemical stripping with very concentrated acids in close proximity to potable water sources	Toxic gases released into the environment kill fish and plants.
Printed circuit boards	Removing solder from chips	Toxic fumes are released into the air
Computer plastics and other external computer hardware, such as printers	Cutting into pieces and low temperature dissolving	Brominated dioxins, heavy metals, and hydrocarbons in the atmosphere
Disassembled printed circuit sheets processing	Open incineration of e-waste circuit sheets	Tin and lead pollution in the surrounding surroundings
Cables	Recover copper by burning in the open	Hydrocarbons and wastes, especially PAHs, released into the atmosphere, water, and soil

Table 4. E-waste sources and their health effects

Source of e-wastes	Constituent	Health effects
Solder is used in printed circuit boards, glass panels, and computer monitor gaskets.	Lead (PB)	Central and peripheral nerve system damage, blood system damage, and kidney damage Affects children's brain development.
Chip resistors and semiconductors	Cadmium (CD)	Toxic impacts on human health are irreversible. It builds up in the kidneys and liver. Damages the nervous system. Teratogenic.
Relays and switches, printed circuit boards	Mercury (Hg)	Brain injury that is sustained over time. Fish bioaccumulation causes respiratory and skin problems.
Untreated and galvanized steel plate corrosion prevention, decorator or hardner for steel housings	Hexavalent chromium (Cr) VI	Asthmatic bronchitis. DNA damage.
Cabling and computer housing	Plastics including PVC	Dioxin is produced during combustion. It results in Reproductive and developmental issues Immune system dysfunction Interfere with hormone regulation
Electronic equipment and circuit boards are housed in plastic.	Brominated flame retardants (BFR)	Disrupts endocrine system functions
Front panel of CRTs	Barium (Ba)	Short term exposure causes: Muscle weakness Damage to heart, liver and spleen.
Motherboard	Beryllium (Be)	Carcinogenic (lung cancer) (lung cancer) Fume and dust inhalation Causes chronic beryllium illness, often known as beryllicosis. Warts are a type of skin disease.

Large home appliances made from e-waste are used in concrete. As a result, the only coarse aggregate used in the concrete was the plastic TV frame. The composition of this substance was tested and examined using SEM scans. Shredders are used in the plastics industry to crush collected electronic waste into the required sizes. The size of the E waste ranged from 10 to 12.5 mm.

E-Waste's Characteristics

The number of materials that can be recovered from e-waste can be widely classified. E-waste is made up of tens of thousands of different materials, of which steel makes up 50%, copper, aluminum, and other metals make up 13%, and plastics make up 21% [ETC/SCP]. Harmful and toxic metals make up about 1% of the overall weight,

Figure 1. Types of e-waste

Figure 2. Collection of electronic waste from local scrap yard

while the rest of the substance is made of valuable metals like gold, silver, and palladium (Ali et al. 2019).

The Quantity of E Waste

Electronic waste generates substantially more garbage than other consumer items due to its rapid obsolescence. It could make up to 5% of municipal solid trash disposed of (E Waste in India, Rajya Sabha Report, 2011)(Ali et al. 2019; Priya Jeyaraj 2021). It's vital to remember that less than 5% of the nation's total amount of electronic waste is handled by the informal sector in India, while the other 95% is sent to the formal sector for recycling. That is more than all the plastic packaging, beverage cans, and disposable diapers put together. It is hard to know with precision how much electronic garbage there is in the world and in our country, though, as millions of tons of it keep coming every day. Numerous methodologies have been proposed and implemented in industrialized nations to assess the possible global volumes of e-waste. The Dutch use the consumption and use method to estimate the potential amount of e-waste, the German Electrical and Electronic Industries Association uses the market supply method and estimates from the Swiss Environmental Agency are based on the assumption that private households are already at capacity. It is difficult to compare the findings of e-waste estimating studies since they provide such a wide range of results due to variations in the technique used and the underlying assumptions mentioned in each investigation.

In industrialized countries, when the market for electrical and electronic devices is saturated, assessing the amount of e-waste is more difficult and complex. But every year, the quality of millions of tons of electronic devices, including toys, televisions, stereos, mobile phones, and other gadgets, decreases (Franklin and Associates, 1998). In India, there are 30 million installed computers, 80% of which are desktop computers that are soon to become obsolete as low-cost laptops and tablets grow in popularity. According to study, 24 million PCs contain about 2,872,000 tons of plastic, 718,000 tons of lead, 1,363 tons of cadmium, and 287 tons of mercury (Puckett & Smith, 2012). The average PC lifespan is rapidly decreasing, yet the worldwide PC industry is still far from being fully saturated. For instance, CPU lifespans decreased from 4-6 years in 1997 to just two years in 2012. Old computers account for a very small percentage of all electronic garbage (Widmer et al., 2005).

All types of portable electronic gadgets are anticipated to have similar levels of electronic waste. Most of this e-waste will be disposed of in landfills and rubbish, with just a small amount being appropriately recycled. The United States and other industrialized countries today use most the world's technology commodities, and their electrical and electronic equipment markets are extremely saturated. According to experts, more than 500 million computers will become outdated in the United

States alone between 2003 and 2013. E-waste has seen a tremendous rise in volume as the electronics sector has developed, which is an unexpected outcome of the technological revolution. Over 3.2 million tons of electronic garbage were dumped in US landfills in 2011, according to the US Environmental Protection Agency (EPA). Californians Against Waste and the Computer Take Back Campaign, both in 2010. According to European studies, the volume of e-waste is expanding at a pace of 3% to 5% per year, which is almost three times faster than the rate at which municipal waste is created. The former 15 nations of the European Union (EU 15) produced between 3.3 and 3.6 kilograms of electronic waste per person between 2000 and 2009, and it was anticipated that starting in 2010, that quantity would rise by more than 5 kilos per person (EEA, 2011)(Streicher-Porte et al., 2005).

Nature of E Waste

E-waste consists of a wide range of products from many categories. There are more than a thousand different substances, both dangerous and benign. Ferrous and nonferrous metals, polymers, glass, printed circuit boards, ceramics, rubber, and other materials make up the majority of its composition. Over half of all e-waste is made up of iron and steel, with the remainder being made up of plastics (21%), nonferrous metals (13%), and miscellaneous materials (Cherukuri et al., 2018). Copper, aluminum, as well as priceless metals like palladium, platinum, gold, and silver, are examples of nonferrous metals. E-waste is labeled as hazardous waste when its lead, mercury, arsenic, cadmium, selenium, and hexavalent chromium contents surpass threshold values. Over the past century, the use of metals in technological development has increased to the point that guaranteeing the supply of key metals, like lead, is viewed as essential (UNEP, 2010). Electrical and electronic equipment is one of the greatest consumers of metals, and WEEE comprises a variety of precious and rare metals. WEEE is currently being given increasing consideration as a secondary source of metals. Many metal elements and organic substances can be hazardous to the environment and to people's health if they are released. Safe handling, storage, and disposal of E-waste are therefore of the utmost importance. One way to handle e-waste is to lower the concentration of these harmful chemicals and components through recycling, resource recovery, and ecologically responsible disposal (Julander et al., 2014).

Composition of E Waste

Waste electrical and electronic equipment (WEEE) is complicated and non-homogeneous in terms of its materials and components. To create a recycling system that is both affordable and environmentally friendly, it is essential to identify and quantify valuable materials, dangerous compounds, non-hazardous components, and physical properties of this waste stream. E-waste goods come in a wide range of categories and have a very diverse material composition. Iron and non-iron metals, polymers, glass, wood, plywood, printed circuit boards, concrete, ceramics, rubber, and other materials make up the bulk of it (Streicher-Porte et al., 2005). About half of all e-waste is made up of iron and steel, with plastics making up 21%, nonferrous metals 13%, and other materials making up the remaining components. Non-ferrous metals include both common metals like aluminum and copper as well as valuable metals like silver, gold, platinum, palladium, and copper. Lead, mercury, arsenic, cadmium, selenium, and hexavalent chromium threshold levels are exceeded in hazardous wastes, along with flame retardants. The effective separation of these components based on the variations in their physical features is the key to developing a mechanical recycling system.

Non-Hazardous Wastes

Heavy metals like iron, steel, copper, aluminum, gold, silver, platinum, and rubber are non-hazardous wastes that can be recycled and reused to make new machinery or for other purposes.

Hazardous Waste

Hazardous trash contains dangerous materials like lead, mercury, arsenic, cadmium, chromium, selenium, flame retardants, and acids that contribute to pollution. These can harm the environment and people's health if they are thrown away or burned because they cannot be recycled (Okunola et al., 2019). The following are some of the potential threats that e-waste poses to the environment and public health.

Lead

People's reproductive organs, circulatory systems, central and peripheral nervous systems, and kidneys are all harmed. It also stops children's brain development and harms animals, plants, and bacteria over the long run. It is used in glass panels, electronics, computer monitors, and soldering.

Mercury

Mercury is present in the majority of electronic devices. The most obvious CFL bulbs are the most recent energy-saving models. Devices that employ it include relays, switches, medical equipment, lighting, and mobile phones (Debnath et al., 2019). It is used in flat panel screens. Electrical and electronic gadgets use an estimated 22% of the mercury consumed globally each year. The brain, kidneys, and other organs of a fetus may be harmed. Methylated mercury, which is created when inorganic mercury enters the water, accumulates in the food chain, particularly through fish.

Cadmium and Arsenic

Both carcinogens, which have an effect on both soil and air, are released into the atmosphere when electronic debris is burned. Toxic cadmium compounds are produced by the human body, especially in the kidneys.

Beryllium

Berylliosis, a chronic lung illness, can be caused by inhaling beryllium or beryllium-containing dust, mist, or fume. Beryllium is also possibly carcinogenic to humans. On motherboards and finger clips, it is often utilized to make connectors stronger.

Plastics and PVC

Hazardous chemical additives (such phthalates) may leak out of PVC components of electronic items when they are dumped in a landfill. Dioxin is released when PVC is burned (Blinová & Godovčin 2021). The majority of polymers (26%) used in 16 electronics are PVC, which is utilized for computer casing and cabling. These days, a lot of computer moldings are made from safe polymers like ABS plastics.

Barium

Short-term exposure to barium weakens muscles, harms the heart, liver, and spleen, and results in brain enlargement.

Barium is a soft silvery white metal, also utilized in computers, is used in CRT front panels to protect users from radiation. Brominated flame retardants (BFRs) are used to reduce flammability in electronic device plastic housings and circuit boards.

METHODS OF DISPOSAL OF E WASTE

Electronic garbage contains lead, which, when combined with groundwater, soon renders it unsafe for direct ingestion (Gupta & Nath, 2020). There are still issues with e-waste disposal. The generated E-waste is normally disposed of using the following techniques (Figure 3):

Figure 3. Disposal of E-waste

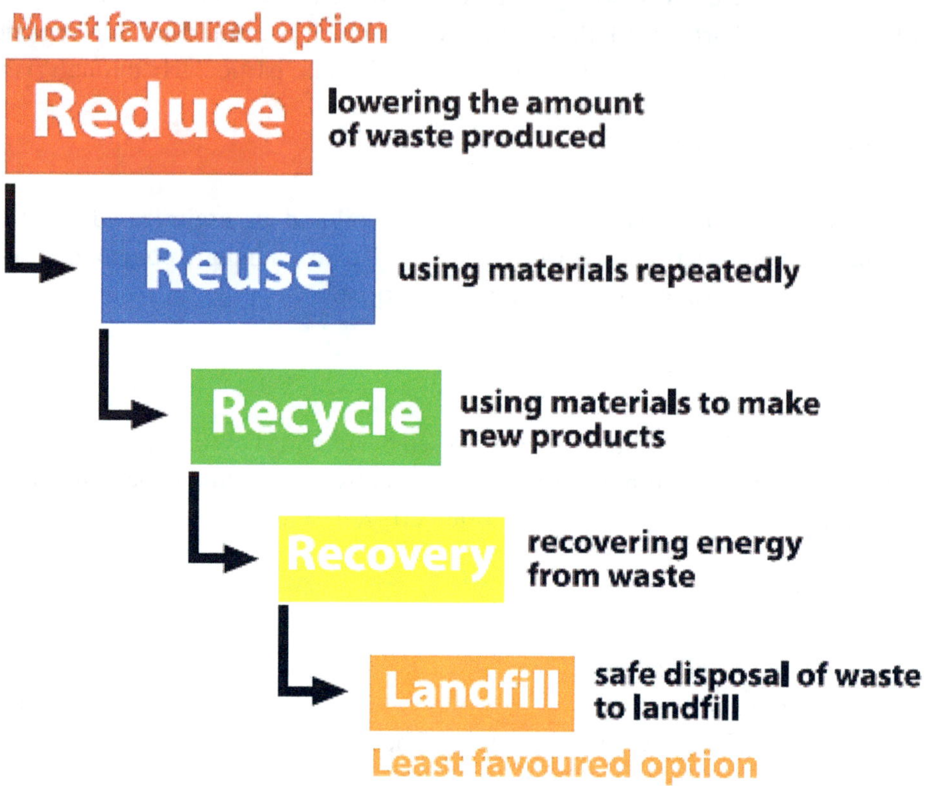

Reduce

The first and most crucial step in the waste hierarchy is waste reduction. Customers are urged to buy in bulk, pick things with fewer packing, and switch to reusable items rather than single-use items in order to reduce their waste. Production methods that consume fewer resources and generate less waste will be employed by businesses. These efforts also generally provide customers and businesses with financial incentives in the form of lower buying costs in order to improve the environment.

Reuse

Reusing anything involves putting it to use in its original form more than once. It's a process where items or resources that don't seem wasteful be employed once more for the original intent behind them. It involves looking over, keeping up with, fixing, and refinishing furniture, appliances, and other household objects.

Recycling

It is a method that entails accumulating new, used, or recycled objects that are unquestionably rubbish. Recycling is the process of separating recyclables into raw materials, processing those raw materials into new products.

E-waste Recycling Practices in India

There are a large number of active recyclers in India who manage e-waste through recycling. They carefully collect, move, disassemble, classify, handle, and dispose of e-waste in order to manage it. One of them, who registered with MOEF/CPCB as a recycler or preprocessor, with environmentally sound management facilities and e-waste preprocessors, is exemplified by the person named below. Examples include K.G. Nandani Enterprises, E- Parisara Pvt. Ltd., Ash Recyclers, Unit II, New Port Computer Services (India) Private Limited, (Karnataka) Eco Recycling Limited, and Earth Sense Recycle Pvt. Ltd. (Maharashtra). Additionally, the Rotary Club of Bangalore helps big city IT enterprises by recommending good strategies for lowering e-waste. It compels IT corporations to donate their usable computers to educational institutions and nonprofit organizations so that they can be used by those institutions and help minimize e-waste (Saran & Yadav, 2019)

Specific Suggestions to Reduce the Volume of Electronic Waste

Product lifespan should be verified. When you're ready to buy anything new, do your homework. Check to make sure it won't break quickly or sustain damage right away after purchase. To avoid having to replace items after a few years or even just a few months, buy items that are likely to endure a lot longer. The electronics industry is widely recognized for producing things with shorter lifespans so that more money may be gained later when they fail or break.

The Power of Knowledge

Learn about the ingredients that are used in your electronics. You may learn how toxic such substances can be if they are thrown into a landfill by doing some research on the basic ingredients used to make your laptop or mobile phone. The better informed you are, the more goods you can buy that won't damage the environment.

Look for a Label That Says It Is Eco-Friendly

Consider checking the Energy Star or Electronic Product Environmental Assessment Tool labels on the items you purchase.

Limit the Number of Electronics You Own

Look for devices with multiple functions if you don't really need an extra gadget.

Educate Children on E-Waste

Children are our future, so it's important to help them develop an early commitment to recycling e-waste.

Constantly Recycle

Whatever the case, recycle, recycle, recycle. Whatever you have, it's crucial to always properly dispose of your e-waste. That involves recycling all your old gadgets, keeping in mind that it's dangerous to dispose of e-waste illegally, especially given how much of it there is now.

Recognize Security Concerns

Another justification for not discarding your electronic devices is the fact that all of your personal data is still there on them even after deletion. Before recycling, a recycling company like Great Lakes Electronics Corporation may clean your device, eliminating the chance for cybercriminals to obtain that information.

Maintain What You Have

Small adjustments extend the life of what you already have. Keep your computer clean on a regular basis and avoid overcharging it to extend the life of your battery.

Current Trends in Recycling E-Waste

Due to our increased reliance on electronics, phones, and computer equipment as well as the continuous introduction of new products, the globe currently has a rising e-waste problem. In addition to the numerous problems with excessive waste and damaging pollutants harming the environment today, these materials raise additional challenges that cannot be easily addressed by current methods. The most recent e-waste recycling trends aim to solve problems by streamlining the entire production to disposal cycle.

New Solutions for E-Waste Excess

How can the issue of e-waste be addressed while simultaneously making the recycling of valuable materials more ecologically and socially responsible? The National Strategy for Electronics Stewardship, released by the EPA, serves as the foundation for new rules and initiatives for the manufacture and recycling of electronic items.

A focus on a zero-waste, circular economy, increased cyber security awareness to improve hardware destruction techniques, and a general focus on proper e-waste recycling certifications, workers' rights, and environmental oversight for current service providers are some of the changes that have been proposed or already implemented.

The world will ultimately rely more and more on automation and technology in the years to come, and people won't give up their own devices any time soon. As the demand for recycling e-waste rises, environmental and economic authorities throughout the world must rebuild the entire production-to-consumer pipeline with an emphasis on sustainability rather than just profitability.

SCOPE FOR FURTHER INVESTIGATION FOR MORE UTILIZATION OF E-WASTE IN CONCRETE

The current research looked on the use of E-waste polymer components as coarse aggregate. Additional research may be undertaken to examine the impact of utilizing e-waste in foundation, footing, and fly ash activities in order to find a sustainable solution to handle the vast volume of garbage (A Gupta et al. 2008). It is possible to research the usage of electronic waste in higher grade concrete. Carefully chosen chemical admixtures and minerals can be utilized to give e-waste concrete the desired greater strength so that it will meet the durability requirements.

- Mineralogical and micrographic research can be done to comprehend the structure and makeup of concrete materials made from electronic trash.
- E-waste can also be proposed as a substitute for materials used in the manufacture of concrete blocks.
- E-waste concrete's behavior at high temperatures can be studied.
- It is possible to conduct studies of NDT testing and analyses of e-waste in frame structures. You may do a quick chloride permeability test on concrete made from e-waste.

REFERENCES

Ahirwar R, & Tripathi AK (2021) E-waste management: A review of recycling process, environmental and occupational health hazards, and potential solutions. *Environ. Nanotechnology, Monit. Manag. 15*. doi:10.1016/j.enmm.2020.100409

Ali H, Khan E, & Ilahi I (2019) Environmental chemistry and ecotoxicology of hazardous heavy metals: Environmental persistence, toxicity, and bioaccumulation. *J. Chem. 2019*(Cd): doi:10.1155/2019/6730305

Ankit, S. L., Saha, L., Kumar, V., Tiwari, J., Sweta, Rawat, S., Singh, J., & Bauddh, K. (2021). Electronic waste and their leachates impact on human health and environment: Global ecological threat and management. *Environ. Technol. Innov., 24*, 102049. doi:10.1016/j.eti.2021.102049

Blinová, L., & Godovčin, P. (2021). Importance of Recycling the Waste-Cables Containing Copper and PVC. *Res. Pap. Fac. Mater. Sci. Technol. Slovak Univ. Technol., 29*(48), 1–21. doi:10.2478/rput-2021-0001

Cherukuri, I., Sultana, N., & Podila, S. P. (2018). Status Of E-Waste In India- A Review. *J. Environ. Sci. Toxicol. Food Technol., 12*(11), 8–16. doi:10.9790/2402-1211010816

Debnath, B., Dey, R., & Roy, S. (2019) Smart Switching System Using Bluetooth Technology. *Proc. – 2019 Amity Int. Conf. Artif. Intell. AICAI 2019* (pp. 760–763). IEEE. 10.1109/AICAI.2019.8701298

Elleuch, B., Bouhamed, F., Elloussaief, M., & Jaghbir, M. (2018). Environmental sustainability and pollution prevention. *Environmental Science and Pollution Research International, 25*(19), 18223–18225. doi:10.100711356-017-0619-5 PMID:29971753

Gupta, A., Singh, A., Singh, U., Singh, A., & Tomar, C. (2008) A Review Paper on Use of E-Waste in Concrete. *Int. Res. J. Eng. Technol, 2762*. www.irjet.net

Gupta, N., & Nath, M. (2020). Groundwater Contamination by E-Waste and Its Remedial Measure - A Literature Review. *Journal of Physics: Conference Series, 1531*(1), 012023. doi:10.1088/1742-6596/1531/1/012023

Jeyaraj, P. (2021). Management of E-waste in India – Challenges and recommendations. *World J. Adv. Res. Rev., 11*(2), 193–218. doi:10.30574/wjarr.2021.11.2.0329

Julander, A., Lundgren, L., Skare, L., Grandér, M., Palm, B., Vahter, M., & Lidén, C. (2014). Formal recycling of e-waste leads to increased exposure to toxic metals: AN occupational exposure study from Sweden. *Environment International*, *73*, 243–251. doi:10.1016/j.envint.2014.07.006 PMID:25300751

Kirthika, S. K., Singh, S. K., & Chourasia, A. (2020). Alternative fine aggregates in production of sustainable concrete- A review. *Journal of Cleaner Production*, *268*, 122089. doi:10.1016/j.jclepro.2020.122089

Mohabuth, N., & Miles, N. (2005). The recovery of recyclable materials from Waste Electrical and Electronic Equipment (WEEE) by using vertical vibration separation. *Resources, Conservation and Recycling*, *45*(1), 60–69. doi:10.1016/j.resconrec.2005.02.001

Needhidasan, S., Samuel, M., & Chidambaram, R. (2014). Electronic waste – an emerging threat to the environment of urban India. *Journal of Environmental Health Science & Engineering*, *12*(36), 1–9. doi:10.1186/2052-336X-12-36 PMID:24444377

Okunola, A. A., Kehinde, I. O., Oluwaseun, A., & Olufiropo, E. A. (2019). Public and Environmental Health Effects of Plastic Wastes Disposal: A Review. *J. Toxicol. Risk Assess.*, *5*(2). doi:10.23937/2572-4061.1510021

Saran, R. K., & Yadav, S. (2019) Significance of Electronic Waste Management for Sustainable Industrial Production Rishabh. In Operations Management and Systems Engineering, Lecture Lecture Notes on Multidisciplinary Industrial Engineering (pp. 241–249). Springer Singapore doi:10.1007/978-981-13-6476-1

Streicher-Porte M, Widmer R, Jain A, Bader HP, Scheidegger R, Kytzia S (2005) Key drivers of the e-waste recycling system: Assessing and modelling e-waste processing in the informal sector in Delhi. *Environ. Impact Assess. Rev.*, *25*(5 SPEC. ISS.), 472–491. doi:10.1016/j.eiar.2005.04.004

Turaga, R. M. R., Bhaskar, K., Sinha, S., Hinchliffe, D., Hemkhaus, M., Arora, R., Chatterjee, S., Khetriwal, D. S., Radulovic, V., Singhal, P., & Sharma, H. (2019). E-Waste Management in India: Issues and Strategies. *Vikalpa*, *44*(3), 127–162. doi:10.1177/0256090919880655

Ullah, S., Qureshi, M. I., Joyklad, P., Suparp, S., Hussain, Q., Chaiyasarn, K., & Yooprasertchai, E. (2022). Effect of partial replacement of E-waste as a fine aggregate on compressive behavior of concrete specimens having different geometry with and without CFRP confinement. *Journal of Building Engineering*, *50*(January), 104151. doi:10.1016/j.jobe.2022.104151

Widmer R, Oswald-Krapf H, Sinha-Khetriwal D, Schnellmann M, Böni H (2005) Global perspectives on e-waste. *Environ. Impact Assess. Rev., 25*(5 SPEC. ISS.), 436–458. doi:10.1016/j.eiar.2005.04.001

Chapter 4
External Barriers Affecting E-Waste Remanufacturing in the Indian Context

Swatantra Kumar Jaiswal
National Institute of Technology, Raipur, India

Suraj Kumar Mukti
National Institute of Technology, Raipur, India

ABSTRACT

In the last few decades, there has been massive growth in the electrical and electronic industries. The demand for electronic products is growing enormously, resulting in a greater amount of e-waste being produced. Remanufacturing is a method of transforming end-of-use (EOU) or end-of-life (EOL) products to the given standards or specifications of the original product. In India, remanufacturing is at a nascent stage. In this study, critical barriers that will affect e-waste remanufacturing in the Indian context are identified, and the logical relationship between them is found using ISM and MICMAC. The result shows that lack of collection centres, customer eagerness to return the product, and uncertainty in demand are major driving barriers; and incomplete recycling systems, sorting policies, consumption attitudes, and uncertainty in quality are major dependent barriers. To increase customer eagerness to return used products, organise proper awareness programs so that people will know the effects of e-waste. It will also increase the demand for used products.

DOI: 10.4018/978-1-6684-7573-7.ch004

INTRODUCTION

In last few decades, massive growth in electronic industry in all over the world as a result of this large number of electronic products are getting obsolete or the intended life span of the electronic product are getting less for consumers, hence there is enormous growing of electrical and electronic waste stream also known as E-waste. E-waste is defined as discarded, surplus, obsolete, broken, electrical, or electronic devices that have lost their intended use. Electronics devices such as refrigerators, cameras, printers, cell phones, air conditioners, consumer electronics, and laptops are common components of e-waste. According to EU directives, this electronic waste stream is acknowledged as e-waste.

The world produces as much as 50 million tons of e-waste a year, weighing more than all of the commercial airliners ever made (UN Report, 2019). In India, 2 million tons per annum E-waste is produced (ASSOCHAM-NEC 2018). E-waste encompasses toxic substances that will directly or indirectly affect the surroundings or have an adverse effect on human health. It contains thousands of poisonous substances like nickel (Ni), polyvinyl chloride (PVC), antimony (Sb), cadmium (Cd), chlorofluorocarbons (CFCs), mercury (Hg), and lead (Pb), which are carcinogenic in nature and have an adverse effect on the environment (Kiddee, Naidu, & Wong 2013).

In India 80% of E-waste is recycled or handled by informal sectors, after recovering precious metal they dumped the remaining part in land, only 20% of the E-waste is recycled by the formal sector or the registered company. Refurbishing, recycling, and remanufacturing of used products are certain options to decrease waste generation, in India remanufacturing is at the emerging stage and has done marginal growth in the last few decades.

Remanufacturing is a closed-loop supply chain approach that can change used products into like-new products or a worn-out product that completes its life span to as good as a new product. In remanufacturing, the first step is the collection of various used electronic products through a collection point or a collection center, there is uncertainty in quality, quantity, and timing of the return used product or waste. The next step is the disassembly of E-waste, which can be done by various skilled labors, after that sorting of waste is accomplished according to parts and materials, cleaning would be done by using some cleaning agents. Further, inspection of parts, it will be accomplished through skilled labors, decisions may be taken either part is in able to reuse or it would be not according to given standards, then assembling and testing would happen. Based on previous literature, it is found that very few studies have been conducted regarding barriers to the remanufacturing of e-waste in the Indian context.

Objective of this study is to find out the external barriers which affect the e-waste remanufacturing in India and determine the driving and dependence power of barriers by using contextual relationship between them by Interpretative Structural Modelling (ISM) and Matriced' Impacts Croises-Multipication Appliqué and Classement (MICMAC) analysis.

LITERATURE REVIEW

In worldwide electronic waste (e-waste) is one of the fast growing waste stream, which contains range of toxic substances which affects the social health and contaminate the surroundings if proper disposal protocol are not managed (Kiddee et al., 2013). In India, green consciousness, government incentive, return intent, and organization vision are the key persuasive factors for E-waste remanufacturing (Singhal, Tripathy, & Kumar 2018). In Malaysia, the potential challenges of remanufacturing are (a) Marketing and Competition (b) raw material collection (c) skill manpower and expert (d) product design (e) environmental and government (f) technology and method (Shamee & Shamsuddin 2019).

The prominent external barriers to remanufacturing are the absence of channels to collect the used product and consumer opposition to returning the product, the casual barriers are the timing of return of the product and uncertainty in quantity (Kumar, Agrawal, & Sharma 2015). Uncertain government policy and regulation, a lack of management foresight, and negligence of the environment are the main hurdles of remanufacturing (Singhal et al., 2018).

In worldwide electronic waste (E-waste) is one of the fast-growing waste streams contains a range of toxic substances that affects the social health and contaminate the surroundings if proper disposal protocol is not managed (Jayaraman et al., 2019). In India the major drivers of remanufacturing can be categories into three parts- (a) economic drivers (b) social drivers (c) environmental drivers, competitive pricing (Sharma, Garg, & Sharma 2016) is the potential economic driver of remanufacturing, the social driver of remanufacturing is more job opportunities, and the best way of product recovery is the important environmental driver of remanufacturing (Singh & Srivastava 2018). Xu et al. discuss the take bake regulation for e-waste, identify the factors affecting the remanufacturing of e-waste in developing countries, and determine the interrelationship between them (Xu et al., 2022). Ni et al. develop a framework for successful implementation of reverse logistics for e-waste to reduce the uncertainty in quality, uncertainty in quantity, and uncertainty in timing of waste material (Ni et al., 2021).

In 1973, J. N. Warfield proposed interpretative structural modelling (ISM), which is used for finding out the circumstantial correlation among variables, gives a basic idea about the different levels, and is also used to determine the driving or dependence power among the factors. ISM is used in supplier selection (Raut et al., 2017), knowledge management (Agrawal & Mukti 2019) (Agrawal and Mukti 2020), and ERP (Mukti & Rawani, 2016) (Mukti et al., 2014). ISM is a statistical tool that uses Structural Self-Interaction Matrix (SSIM) and reachability matrices to find out the driving and dependence powers of factors. ISM requires less expertise than other Delphi or brainstorming techniques.

METHODOLOGY

Firstly, this study involved identification of critical external barriers which mainly affect the remanufacturing of E-waste in India, then develop an appropriate relationship with the help of experts among barriers, after that self-structured interaction matrix (SSIM) was developed between the barriers, next step is formation of reachability matrix by using binary code, then different levels are assigned to barriers by performing multiple iterations after that authors have prepared the ISM based model according to assigned levels and then MICMAC analysis is done by using reachability matrix.

Identification of Barriers

First, authors have done a literature survey on the topics like E-waste, E-waste management, remanufacturing, e-waste remanufacturing, enablers of remanufacturing, challenges related to remanufacturing, recycling, refurbishing and find out the critical barriers which are responsible for e-waste remanufacturing in Indian context, Finally, with the help of previous literature and experts, ten barriers (Table 1) have been identified, such as lack of collection centre (B1), imperfect law (B2), consumption attitude (B3), customer eagerness to return the product (B4), uncertainty in quality (B5), uncertainty in timing (B6), uncertainty in quantity (B7), uncertainty in demand (B8), incomplete recycling system (B9) and sorting policies (B10) which affect the remanufacturing practices in India. A set of the questionnaire was prepared according to ISM and response were taken from experts.

Interpretative Structural Modelling (ISM)

ISM is a robust method in which directly and indirectly related barriers are grouped together to determine the contextual relationship between barriers, ISM is a statistical tool that is used for determining the driving and dependence power of barriers by developing a hierarchical model for barriers. In ISM, the first step is the identification of barriers, In this study, the authors have identified the external barriers of remanufacturing, a total of ten barriers have been identified, as shown in Table 1. Next step is the formation of self-structured interaction matrix (SSIM), in

Table 1. External barriers of remanufacturing

Code	Barriers	Description	Source
B1	Lack of Collection Centre	There is a lack of collection points so that consumers don't know where to dump the used product	(Singhal, Tripathy, and Kumar 2019), (Sharma et al. 2016)
B2	Imperfect Law	In India, there is less law related to manufactures or very few law related to remanufacturing	(Singh and Srivastava 2018), (Xiang and Ming 2011)
B3	Consumption Attitude	Generally, customers are having an attitude of one-time consumption	(Singhal et al. 2018),(Singhal et al. 2019)
B4	Customer Eagerness to Return the Product	Customers are not willing to give the used product to the collection points	(Sharma et al. 2016), (Singh and Srivastava 2018)
B5	Uncertainty in Quality	There is variation in the quality of the used product, so an additional number of operations are needed for remanufacturing activities	(Singh and Srivastava 2018), (Kiddee et al. 2013), (El, Jaber, and Bonney 2013)
B6	Uncertainty in Timing	Timing of return of a used product is not known, it would differ for a different product or the same product	(Singh and Srivastava 2018), (Mukhopadhyay and Ma 2009), (Guide Jr 2000)
B7	Uncertainty in Quantity	Quantity of used product is not certain so proper planning of the product would be difficult	(Singh and Srivastava 2018), (Galbreth and Blackburn 2010), (Robotis, Boyaci, and Verter 2012), (Mukhopadhyay and Ma 2009)
B8	Uncertainty in Demand	Estimation of the demand for the remanufactured product would be difficult and so proper planning is not possible	(Teunter and Flapper 2011), (Inderfurth 2005), (Van Der Laan et al. 1999)
B9	Incomplete Recycling System	Proper recycling system would be needed for used product. Incomplete recycling directly or indirectly affects the environment.	(Singh and Srivastava 2018),(Singhal et al. 2018)
B10	Sorting Policies	Sorting would be one of the important steps in remanufacturing proper rules and regulation would be necessary for sorting purpose	(Singhal et al. 2018), (Singhal et al. 2019), (Kumar and Dixit 2018)

this matrix four symbols have been taken V, A, X, and O with the help of expert's responses are taken, if barrier i having impact on j then symbol V is used, if barrier j is having impact on barrier i then symbol A, if symbol X is taken then both barrier

Table 2. SSIM Matrix

	B1	B2	B3	B4	B5	B6	B7	B8	B9	B10
B1		A	X	V	V	V	V	A	O	V
B2			V	X	O	V	V	X	V	V
B3				V	O	O	O	X	V	V
B4					V	V	V	V	V	V
B5						X	X	A	V	V
B6							X	X	V	V
B7								X	V	V
B8									V	V
B9										X
B10										

i and j having impact on each other, and if both the barrier have no relation with each other than symbol O.

Further, reachability matrix is developed by using SSIM matrix, in this matrix symbols of SSIM matrix is replaced by binary code 0 and 1, let the barrier will be i and j, (a) For symbol "V",if position (i,j) then '1' and if position (j,i) then '0', (b) For symbol "A", if position (i,j) then '0' and if position (j,i) then '1', (c) For symbol "X", if position (i,j) then '1' and if position (j,i) then '1', (d) For symbol "O", if position (i,j) then '0' and if position (j,i) then '0'.

Next step is assignment of level to the barriers, by using reachability matrix, in level partition reachability set and antecedent set is determined by using row and column values of reachability matrix, then intersection set is formed by considering the common values in reachability set and antecedent set. If the value of reachability set and intersection set is same then first level is assigned to the respective barrier as shown in table 4, multiple iterations are performed to assign level to the different barriers. A total of five iterations were performed to identify the levels of barriers as shown in table 5, table 6, table 7 and table 8. Number of iterations in ISM depends on the responses of the experts and the assignment of level to barriers, so for this study only five iterations are possible. It can be observed that all the ten barriers are divided into six levels where barrier 1 lack of collection centres are at level VI and incomplete recycling system, sorting policies are at a level I, all the other barriers lie between level I to level VI. Finally, ISM based model is constructed by using the values of level partition.

Matrices' Impacts Croises-Multipication Appliqué and Casement (MICMAC) is also known as Cross-Impact Matrix Multiplication Applied to Classification, this

Table 3. Reachability Matrix

	B1	B2	B3	B4	B5	B6	B7	B8	B9	B10	DRP
B1	1	1	1	1	1	1	1	1	0	1	9
B2	1	1	1	1	0	1	1	1	1	1	9
B3	0	0	1	1	0	0	0	1	1	1	5
B4	0	1	1	1	1	1	1	1	1	1	9
B5	0	0	0	0	1	1	1	0	1	1	5
B6	0	0	0	0	1	1	1	1	1	1	6
B7	0	0	0	0	1	1	1	1	1	1	6
B8	1	1	1	0	1	1	1	1	1	1	9
B9	0	0	0	0	0	0	0	0	1	1	2
B10	0	0	0	0	0	0	0	0	1	1	2
DEP	3	4	5	4	6	7	7	7	9	10	

analysis is done to determine the driving and dependence power of barriers. Based on the driving and dependence power of each reagent, the barriers are classified as either (a) autonomous, representing barriers with low driving and dependence power; (b) dependent, representing barriers with low driving and high dependence power; (c) linkage, representing barriers with high driving and high dependence power; and (d) independent, representing barriers with high driving and low dependence power.

RESULT AND DISCUSSION

Through a literature review, consultation with experts, etc., this research attempts to identify the ten external barriers to e-waste remanufacturing in the Indian context, and assess the relative importance of these factors. Table 2 displays the logical relationship established by the experts after the barriers were identified, as shown in Table 1, by means of the symbols V, A, X, and O. Then, following the guidelines in Table 3, a reachability matrix is constructed by translating the symbols into binary codes (1 and 0). A total of 5 iterations were performed to assign a level to each barrier (Table 4), and then an ISM-based model (Figure. 1) was prepared making use of the levels from Table 9. ISM model divides the factors into six levels, strong driving power is shown at the bottom level, viz., barrier B1 (lack of collection centre), barrier B4 (customer eagerness to return the product), which have strong influence on the other barriers in the hierarchy, and strong dependence power is shown in the top level of the ISM model, viz., barrier B9 (incomplete recycling system) and barrier

Table 4. Level Partition Iteration I

Barriers	Reachability set	Antecedent sets	Intersection set	Level
B1	1, 2, 3, 4, 5, 6, 7, 8, 10	1, 2, 8	1, 2, 8	
B2	1, 2, 3, 4, 6, 7, 8, 9, 10	1, 2, 4, 8	1, 2, 4, 8	
B3	3, 4, 8, 9, 10	1, 2, 3, 4, 8,	3, 4, 8	
B4	2, 3, 4, 5, 6, 7, 8, 9, 10	1, 2, 3, 4	2, 3, 4	
B5	5, 6, 7, 9, 10	1, 4, 5, 6, 7, 8	5, 6, 7	
B6	5, 6, 7, 8, 9, 10	1, 2, 4, 5, 6, 7, 8	5, 6, 7, 8	
B7	5, 6, 7, 8, 9, 10	1, 2, 4, 5, 6, 7, 8	5, 6, 7, 8	
B8	1, 2, 3, 5, 6, 7, 8, 9, 10	1, 2, 3, 4, 6, 7, 8	1, 2, 3, 6, 7, 8	
B9	9, 10	2, 3, 4, 5, 6, 7, 8, 9, 10	9, 10	**I**
B10	9, 10	1, 2, 3, 4, 5, 6, 7, 8, 9, 10	9, 10	**I**

B10 (sorting policies), all other barriers lie between level I and level VI. Thereafter, figure 2 shows the MICMAC analysis, barrier B1 (lack of collection centre), barrier B2 (imperfect law), and barrier B4 (customer eagerness to return the product) are

Table 5. Level Partition Iteration II

Barriers	Reachability set	Antecedent sets	Intersection set	Level
B1	1, 2, 3, 4, 5, 6, 7, 8	1, 2, 8	1, 2, 8	
B2	1, 2, 3, 4, 6, 7, 8	1, 2, 4, 8	1, 2, 4, 8	
B3	3, 4, 8	1, 2, 3, 4, 8,	3, 4, 8	**II**
B4	2, 3, 4, 5, 6, 7, 8	1, 2, 3, 4	2, 3, 4	
B5	5, 6, 7	1, 4, 5, 6, 7, 8	5, 6, 7	**II**
B6	5, 6, 7, 8	1, 2, 4, 5, 6, 7, 8	5, 6, 7, 8	**II**
B7	5, 6, 7, 8	1, 2, 4, 5, 6, 7, 8	5, 6, 7, 8	**II**
B8	1, 2, 3, 5, 6, 7, 8	1, 2, 3, 4, 6, 7, 8	1, 2, 3, 6, 7, 8	

Table 6. Level Partition Iteration III

Barriers	Reachability set	Antecedent sets	Intersection set	Level
B1	1, 2, 4, 8	1, 2, 8	1, 2, 8	
B2	1, 2, 4, 8	1, 2, 4, 8	1, 2, 4, 8	**III**
B4	2, 4, 8	1, 2, 4	2, 4	
B8	1, 2, 8	1, 2, 4, 8	1, 2, 8	

Table 7. Level Partition Iteration IV

Barriers	Reachability set	Antecedent sets	Intersection set	Level
B1	1, 4, 8	1, 8	1, 8	**IV**
B4	4, 8	1, 4	4	
B8	1, 8	1, 4, 8	1, 8	

situated in independent barriers, these barriers are the critical barriers and have a direct impact on other barriers. In linkage barriers, barrier B6 (uncertainty in timing), barrier B7 (uncertainty in quantity), and barrier B8 (uncertainty in demand) are situated, which have strong driving and dependence powers; moreover, barrier B5 (uncertainty in quality), barrier B9 (incomplete recycling system), and barrier B10 (sorting policies) are situated in a dependent barrier cluster having high dependency on other barriers, in the autonomous category, barrier B3 (consumption attitude) is situated, which has low driving and dependence power.

CONCLUSION

Remanufacturing is a method of transforming an end-of use product (EOU) or end-of-life product (EOL) to the given standards or specifications of the original product. This research aims to identify the critical or influential barriers that will affect e-waste remanufacturing in India and find out the logical relationship between them by using interpretative structural modelling (ISM). A total of ten barriers have been taken from the literature and experts in the respective fields.

Further, with the help of experts, these ten factors are divided into six levels based on logical relationships among them or by using the ISM model. The result shows that lack of collection centres (B1), customer eagerness to return the product (B4) and uncertainty in demand (B7) are major drivers of remanufacturing of E-waste and incomplete recycling system (B9), sorting policies (B10), consumption attitude (B2), uncertainty in quality (B5) are the dependent external barriers for e-waste remanufacturing in India. To overcome these driving barriers, we can establish a

Table 8. Level Partition Iteration V

Barriers	Reachability set	Antecedent sets	Intersection set	Level
B4	4, 8	4	4	**V**
B8	8	4, 8	8	**VI**

Table 9. Final Level

Code	Barriers	Level
B1	Lack of Collection Centre	VI
B2	Imperfect Law	III
B3	Consumption Attitude	II
B4	Customer Eagerness to Return the Product	V
B5	Uncertainty in Quality	II
B6	Uncertainty in Timing	II
B7	Uncertainty in Quantity	II
B8	Uncertainty in Demand	IV
B9	Incomplete Recycling System	I
B10	Sorting Policies	I

Figure 1.

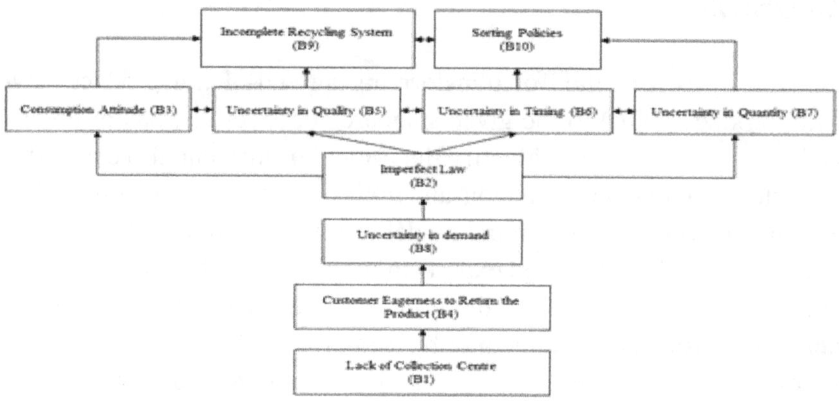

proper channel of collection points that will be easily assessable to customers, and by doing this, the uncertainty in quantity will decrease to a certain extent, as will the uncertainty in quality and time. To increase customer eagerness to return used product, we have toracticee proper awareness programs, so that people will know the effects of e-waste on our environment. This will decrease the uncertainty in quantity, quality, and timing of return, and increase the demand for used product, which will be beneficial for the manufacturers who perform remanufacturingracticess in India.

Figure 2.

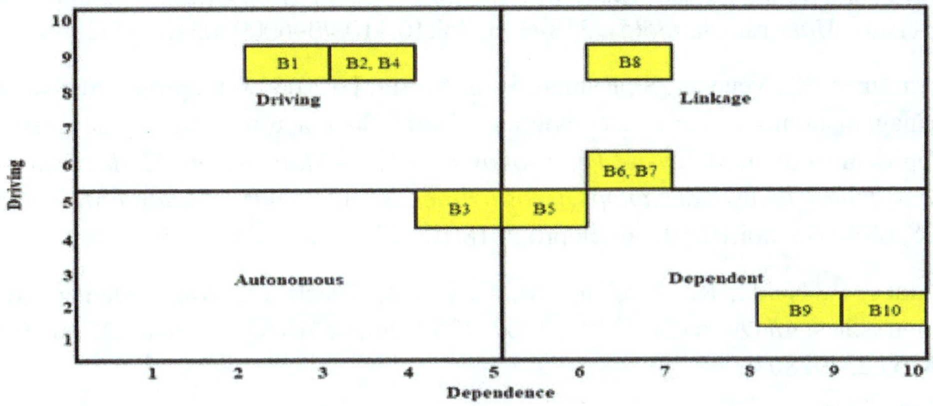

REFERENCES

Agrawal, A., & Mukti, S. K. (2020). Knowledge Management & It's Origin, Success Factors, Planning, Tools, Applications, Barriers and Enablers: A Review. [IJKM]. *International Journal of Knowledge Management*, *16*(1), 43–82. doi:10.4018/IJKM.2020010103

Animesh, A., & Mukti, S. K. (2019). Case Study of Critical Success Factors Affecting Knowledge Management in Small-and Medium-Sized Enterprises in Developing State: Steel Sector. In Advances in Industrial and Production Engineering, 825–31. Springer. doi:10.1007/978-981-13-6412-9_76

ASSOCHAM-NEC. (2018). *India among the Top ve Countries in E-Waste Generation*, 1–4. ASSOCHAM-NEC Study.

El, A. M. A., Jaber, M. Y., & Bonney, M. (2013). Int. J. Production Economics How Many Times to Remanufacture? *International Journal of Production Economics*, *143*(2), 598–604. doi:10.1016/j.ijpe.2011.11.017

Galbreth, M. R., & Blackburn, J. D. (2010). Optimal Acquisition Quantities in Remanufacturing with Condition Uncertainty. *Production and Operations Management*, *19*(1), 61–69. doi:10.1111/j.1937-5956.2009.01067.x

Guide, V. Jr, & Daniel, R. (2000). Production Planning and Control for Remanufacturing: Industry Practice and Research Needs. *Journal of Operations Management*, *18*(4), 467–483. doi:10.1016/S0272-6963(00)00034-6

Inderfurth, K. (2005). Impact of Uncertainties on Recovery Behavior in a Remanufacturing Environment. *International Journal of Physical Distribution & Logistics Management*, *35*(5), 318–336. doi:10.1108/09600030510607328

Jayaraman, K., Vejayon, S., Raman, S., & Mosta, I. (2019). The proposed e-waste management model from the conviction of individual laptop disposal practices-An empirical study in Malaysia. *The Proposed E-Waste Management Model from the Conviction of Individual Laptop Disposal Practices-An Empirical Study in Malaysia.*, *208*, 688–696. doi:10.1016/j.jclepro.2018.10.125

Kiddee, P., Naidu, R., & Wong, M. H. (2013). Electronic Waste Management Approaches. *An Overview.*, *33*(5), 1237–1250. doi:10.1016/j.wasman.2013.01.006 PMID:23402807

Kumar, A., & Dixit, G. (2018). An Analysis of Barriers Affecting the Implementation of E-Waste Management Practices in India : A Novel ISM-DEMATEL Approach. *Sustainable Production and Consumption*, *14*, 36–52. doi:10.1016/j.spc.2018.01.002

Kumar, D., Agrawal, R., & Sharma, V. (2015). Enablers for Competitiveness of Indian Manufacturing Sector : An ISM-Fuzzy MICMAC Analysis. *Procedia: Social and Behavioral Sciences*, *189*, 416–432. doi:10.1016/j.sbspro.2015.03.200

Laan, V. D., Erwin, M. S., Dekker, R., & Van Wassenhove, L. (1999). Inventory Control in Hybrid Systems with Remanufacturing. *Management Science*, *45*(5), 733–747. doi:10.1287/mnsc.45.5.733

Mukhopadhyay, S. K., & Ma, H. (2009). Joint Procurement and Production Decisions in Remanufacturing under Quality and Demand Uncertainty. *International Journal of Production Economics*, *120*(1), 5–17. doi:10.1016/j.ijpe.2008.07.032

Mukti, S. K., & Rawani, A. M. (2016). ERP System Success Models: A Literature Review. *Journal of Engineering and Applied Sciences (Asian Research Publishing Network)*, *11*(3), 1861–1875.

Mukti, S. K., Tripathi, P., & Rawani, A. M. (2014). Identification of Factors and Indicators for Success Measurement of ERP System. *International Proceedings of Economics Development and Research*, *75*, 117.

Ni, Z., Chan, H. K., & Tan, Z. (2021). Systematic Literature Review of Reverse Logistics for E-Waste: Overview, Analysis, and Future Research Agenda. *International Journal of Logistics Research and Applications*, *0*(0), 1–29. doi:10.1080/13675567.2021.1993159

Raut, R. D., Narkhede, B., & Gardas, B. B. (2017). To Identify the Critical Success Factors of Sustainable Supply Chain Management Practices in the Context of Oil and Gas Industries. *Renewable & Sustainable Energy Reviews*, *68*, 33–47. doi:10.1016/j.rser.2016.09.067

Robotis, A., Boyaci, T., & Verter, V. (2012). Investing in Reusability of Products of Uncertain Remanufacturing Cost: The Role of Inspection Capabilities. *International Journal of Production Economics*, *140*(1), 385–395. doi:10.1016/j.ijpe.2012.04.017

Shamee, A. & Shamsuddin, A. (2019). *End-of-Life Electrical and Electronic Equipment Remanufacturing Prospects in Malaysia End-of-Life Electrical and Electronic Remanufacturing Prospects in Malaysia*. IOP Science.. doi:10.1088/1757-899X/530/1/012033

Sharma, V., Garg, S. K., & Sharma, P. B. (2016). Identi Fi Cation of Major Drivers and Roadblocks for Remanufacturing in India. *Journal of Cleaner Production*, *112*, 1882–1892. doi:10.1016/j.jclepro.2014.11.082

Singh, M., & Srivastava, R. K. (2018). Resources, Conservation & Recycling Analysis of External Barriers to Remanufacturing Using Grey-DEMATEL Approach : An Indian Perspective. *Resources, Conservation and Recycling*, *136*(February), 79–87. doi:10.1016/j.resconrec.2018.03.021

Singhal, D., Tripathy, S., & Kumar, S. (2018). ScienceDirect ScienceDirect ScienceDirect ScienceDirect School for Costing Models Capacity Optimization between Capacity and Operational Efficiency. *Procedia Manufacturing*, *20*, 452–457. doi:10.1016/j.promfg.2018.02.066

Singhal, D., Tripathy, S., & Kumar, S. (2019). Sustainability through Remanufacturing of E-Waste : Examination of Critical Factors in the Indian Context. *Sustainable Production and Consumption*, *20*, 128–139. doi:10.1016/j.spc.2019.06.001

Teunter, R. H., & Simme Douwe, P. F. (2011). Optimal Core Acquisition and Remanufacturing Policies under Uncertain Core Quality Fractions. *European Journal of Operational Research*, *210*(2), 241–248. doi:10.1016/j.ejor.2010.06.015

Xiang, W., & Ming, C. (2011). Implementing Extended Producer Responsibility: Vehicle Remanufacturing in China. *Journal of Cleaner Production*, *19*(6–7), 680–686. doi:10.1016/j.jclepro.2010.11.016

Xu, J., Luo, C., Ng, C. T., Feng, X., & Cheng, T. C. E. (2022). Remanufacturing with Random Yield in the Presence of the Take-Back Regulation. *Computers & Industrial Engineering*, *168*(March), 1–9. doi:10.1016/j.cie.2022.108097

Chapter 5
Impact of E-Waste on the Environment

Shilpa R. Kalambe
Dr. Babasaheb Ambedkar College of Engineering and Research, India

Bhojraj Natthuji Kale
https://orcid.org/0000-0003-4084-4796
Dr. Babasaheb Ambedkar College of Engineering and Research, India

Sanjay Jain
RKDF University, India

Anuprita Sandeep Mishra
IES College of Technology, India

ABSTRACT

e-waste is some of the most hazardous waste on the globe. It is currently at its highest degree of distress, if accounting its production of 50 million tons per year, which will eventually be increasing even more as electronics become a key factor in machines used by mankind. e-waste constitutes numerous chemicals, like lead, mercury, brominated flame retardants, beryllium, and cadmium, that may hamper all living things as well as the environment. The mishandled electronics disposal ends up in our air, water, and soil. This chapter presents the significant impacts of e-waste on the environment, and lists out the possible ways which may help to get rid of it to a considerable extent.

DOI: 10.4018/978-1-6684-7573-7.ch005

INTRODUCTION

Waste generation and its management has always been a topic rendered among environmentalists. Preventing globe and keeping it striking is a stance handled by ecologist with tedious efforts which leads to safe and healthy livelihood. There is a huge quantity of waste that ends up in landfills throughout the world. This waste has caused detrimental impacts on ecology and human health. To alleviate those concerns, households reprocess, recycle their waste or manure food scraps. It is the upmost way out to save the planet. One of the types of waste which is often misunderstood as not to be going to harm landfills is electronic waste, also known as e-waste.

Electronic, microchip technology, Integrated circuit, Semiconductor technology are the sphere's chief and wildest growing industrial sectors. Speedy growth accompanied with fast product uselessness often results in thrown away or discarded electronics devices or elements which is now the wildest rising waste stream in the industrial ecosphere. The rising amount of e-waste from electronic industry is commencing to reach calamitous magnitudes. e-waste is the term used to define timeworn, damaged, or ruined electronic equipment like laptops, music systems, cell phones, geysers, telephones, televisions, old radios, refrigerators, computers, tube lights, lightning gadgets, toasters etc. which have been disposed of by their owners due to main two reasons either desire for upgraded version or damage to older appliance. Therefore, e-waste is engendered as comparatively costly and principally long-lasting products used for telecommunications, data processing, or amusement in domestic, households and businesses.

The issues, concerns and aftereffects of e-waste are observed, experienced, examined and analysed from last few decades. Industrialization, automation, and luxury of life laid down the generation of huge e-waste for future. If not taken seriously it may cause serious impacts on the planet. Electronics is rendering its path everywhere from villages to metros. Individuals use electronics every moment. After using them to some extent and after getting issues in their normal working people often either replace of through them. This trash is reaching to its gigantic shape gradually and its rate is increasing even with a dangerous speed which if not taken care of will create an unbelievable impact harmful not only to society but to the planet even. So, it's a growing problem that needs to be further addressed.

This chapter deals with the introduction, information, and investigations of the impacts of e-wastes due to three major concerns of it such as 1) it is the major copper-bearing waste type, 2) its steep increase in generation globally and 3) it is highly enriching in major metals acknowledged by the World Bank.

e-waste is technically known as waste of electrical and electronic equipment (WEEE).

Figure 1. Resources of e-waste

e-waste is the type of waste which is generated in huge quantity as an output of steeply increasing reliance on electronic gadgets and tremendous rate in exploring the innovations in technologies. Figure 1 shows some of the resources of e-waste generation. Most of e-waste pollutants leached into the land and water and release toxic substances It is a worldwide problem that is rendering stress on all income-range countries. Those lower to intermediate profile countries are the highest receivers of huge bulks of e-waste per year from developed countries. In certain places, valuable elements and metals are hauled out from this disposed-off e-waste. Nevertheless, it can slightly reduce the hordes of waste accumulations, keeping much of it useless. The toxic components of e-waste sat global eco as well as health risks; it contains valuable recoverable components which may generate revenue leading to economic paybacks (Andrade, Soares, Solé, Pereira and Freitas, 2022).

The reformatting of the earth by mankind is revealing noticeable concerns for the humanity and their health concerns and environment as well. Although the deprivation is perceptible and persistent maximum of the distress caused by humans found to exterior to the monarchy justice systems. Such harms must be regulated, and such regulation shall lead to the complete preservation of the environment and the entire life on the sphere (Crutzen & Stoermer, 2000).

Effective e-waste reparation, recycling and reprocessing is said to be optimally exhibited if its elements are segregated and effectively processed by advanced technological and chemical methodologies. If waste component collection, separation, segregation, and split-up is manually executed, the probable chances of their health at risk could not be deniable which may cause wastage of time also. To overcome this issue automation and robotics involvement in waste handling, processing and disposal are essential (Bisschop, 2016)

WHAT IS E-WASTE?

Anything from washing machine to telephone instrument, which is outmoded, not required or damaged electronic/electrical equipment is accounted as e-waste or electronic waste. It can also be defined as anything that operates on electricity and required to be discarded off.

Mostly the hazards of this waste were often overlooked. It is found that globally around 10% of e-waste is recycled and remaining 90% is getting landfilled, burned, or illicitly transacted. These numbers are annoying as well as disappointing too (Rehamn, Mack, & Alam, 2021).

BACKGROUND

The tremendous upsurge in worlds' population is going with the exponential development in huge accumulations of waste of e-waste) generation due to the rigorous technological implementations. Nowadays e-waste is of great concern of ecologists and considered as incipient issue to be taken care seriously. It is also accounted as of having a huge business prospect which may gain significant attention as the volumes of e-waste being generated found to have content of both toxic, nontoxic and sources of valuable metals in them. It has been observed that over 60% fractions include copper, iron, aluminium, gold and many other valuable metals whereas pollutants contains approximately 3% (Davies, Hernandez, & Wyatt, 2019).

Davies, P (Davies, Hernandez, & Wyatt, 2019), presented the previous and present socio-economic and ecological backgrounds of a destitute communal intertwined in worldwide hard work to guard and mend the atmosphere. It divulges an address from companies endorsing their societal and ecological accountability in a civic that undergoes socioeconomic demotion and ecological contests.

The incongruous and incompatible handling, processing and disposal of e-waste incorporating Rare-Earth Elements (REEs) such as gadolinium (Gd), is continually enhancing its occurrence on the earth. In specific, the Gd content in aquatic systems may create ecological risks which were still unidentified, particularly considering expected weather variations like water salinity changes due to extreme weather conditions (Widmer, Oswald, Sinha, Schnellmann, & Böni, 2005). In (Widmer, Oswald, Sinha, Schnellmann, & Böni, 2005), the assortment of information presented by author indicated the possibility of recycling and reprocessing e-waste probable way out for effective waste management through robotic application implementation. Their methods involved finding e-waste, classifying and segregating diverse working fragments from the discarded electronic discharges and reusing or reprocessing those into stretchy components with entire, joined, translational and rotating freedom. This work shows the probability in reprocessing e-waste to produce practical electro-mechanical devices.

In (Rehamn et al., 2021) authors highlight the hazards of e-wastes, the need for its appropriate management and options that can be implemented.

In it is nicely explained that a crossover class of climate change solutions (which we term “technological solutions”) may disproportionately and adversely impact some populations over others. Suggest that many of the most attractive technological solutions to climate change, such as solar energy and electric car batteries, will likely add to the rapidly growing stream of electronic waste (“e-waste”). This e-waste may have negative downstream effects on otherwise disenfranchised populations. We argue that e-waste burdens women unfairly and disproportionately, affecting their mortality/morbidity and fertility, as well as the development of their children.

Various authors report the vast insight into this field. The thorough study over it narrates various awful bullets as well as their remedies which are listed in the Table 1 Given below.

IMPACTS OF E-WASTE

Among all other types of wastes E-Waste is the less attended category due to either scarcity in measures to rectify their impacts of lack of awareness regarding its worst impacts. So the upmost priority should be given to gather the knowledge about all possible impacts of it.

Following are the noteworthy impacts of e-waste worth mentioning while discussing about it which are listed after thorough study and analysis of available literature in this field.

- It can be toxic,

Table 1. Noteworthy contribution of e-waste and its information

SN	Reference	Noteworthy contribution	Type of waste discussed	Highlights	Concluding remark	Applicable to (Specfic Country / Global)
1.	(Bassiouny, Farhan, Maged and Awaad, 2021)	Recycle and reuse only temporary transfer the e-waste to other places and after sometimes, the e-waste still landed on the landfill	E Waste	Reducing the use of toxic materials is a better way of managing e-waste, but for time being, it is difficult and not cost effective to replace and reduce all these materials	College of Information Technology (COIT), UNITEN in order to identify whether the students are aware of the essentiality of e-waste management. Based on the study results, it shows that COIT students are not ready for e-waste management and the factor of not offering of Green IT or e-waste management subject is having the main factors of the unawareness of e-waste management	
2.	(Chen and Yee, 2011)	provides a critical review of the technological details of different metal recycling processes, pre-treatment methods, and the advancements made in these techniques.		Critical evaluation of different metal recovery techniques has also been presented based on the available life cycle assessment (LCA), techno-economic, and industrial-scale studies.	the integrated metal recovery techniques serve better in terms of recovery efficiency and environmental performance than any single recovery technique. Also, scaling up of bio metallurgical, electrochemical and super critical fluid extraction methods needs to be promoted due to their better environmental performances.	Global
3.	(Prasad, Vithanage and Borthakur,2019)	readers an overview on the e-waste topic—how e-waste is defined, what it is composed of and which methods can be applied to estimate the quantity of e-waste generated.	E waste	The three countries are compared using an assessment indicator system which takes into account the structural framework, the recycling system and its various impacts of e-waste	a) e-waste recycling has developed in all countries as a market based activity, b) in China and India it is based on small to medium-sized enterprises (SME) in the informal sector, whereas in South Africa it is in the formal sector, and c) each country is trying to overcome shortcomings in the current system by developing strategies for improvement.	China, India, South Africa
4.	(Monika and Kishore, 2010)	Presents e-waste composition, categorization, Global and Indian e-waste scenarios, prospects of recoverable, recyclable, and hazardous materials found in the e-waste,	e-waste	Best Available Practices, recycling, and recovery processes followed, and their environmental and occupational hazards.	Unlike the European e-waste system where consumers need to pay for the disposal of their e-waste or return them free of charge to recycler, consumers in India, however, expect to receive payment for their e-waste, which is viewed as a potentially valuable resource	India
5.	(Ramachandran, Dharmaraj and Natarajan, 2022)	focuses on the detailed explanation of the e-waste management system includes the recycling process and its effects in India	E waste	Discussion of legal instruments and awareness programs in India at various periods of time. Also, it provides sufficient knowledge to the readers in various aspects of increasing e-waste and its controlling methods	gives adequate information for reducing the utilization of e-product in consumer-side and control measures on the manufacturer-side	India
6.	(Rima and Sukha, 2022)	Detailed presentation of e-waste handling methods that also includes examples to further demonstrate how they work in the real world	e-waste	Presented the data that reveals the geographies of e-waste flows at global, national and subnational levels.	a detailed presentation of e-waste estimation methods that also addresses both the handling of e-waste and their hazardous effect on the surrounding environment.	Global
7.	(Robinson, 2009)	Thorough analysis of e-waste management in India	e-waste	Burden of e-waste, its health impacts, Current status of E-management and initiatives taken for it.	For e-waste management many technical solutions are available, but to be adopted in the management system, prerequisite conditions such as legislation, collection system, logistics, and manpower should be prepared.	India, China
8.	(Tabelin, Park, Phengsaart, Jeon,Villacorte, Alonzo,Yoo, Ito and Hiroyoshi 2021)	Presented a critical analysis of ore distribution/processing, metal extraction, e-waste generation and e-waste recycling is	Copper and critical metals production from porphyry ores and e-wastes	provide valuable information about the current status of copper and critical metal resources from porphyry deposits and e-wastes, advances and challenges in their processing/recycling, and sustainable strategies to limit impacts of their exploitation on society and the environment.	more sustainable mine waste management strategies, including better Life Cycle Impact Analysis tools with spatial and temporal dimensions, are needed to limit socio-environmental impacts of resources exploitation and maintain the sector's Social Licence to Operate	
9.	(Rehamn, Mack and Alam, 2021)	It assess the global production of e-waste, the contaminants and contaminant fluxes associated with e-waste, and the likely environmental impact of e-waste associated contaminants	E waste	focuses on the physical, chemical and environmental effects of e-waste, rather than e-waste associated politics, which are discussed extensively in publications from the aforementioned NGOs	The negative effects of the contaminants at these sites may be reduced using standard remediation technologies.	Global
10.	(Wath, Dutt and Chakrabarti, 2011)	Discusses case study conducted into the upcycling of e-waste	e-waste	highlights some of the impracticalities and challenges involved in e-waste upcycling	there is feasibility in re-engineering e-waste to manufacture functional electro-mechanical products and devices, and that upcycling e-waste in this way is a step towards responsible e-waste management through engineering knowledge and practice	Malaysia
11.	(Widmer, Oswald, Sinha, Schnellmann and Böni, 2005)	The presence of Gd in marine systems may lead to environmental risks which are still unknown, especially considering foreseen climate modifications such as water salinity shifts due to extreme weather events.	e-waste	the present study intended to assess the combined effects of Gd at variable salinities.	the need to investigate the risk of emerging contaminants such as rare-earth elements at present and forecasted climate change scenarios, thus providing a more realistic environmental risk assessment.	

- It is not biodegradable accumulates in the environment, in the soil, air, water and living things.
- Its impact on disposal on lead concentrations in landfill leachate
- Electronic waste leading to environmental pollution
- Health issues
- e-waste management

e-waste disposes plenty of toxic components which are harmful to health of humans. These toxic components are listed as follows.

- mercury,
- lead,
- cadmium,
- polybrominated flame retardants,
- barium
- lithium.

The adverse health impacts of these toxins on human body may include damage to heart, liver, skeletal system, kidney and even to brain.

SOLUTION TO DISCARD E-WASTE

Comprehensive climate alteration poses threat to flora and fauna and to the vivacity of the globe's biomes. It is the topic to be argued that the globe is now in a new geological era, the "Anthropocene" (Crutzen et al., 2000), in which the impact that humans create will be the defining feature of the planet (Verma & Prakash, 2020).

Humans ominously reformed the globe. This reforming in overriding the phase of progress and attaining the high prosperity for many societies has definitely been leading to vast wreckage to the sphere, its ecologies and obviously to a great extent to non-anthropoid. Nonetheless this damage is noticeable and persistent much of the harm is even now not noticed by common people and the lacks in taking preventive measures at huge levels which is the prime requirement of today's era (Crutzen et al., 2000). Such destructions have a tendency of not to be outlawed whereas it can be regulated which has not led to the overall preservation of the environment or to the protection of fauna.

Undeniably, pathetic and erratic ecological protocols through the world have empowered global establishments to upsurge their revenues without bothering ecological balance and benefits to local societies. Figure 2 shows the chief factors which are responsible for e-waste generation.

Figure 2. Factors responsible for generation of hazardous e-waste

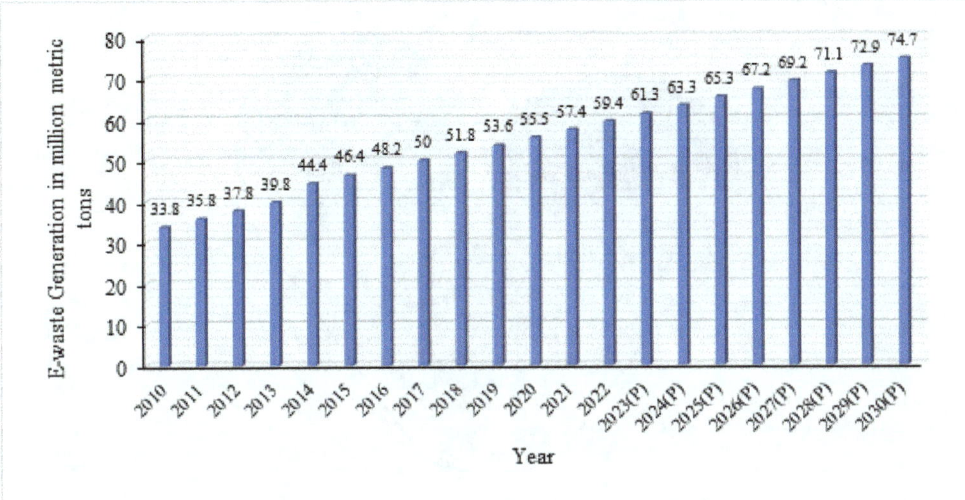

This chapter surveys how e-waste that is the waste generated by electronic and electrical tools which presents a challenge for ecological authority. A casual e-waste recycling division pays many households in metropolitan areas to accumulate, sort, patch-up, mending, renovate and dismantle trash electrical/ electronic devices. It is projected that the quantity of e-waste engendered worldwide per year will touch the dreadful number of 100 billion tones by 2025. Looking in the severity of its existence and the overlooked outcomes of its deadlier aftereffects this chapter also extends its study in discussing how e-waste is a challenge for controlling the banned trade and stoppage of ecological destruction. A steep focus on the role of corporate, state as well as civil culture, insights are gained into the fortes and confines of the supremacy context. It strongly recommends the essential for echo about both practical and theoretical insinuations that ascend for ecological domination (Wath et al., 2011).

Looking into the severe impact of this waste environmentalist strongly demands the impactful remedies for overcoming this issue. Figure 3 shows some of the way outs to get rid of this waste.

Continues efforts of researchers and economist have searched for four remedies or R's which may be implemented to get rid of vulnerabilities of E-Waste are listed as.

- Reduction
- Reparation
- Reprocessing
- Reutilization

Figure 3. Possible way-outs to get rid of e-waste

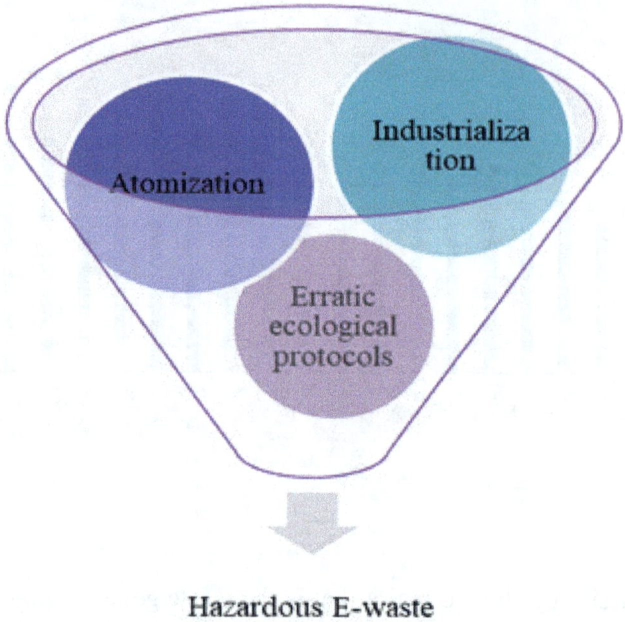

These R's were studied, analyzed, implemented, executed, and thereafter found impactful and worthy of execution to find an effective way out to get rid of most of the impacts of e-waste. It is observed that with slight changes as per the mode of requirement these R's may also be implemented for discarding other types of wastes also.

REDUCTION

Nonetheless it is easier said than done, the simplest strategy to solve the e-waste crunch is to diminish rate of its production. Enterprises and firms are frequently proposing updated products which is attracts to the consumers for switching to updated version by discarding the older one. If consumers stick to need rather than new features the generation of e-waste can be reduced to much extent.

REPROCESSING

Now day sophistication urges people to discard their electronic gadgets rather than replacing or repairing it. Reprocessing is often found ignored by mass utilizers.

REPARATION

Instead of throwing out those old gadgets they may be considered for reusing after altering the old form, reselling, or offering it to the needy. Else or could also hold on to the same with little editing which may found useful any day in future.

REUTILIZATION

As an upmost possibility e-waste can always be reutilized or recycled with precise methodology. Often e-waste recycling centers and drop-off depots to handle the gadgets for further possible process are available. The deficiency of a consistent e-waste cataloging and legislature, including deprived assemblage rates, leftovers stern difficulties in e-waste recycling.

IMPACTS OF E-WASTE

Electronic waste is an unintended and inconceivable waste which may develop adversative impacts on the human, nonhuman and atmosphere by contaminating the natural resources like air, soil and water. Amassing and adulteration of the e-waste for an extended duration may severely disturb the ecological assets. The current universal assembly of e-waste is assessed to be 20–25 million tonnes per annum and it has been observed that maximum of the e-waste is generated in The United States, Europe, and Australasia. Analysts predicts that China, Eastern Europe and Latin America may be the major e-waste producers in the next ten years. All those countries were found to be the major consumers of electronic gadgets so often blamed responsible for an increase in the e-waste (Ramachandran et.al., 2022). In 2015, 193 governments approved seventeen Sustainable Development Goals (SDGs) directing to eliminate scarcity, guard the earth and confirm peaceable and flourishing survives for everybody by 2030 (Tabelin et al., 2021) 13th point of the SDGs regarding Climate action envisages severe carbon dioxide (CO_2) emission diminutions to fight climate alteration and attain a carbon-free planet by 2030. According to the Paris Agreement, "Climate Action" explicitly goals the power generation and transportation segments and sectors claiming about 70% approximately 22 Mega tones of CO_2 emissions from fuel combustion in 2018 which utters the replacement of fossil fuels with renewable sources and transition of automobile fields to electric-vehicles (EVs). Unfortunately, most of the low-carbon tools are more mineral, material and metal comprehensive than traditional fossil fuel technologies. For example, a specific 5 Megawatt wind generator entails approximately 5 tons of copper, 3.5 tons of aluminum and 2.1 tons of Rare Earth Elements (REEs) and considerable amounts of molybdenum and zinc whereas PV (photovoltaic) cells need approximately 1 percent copper and 0.1-0.2 percent of silver, lead and tin. Even electric vehicles require more 1.5 to 12 times more copper than conventional vehicles rather it also demands crucial elements like cobalt, manganese, aluminum, nickel and lithium. These numbers ask for persistent supplies of approximately seventeen types of metals and elements like materials—

- zinc
- chromium,
- nickel
- lithium
- aluminium,
- vanadium
- manganese,
- cobalt
- titanium
- lead
- indium,
- graphite,
- iron
- molybdenum
- silver
- neodymium and
- copper

These elements are acute in the next 20–40 years for the clean global scenario. Figure 4 shows the possible extracts of e-waste which indicates the benefits in recycling or reparation it.

e-waste contains many valued elements and metals such as copper and of platinum group. It is also found to behaving high potential ecological impurities and toxic gases. Few of them are listed as

- Lead i.e., Pb,
- Antimony i.e., Sb,

- Mercury i.e., Hg,
- Cadmium i.e., Cd,
- Nickel i.e., Ni,
- polybrominated diphenyl ethers i.e., PBDEs
- polychlorinated biphenyls i.e., PCBs.
- Apart from this burning e-waste may generate.
- dioxins,
- furans,
- polycyclic aromatic hydrocarbons (PAHs),
- polyhalogenated aromatic hydrocarbons (PHAHs), and
- hydrogen chloride.

Figure 4. Possible extracts of e-waste

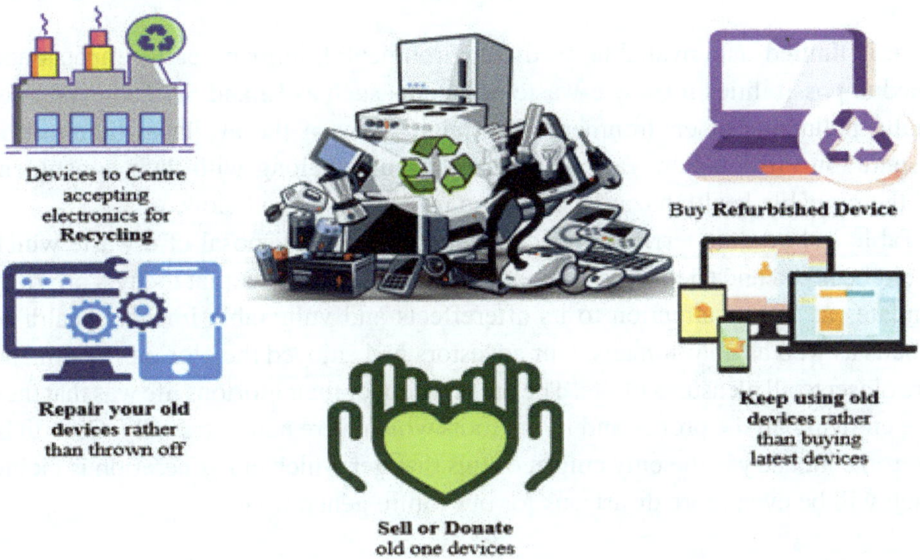

The worst part of this disaster is that the chemical composition of e-waste alters with the alteration, updating or improvement of new technologies. The pressure from ecological societies, environmentalists on high ranked e-waste generating countries, on electronics companies to explore, investigate and implement fruitful solution to e-waste and its impacts renders speed to pay attention towards it. But it is found that instead of finding beneficial solution those culprits dispose e-waste in landfills as efficient recycling and reprocessing techniques which can recover the valuable metals and elements with trifling ecological influence are complicated, time consuming and costly too. Subsequently, though it is illegitimate as per the Basel Convention, developed countries often export an indefinite amount of e-waste to deprived countries, where reprocessing methods comprises of burning and dissolution in strong acids with few procedures to protect health hazards and the atmosphere. Such recycling processes at the start consequence extreme local adulteration which is trailed by movement of the pollutants into water passages and food cycles. e-waste personnel agonize adverse health impacts due to the skin contact and breathing, although the broader communal are highly wide-open to the toxins and impurities due to dust, smoke, contaminated drinking water and adulterated food. There is indication that e-waste allied pollutants may be existing in few of farming or manufactured products for export (Rehamn, 2021).

HEALTH IMPACTS DUE TO E-WASTE

There is limited data available on the environmental, human health impacts and remedial possibilities for few e-waste pollutants such as Li and Sb as they are not usually pollutants of environment. Thorough study of the available literature is presented in this chapter. Various e-waste resources along with their constituent and the probable health hazards and given in Table 2 given below.

Table 2 shows the terrible impacts of inappropriate disposal of e-waste which should be noted and tried to be avoided. It is unavoidable fact that today's progress mandates us to pay attention to its aftereffects and vulnerable impacts health of humans as well as non-humans. Our ancestors had enjoyed the glorious health and were closer to all pleasures of life. The only reason for their glorious life was that they were enjoying all the probes and living tools which were nature friendly. Growth in electronic industry is the only culprit of this disaster which our generation is facing which will be even more disastrous for our future generation.

CONCLUSION

In this era of rapid industrialization and automation e-waste is counted as one of the significant environmental threats which required keen attention to be paid off. Now days many researchers and environmentalist are working hard to pave out and find the way to get rid of this calamity which arouse as a side effect of mankind progress. This chapter focuses the most severe impact of it which indirectly pricks the safety of environmental peace of globe.

Table 2. e-waste resources along with their constituent and the probable health hazards

SN	Resources	Components	Health Impacts
1.	Panels of Cathode Ray Tubes (CRTs): A vacuum tube used as a display screen in a computer monitor or TV	• Phosphorus • Barium • heavy metals	• calcification, or hardening, of heart arteries and increase inflammation • Nausea, vomiting, irregular heartbeat, muscle weakness • Lower energy levels and damage the functioning of the brain, lungs, kidney, liver, blood composition and other important organs.
2.	PCB Soldering	Lead	• Causes anemia, weakness, and kidney and brain damage • Adverse impacts on brain growth of children
3.	• Computers, • Televisions • VCRs • stereos, Xerox machines • fax machine	• mercury • lead • beryllium • brominated flame retardants • cadmium	• Causes anemia, weakness, and kidney and brain damage • Adverse effects on brain development of children • Beryllium sensitization, chronic beryllium disease (CBD), and lung cancer. • Breathing high levels of cadmium damages people's lungs and can cause death
4.	• Chip resistors • semi-conductors	Cadmium	• Breathing high levels of cadmium damages people's lungs and can cause death
5.	• Copper wires • PCB tracks.	• Copper • Lead	• cause vomiting, nausea, abdominal pain, and or diarrhea • Causes anemia, weakness, and kidney and brain damage • Adverse impacts on brain growth of children
6.	• Relays and switches • printed circuit boards	Mercury	• difficulty breathing • chest pain, cough • irritability • irritation to the eyes, skin, and stomach • weakness or exhaustion • insomnia • headache • indecision
7.	• Cabling • computer housing	• Plastics • PVC • copper	• Burning produces dioxin that causes reproductive and developmental problems • cause vomiting, nausea, abdominal pain, and or diarrhea
8.	Motherboard	Beryllium	• beryllium sensitization, chronic beryllium disease (CBD), and lung cancer.
9.	Lithium-ion battery	Lithium	• cause loss of appetite, nausea, vomiting, diarrhea and abdominal pain
10.	• Electronic equipment • circuit boards	• Lead • Brominated flame-retardants	• Causes anemia, weakness, and kidney and brain damage • Adverse impacts on brain growth of children • May cause diabetes, neurobehavioral and developmental disorders, • cancer, reproductive health effects and alteration in thyroid function
11.	Nickel–cadmium rechargeable batteries	Nickel–cadmium	• May cause various types of cancers like lung, pancreas, breast, prostate, kidney and nasopharynx
12.	• Galvanized steel plates • Decorator • hardener for steel housing	Chromium	• It may cause stomach problems, headaches, insomnia • It has been found that high doses may cause serious side effects including liver or kidney damage

ACKNOWLEDGMENT

This research received no specific grant from any funding agency in the public, commercial, or not-for-profit sectors.

REFERENCES

Andrade, M., Soares, A. M., Solé, M., Pereira, E., & Freitas, R. (2022). Will climate changes enhance the impacts of e-waste in aquatic systems? *Chemosphere, 288*, 132264. doi:10.1016/j.chemosphere.2021.132264 PMID:34624793

Bassiouny, A. M., Farhan, A. S., Maged, S. A., & Awaad, M. I. (2021). Comparison of Different Computer Vision Approaches for e-waste Components Detection to Automate e-waste Disassembly. *2021 International Mobile Intelligent, and Ubiquitous Computing Conference*, (pp. 17–23). IEEE. doi:10.1109/MIUCC52538.2021.9447637

Bisschop, L. (2016). How E-Waste Challenges Environmental Governance. In: Tanya Wyatt (eds) Hazardous Waste and Pollution, 27-43 . Springer, Cham. doi:10.1007/978-3-319-18081-6_3

Chen, L. F., & Yee, H. W. (2011) E-waste management: Are we ready for it? A study on the awareness of COIT students toward e-waste management. *ICIMU 2011: Proceedings of the 5th international Conference on Information Technology & Multimedia*, (pp. 1-5). IEEE. 10.1109/ICIMU.2011.6122729

Crutzen, P. J., & Stoermer, E. F. (2000). The "Anthropocene". The International Geosphere–Biosphere Program (IGBP): A Study of Global Change of the International Council for Science (ICSU) (*International Geosphere-Biosphere Program*). *Newsletter, 41*(May), 17–18.

Davies, P., Hernandez, M.P. & Wyatt, T. (2019). Economy Versus Environment: How Corporate Actors Harm Both. *Critical Criminology 27*(1), 85–99. . doi:10.1007/s10612-019-09433-z

Kumari, R., & Samadder, S. R. (2022) A critical review of the pre-processing and metals recovery methods from e-wastes, *Journal of Environmental Management, 320*. doi:10.1016/j.jenvman.2022.115887

Monika, K. J. (2010, July). e-waste management: As a challenge to public health in India. *Indian Journal of Community Medicine, 35*(3), 382–385. doi:10.4103/0970-0218.69251 PMID:21031101

Prasad, M. N. V. & Borthakur, A. (2019). Handbook of Electronic Waste Management International Best Practices and Case Studies.

Ramachandran, R., Dharmaraj, K., & Natarajan, P. (2022). Electronic waste: A critical assessment on the unimaginable growing pollutant, legislations and environmental impacts. *Environmental Challenges*, *7*, 100507. doi:10.1016/j.envc.2022.100507

Rehamn, U., Mack, J., & Alam, P. (2021). From e-waste to robots: a case study on e-waste up cycling in low-to-middle income countries. *2021 IEEE International Humanitarian Technology Conference (IHTC)*, (pp. 1-4). IEEE. 10.1109/IHTC53077.2021.9698905

Rehamn, U., Mack, J., & Alam, P. (2021). From e-waste to robots: a case study on e-waste upcycling in low-to-middle income countries. *2021 IEEE International Humanitarian Technology Conference (IHTC)*, (pp. 1-4). 10.1109/IHTC53077.2021.9698905

Robinson, B.H. (2009). e-waste: an assessment of global production and environmental impacts. *Science of the Total Environment*, *408*(2),183-91. doi:.2009.09.044. doi:10.1016/j.scitotenv

Tabelin, C. B., Park, I., Phengsaart, T., Jeon, S., Villacorte-Tabelin, M., Alonzo, D., Yoo, K., Ito, M., & Hiroyoshi, N. (2021). Copper and critical metals production from porphyry ores and e-wastes: A review of resource availability, processing/recycling challenges, socio-environmental aspects, and sustainability issues. *Resources, Conservation and Recycling*, *170*, 105610. doi:10.1016/j.resconrec.2021.105610

Verma, A. K., & Prakash, S. (2020). e-wastes and their impact on environment and public health. *International Journal of Applied Research*, *6*(9), 164–168. doi:10.22271/allresearch.2020.v6.i9c.7111

Wath, S. B., Dutt, P. S., & Chakrabarti, T. (2011). e-waste scenario in India, its management, and implications. *Environmental Monitoring and Assessment*, *172*(1-4), 249–262. doi:10.100710661-010-1331-9 PMID:20151189

Widmer, R., Oswald-Krapf, H., Sinha-Khetriwal, D., Schnellmann, M., & Böni, H. (2005). Global perspectives on e-waste. *Environmental Impact Assessment Review*, *25*(5), 436–458. doi:10.1016/j.eiar.2005.04.001

Chapter 6
A Comparative Study on E-Waste Management Systems in Developed and Developing Countries:
Legislative Compliances and Initiatives

Sanjeeda Atiya
CHRIST University (Deemed), India

Yvonne Josephine Chung
CHRIST University (Deemed), India

Ginu George
CHRIST University (Deemed), India

ABSTRACT

E-waste is an ongoing issue that still lacks a suitable solution, particularly in developing nations. The environment and human health have suffered dramatically as a result of poor recycling practices of waste of electrical and electronic equipment (WEEE), transboundary movement, improper management of e-waste, the lack of environmentally sound management (ESM) programs, and the ineffective EPR (extended producer responsibility) schemes. Although developed nations have implemented efficient legislative frameworks and regulations, emerging nations suffer due to their plans. E-waste management systems differ in developed and developing countries; thus, this study evaluates the differences between the management systems and outlines the areas where the developing nations lack effective e-waste management and the advantages developed countries enjoy. Therefore, the current study results are crucial for comprehending the severe hazard posed by improper management of e-waste and the viability of future research into creating strategies to address these problems of developing nations.

DOI: 10.4018/978-1-6684-7573-7.ch006

INTRODUCTION

The recent developments in information technology and telecommunications have significantly changed the world, yet while these technologies have significantly favorable effects on people's lives, they have also harmed the environment (Osibanjo & Nnorom, 2007). Because of this, used EEE, often referred to as WEEE and E-waste, has emerged as a significant societal issue and environmental danger in numerous nations worldwide (Herat & Pariatamby, 2012). Moreover, during the SARS-CoV-2 virus pandemic, human life became more dependent on Electronic and Electrical Equipment (EEE) while staying at home without commuting due to the immediate need for social isolation and lockdown (Trivedi et al., 2022).

Furthermore, the issue of handling and recycling E-waste affects both industrialized and developing countries, especially the latter. This is because most emerging countries' collected E-waste consists of both domestically produced E-waste and E-waste illegally imported from developed countries (Widmer et al., 2005). There are growing worries that the majority of the electronic waste produced in developed nations is being sent to underdeveloped nations, where there is a lack of infrastructure for Environmentally Sound Management (ESM) of electronic waste, which can have adverse effects on the socioeconomic situation, public health, and the environment (Herat & Pariatamby, 2012).

There is a significant variation in how E-waste is managed in developing countries due to factors like a lack of infrastructure for proper waste management, a lack of legislation that deals specifically with E-waste, and the lack of any framework for the take-back of end-of-life (EoL) products or the implementation of EPR (Osibanjo & Nnorom, 2007). Although the E-waste legislation in each country is designed to address problems unique to that nation, it is typically not comprehensive, which creates a variety of management concerns (Patil & Ramakrishna, 2020). This highlights the significance of efficient management, disposal, and recycling of E-waste, which also compares the E-waste management systems of both developed and developing countries.

OBJECTIVES

It is necessary to realize that as time has evolved and there is evident evolution in our daily lives, technology is an area that has seen immense growth. This growth has not only led to positive advancements, but the excessive increase in technology has led to millions of tonnes of E-waste being generated daily. The objectives of our study are as follows:

1. To understand the various E-waste management systems adopted by developed and developing nations.
2. To understand the laws and regulations in the different countries that have evolved over the years, along with acknowledging the initiatives undertaken by them.
3. To understand the magnitude of initiatives like Producer Responsibility Organisation (PRO) and EPR.
4. To understand the impact of Covid-19 on E-waste production.

E-WASTE LIFE CYCLE

Each piece of EEE goes through a lifespan process roughly divided into three phases. Wath et al. (2010) mentioned that although there are more classifications and it is more organized in developed countries such as Switzerland, the processes could be better defined and more organized in developing countries such as India.

Stage 1: EEE Production

EEE is produced, manufactured, or imported from other nations to meet the demand. Debnath et al. (2015) stated that the countdown to an object's lifecycle begins when it is manufactured or produced and enters the value chain process. Wath et al. (2010) mentioned that this EEE moves through the supply chain and is handled by Distributors, Retailers, or Suppliers of Raw Materials, referred to as "Actors of EEE." Their involvement will end there because they are not final consumers and will not be considered WEEE generators. Since the waste is produced after the equipment is used to its maximum potential, the end users of EEE are regarded as WEEE Generators.

Some countries may import Semi-Knocked-Down (SKD) components or the equipment as a whole or a finished product. Developing nations import SKD primarily so they can be assembled and sold for significantly lower prices (Atlason et al., 2017). These suppliers are referred to as "Assemblers," and their market is far more extensive than branded products. As a result, the grey market thrives in developing countries and aids in meeting demand at a reduced cost, but it also carries consequences (Debnath et al., 2015). The primary drawback of grey market assemblers is a large number of non-identifiable producers, making implementing Extended Producers Responsibility (EPR) extremely difficult (Manomaivibool et al., 2009). However, these assemblers now return their goods as part of an "exchange offer" practice (Debnath et al., 2015).

Figure 1. Comparison of Per Capita Production (Kg) in Developed and Developing Countries

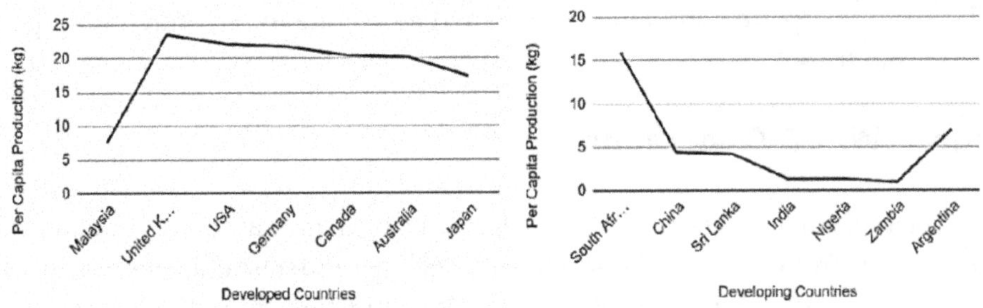

Furthermore, some nations receive donated used EEE from the developed nations. Undocumented EEE is a problem in nations like Nepal since it is distributed through countless unidentified assemblers. Additionally, due to the poor purchasing power of the population in these countries, repair and reuse are highly prevalent, significantly extending the lifespan of the product (Atlason et al., 2017).

Source: E-waste in the international context – A review of trade flows, regulations, hazards, waste management strategies and technologies for value recovery 2018 (Ilankoon et al., 2018); The global E-waste monitor 2017 (Baldé et al., 2017); StEP, E-waste world map (StEP, 2018).

In Figure 2, it is evident that developed countries' per capita production is more than that of developing countries. The reasons for it can be the increased demand for updated EEEs and the high purchasing power of the citizens of developed countries.

Figure 2. Comparison of Total E-waste Production (Kg) in Developed and Developing Countries

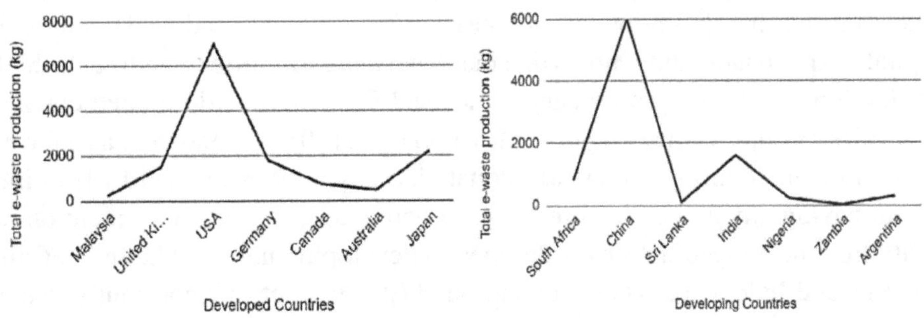

Similarly, in Figure 3, it can be observed that E-waste production is also high in developed countries. The reason being developing countries produce comparatively lesser E-waste even though they are subjected to transboundary movement because the UEEEs are repaired and reused again and again due to less purchasing power of these countries.

Stage 2: WEEE Generation

EEE's final consumers generate WEEE. The majority of these are domestic consumers who buy EEE from distributors or retailers, such as students and people employed in workplaces and institutions. WEEE is also generated by individuals who use electronics like smartphones, televisions, and personal computers. When comparing WEEE producers on a continental scale, the Asia Pacific region generated roughly 25 Mt, followed by America at 13.1 Mt and Europe at 12.1 Mt in 2019 alone (Murthy & Ramakrishna, 2022). However, despite being the second-most populous continent in the world, the African continent generated the least amount of E-waste (Andeobu et al., 2021). According to the Global E-waste Monitor (2019), China is one of the leading producers of E-waste, with 10.1 million metric tonnes produced, and India is third with 3.2 million metric tonnes. At the same time, a developed nation like the United States produces 6.9 million metric tonnes of E-waste, the second largest producer, followed by Japan, which produces 2.5 million metric tonnes, the fourth largest producer of WEEE.

As for the constituents of WEEE, Ongondo et al. (2011) mentioned that iron and steel make up over 50% of the total weight of WEEE and are one of the most common elements in EEE. Plastic follows as the second-largest component, accounting for 21% of WEEE. Finally, 13% of the weight of WEEE is composed of non-ferrous metals such as precious metals, including copper, which is 7%. When it comes to the nations that produce the most WEEE, developed nations with solid EEE demand are the primary producers of WEEE (Hopson & Puckett, 2016).

The leading causes of the high WEEE generation include technical advancements, new models and features, and using the latest equipment while discarding older models. Furthermore, some people donate used EEE to their friends and family and to hospitals, orphanages, and schools that face financial difficulties (Wath et al., 2010). The developed nations also transfer some used EEE to underdeveloped countries as charities exempt from any duties. Mmereki et al. (2016) stated that, as a form of charity, foreign countries frequently donate large quantities of used EEE to India without paying any associated duties. Andeobu et al. (2021) further mention that charity from developed nations is dumped in developing nations like India, China, Malaysia, and Indonesia, whose second-hand EEE markets are constantly rising.

Figure 3. Life cycle of EEE from its production stage to the generation stage of WEEE

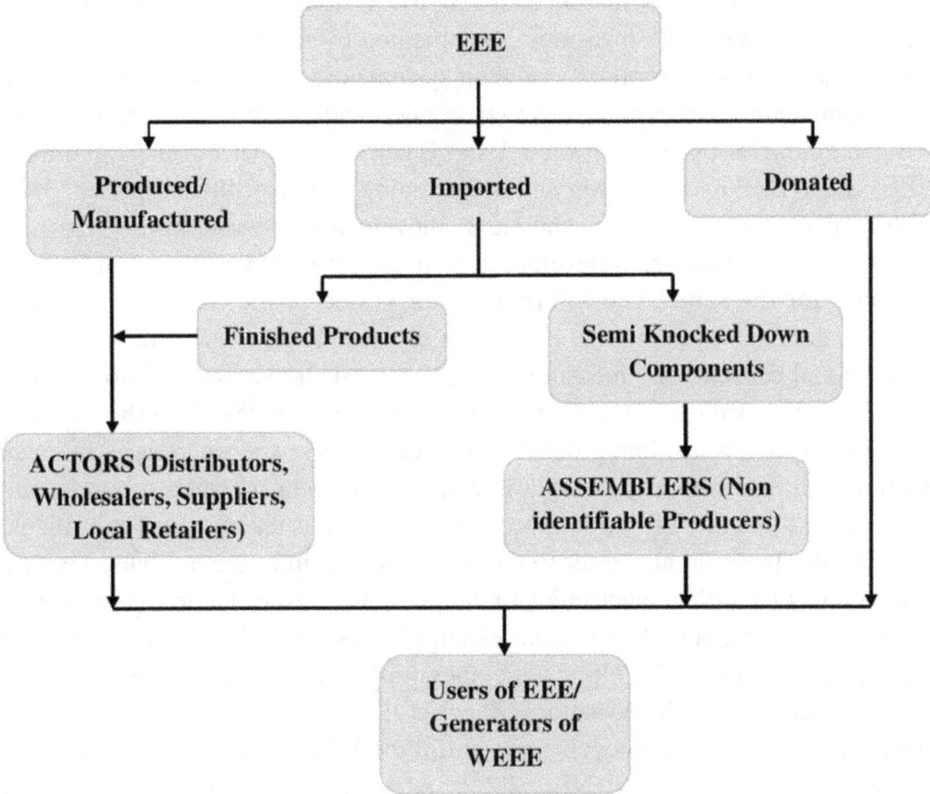

Due to the adoption of newer EEE and the shorter lifespan of older EEE, developed countries produce more WEEE than underdeveloped countries. However, even though used EEE is dumped in developing countries, less WEEE is generated since these countries attempt to repair or recycle the EEE due to their limited purchasing capacity. Batoo et al. (2021) explain it by giving an example- EU citizens generate 15 kg of WEEE per person per year, compared to 1 kg produced annually per person in China and India. As a result, developing nations collect much less electronic waste than developed nations do.

Stage 3: The Disposal of WEEE

The last and final stage of the E-waste life cycle is usually followed by its disposal. Wei and Liu (2012) have discovered that the E-waste generated is generally disposed of both legally and illegally; that is, WEEE is disposed of two private legal disposal entities. However, there is a clear distinction between the disposal systems in developed and developing countries (Mmereki et al., 2016). Although developed nations have strict laws and compliance measures implemented to ensure the safe disposal of E-waste in their nation, several flaws in their system pose a threat to the environment of developing nations. For the wastes to be economically disposed of and recycled, developed countries opt to ship their Used Electrical and Electronic Equipment (UEEE) and WEEE to developing and transitioning nations (Sthiannopkao, 2012).

While keeping in mind that transboundary movement of E-waste is one of the main drivers of the accumulation in developing countries, it is imperative to understand the reasons for the same. The informal sector is one of the most significant in developing nations, revealing that E-waste is discarded improperly due to the lack of professional capacity and measures to undertake such disposal activities (Alias et al., 2014). Agarwal (2012) further mentions that several "leaks" in the informal sector led to the E-waste being discarded through uncrewed operations by open incineration, illegal imports, backyard scraping, dumping in open grounds, and in water bodies that are left unmonitored from hazardous chemicals. These illegal disposal methods would also refer to the waste ending up in second-hand markets and donations of obsolete electronics to the poor. Moreover, due to the high costs incurred in disposing of such hazardous electronic waste, developing countries such as China, Pakistan, and India, along with other third-world countries, are significant dumping grounds for developed nations (Zhang et al., 2012). The activities undertaken by developing countries to upcycle and control the WEEE accumulated are illegal under the Basel Convention (UNEP, 2009). Africa is another nation with a poor E-waste management system imposed with multiple loopholes where E-waste systems and laws have been implemented only since 2017 and later. Being its victim, Ghana has one of the largest landfills in the world, with numerous experiments from their blood samples indicating that the hazardous chemicals affect the workers and people near these dumps with high levels of toxicants in their DNA (Lebbie et al., 2021).

On the other hand, regarding the E-waste disposal and recycling systems in developed countries, research shows that European countries such as Finland, Norway, Sweden, and Switzerland have the highest rate. That is an average of over 42% of systematic E-waste collection with the most advanced techniques adopted to dispose of them (Fadaei, in press). While considering that there are stricter legislative compliances implemented in developed countries, there is a high degree of geographical imbalance between the developed nations with higher income per capita and the developing nations with lower incomes. The E-waste generated by Norway is 140 times higher than in Ghana, Nigeria, and other developing countries alone. This indicates a correlation between the waste generated in developing countries and the amount of E-waste that is collected and required to be disposed of in developing and under-developed nations (Mihai et al., 2019).

In conclusion, due to the unsound infrastructure of E-waste management in developing countries whose environment quality is significantly lower, India, Pakistan, West Africa, Bangladesh, and other parts of Asia are targeted by developed nations as dumping grounds to dispose of their used electronics. The waste generated from it poses a significant threat to those living in the vicinity in the long run (Mohammed, 2022).

E-WASTE MANAGEMENT SYSTEMS

When we distinguish between developing and developed countries, the E-waste management systems of the various countries vary. The formal and informal sectors' systems in developed and developing nations also differ. It is essential to understand that the lack of legislation and improper systems implemented in developing countries affect the regulation of the E-waste generated in the nation, which can simultaneously deteriorate the national health and environment. Although the E-waste legislation implemented in every country is unique depending on the constituents of the composition of their wastes, government policies, and economic position. The laws implemented need to be more holistic to protect the developing nations from being exploited through the transboundary movement of E-waste from developed nations (Doan et al., 2019).

Moreover, the recycling facilities and control facilities to treat E-waste are lacking in developing countries where tiny children are subject to labor who manually scrap the E-waste in small family yards (Greenpeace, 2009; Awasthi et al., 2016). When treating WEEE properly, the developed nations use sophisticated technologies and centralized facilities in a highly regulated environment which helps in the value generation of the product. In contrast, developing countries use labor-intensive methods in an unregulated, hazardous environment. Therefore, value is generated securely in developed countries using measures and precautions. In developing countries, E-waste export and residual waste become a burden on the environment (Ilankoon et al., 2018). This results in a cleaner environment in developed countries at the expense of a dirtier environment in developing countries (Abalansa et al., 2021).

E-Waste Management System of Developed Countries

To understand the differences in the E-waste management systems, we must comprehend the different types of systems followed in developing and developed countries. We will first begin with developed nations, including Japan, Switzerland, and The United States of America.

The United States of America

The residents of the United States of America need more understanding of where they should dispose of their old computers, which is why 75% of these used EEE are still lying inside their houses. It indicates how out of order the proper disposal system still is in America. Since it is uncertain where the used EEE ends up, experts assume that most of it is disposed of in landfills, incinerated, exported, or left in storage (Awasthi et al., 2016). As a result, it is essential to have an appropriate management strategy or system in place to protect public health from hazardous and toxic chemicals in the E-waste stream, as it is apparent that the E-waste management system in the United States is a considerable challenge (Seeberger et al., 2016). The below-mentioned E-waste management systems are followed in the United States.

Circular Economy Concept

The Circular Economy Concept is one of the critical E-waste management systems followed by the United States and other developed nations. As defined by Kirchherr et al. (2017), "circular economy" is an economic system that tries to replace the notion of "end-of-life" with the 3Rs concept, which includes reducing, reusing, and recycling as well as recovering or regenerating the resources during production/distribution and consumption. Therefore, after a product's life cycle is over, it should be retained or sustained in the economy as a resource that may repeatedly be used, thereby increasing its value even further (European Commission, 2008).

When this concept is implemented, it has several beneficial impacts, including increased employment opportunities, a reduction in the production of electronic waste, and increased economic value. By implementing the CE concept, the principle of stock optimization, eco-efficiency, waste prevention, and 4Rs can be achieved where eco-efficiency simply refers to a strategy for reducing the toxicity, volume, and velocity of electronic waste materials; and stock optimization refers to the maximization of the value of resources. These principles or approaches are attained through the application of the primary environment strategy of the 4R's: Reduce, Reuse, Recycle and Recover (Kalmykova et al., 2018; Stahel, 2013).

Refund Deposit System

The Refund Deposit System is one of the traditional methods used to address the problem of disposing of E-waste in the environment. A system called the "bottle bill" system was implemented in the United States of America, and it encouraged customers to return their beverage containers by refunding their deposits. The stakeholders then claimed it should be expanded as a federal policy because it was a successful recycling program (Kahhat et al., 2008). These bottle bills were initially established to address the issue of litter. However, their application was broadened to include other types of containers, tires, lead-acid batteries, electronics, hazardous items, motor oil, and many others. Thus, it became known later as the Deposit-Refund System (Walls, 2011).

The Deposit-Refund concept involves giving a customer a partial or full refund after they return an item to the manufacturer. Thus, this system became preferable over producer take-back schemes and advanced disposal fees (Shinkuma, 2003). When a customer purchases a product, the manufacturer receives the deposit from the retailer and transfers it to the third-party recycling business with some subsidy. The processing capacity and scalability of the companies determine how much subsidy is provided. Later, the unwanted items that customers return are sent to the recycling businesses of their choice (Zhong & Schiller, 2011). This system benefited both parties; it prevented the customers from throwing away their E-waste and helped the producers cut their costs and use the returned product by recycling them and generating their value over time.

Landfill Disposal System

In developed nations, materials accepted for the final disposal of WEEE in landfills are subject to strict and controlled emissions regulations (Morf et al., 2007). While in the US, when only the disposal of desktop and laptop computers was considered, 40 million of the scrap computers were disposed of, eventually entering the end-of-life management sector, of which 17–21% were landfilled in domestic sites in 2010 (Kahhat & Williams, 2012). These statistics are based on the assumption that corporations and public organizations contract with disposal services rather than recycling services to send computers to landfills. It further threatens the environment and poses unimaginable risks, but the exact extent of those risks is still up for debate. Additionally, some properly run landfills assert that the toxicity level of electronic trash is under control, but some studies argue that the problems with toxicity are exaggerated (Williams et al., 2008).

Municipal Solid Waste (MSW) and the vast majority of E-waste are dumped in landfills in the United States (Olsen, 2006). Heavy metals and hazardous substances are both present in E-waste. In the United States, landfills are thought to contain roughly 60% other heavy metals and 40% lead. Because lead is the most dangerous metal, it can be found in Cathode Ray Tubes (CRTs), primarily in non-flat computer monitors and televisions (Wong et al., 2007). Along with other heavy elements like cadmium, batteries can contain significant amounts of lead. Even though some states in the US have effectively banned the disposal of electronic waste containing CRTs and batteries, this is one of the most practical options for consumers to get rid of their electronic waste because they are not aware of the dangers and risks involved with disposing of such toxic metals in landfills. However, it cannot be sustained over the long run as it may raise the danger of exposure for those close to dump sites. Even the most modern landfills emit harmful compounds into the air and leak leachate into nearby groundwater supplies. Safer solutions are, therefore, desperately needed because these risks should not be accepted as the norm (Pak, 2008). Additionally, the government is trying to construct a landfill that can adequately store the remaining or residue electronic waste without harming the environment (Raharjo & Utomo, 2021).

Exportation of E-Waste

Globally, E-waste is increasing due to domestic use and transboundary exports (Thakur & Kumar, 2021). When 200 GPS trackers were used to track them, it was discovered that the United States exports 80% of its electronic waste to Asian countries, 90% of which goes to China alone (Ilankoon et al., 2018; Lee et al., 2018). It was also discovered that in 2007, the US contributed 50,000 tonnes of electronic waste to India, which produced about 4 lakh tonnes on its own, while an additional 20% of E-waste was exported to both India and China as a form of a donation so that these developing nations could recycle them and reap short- term benefits out of them (Zeng et al., 2017; Awasthi et al., 2016).

The US has been exporting its electronic waste to developing nations because they lack the laws and regulations necessary to protect the environment and the workers. Also, since the glass-to-glass recycling of monitors is far more expensive than the developing countries, the US tends to export them to developing countries as recycling there is cheaper (Askari et al., 2014). Furthermore, stringent environmental and social standards are required for recycling operations in the US, which encourages the illegal exportation of E-waste to countries where there is a lack of measures and proper sound environmental management systems (Mmereki et al., 2016). It is legal for the US to export its electronic waste to other nations due to a lack of domestic statutory authority to implement all of the Basel Convention's provisions and the fact that the US has not ratified it yet (UNEP, 2007)

Switzerland

Switzerland was one of the first nations to formally establish and operate an E-waste management system, formally the most technologically advanced nation, and has one of the greatest per capita incomes globally. Additionally, Switzerland has the most meticulously planned and well-organized system for collecting and recycling electronic waste (Khetriwal et al., 2005).

The UEEE are taken back by the retailers, manufacturers, and importers in an environmentally responsible manner free of charge (Fishbein, 2002). However, in addition to recycling, materials that cannot be recycled are also burned for energy, with about 2% being disposed of in landfills (Bhutta et al., 2011). The reason why this country's system of E-waste recycling is effective is that they run two different recycling programs, which are very active- SWICO Recycling Guarantee and Stiftung Entsorgung Schweiz System (SENS). The former looks after or manages the "brown" equipment, such as computers, televisions, radios, etc., while the latter manages the "white" equipment, such as washing machines, refrigerators, and ovens (Gaidajis et al., 2010).

To manage the disposal system of E-waste, Advance Recycling Fee (ARF) was introduced by the Swiss System for consumers of electronic devices to collect a small fee during the time of purchase to cover the future disposal problem of abandoned devices. This fund is charged for covering the cost of collection, transporting, and proper recycling of e wastes. Therefore, ARF is collected to be utilized later to process the existing E-waste. Regarding the export of electronic waste to other developing nations, Swiss law stipulates that the exporter must prove that the waste was handled responsibly from an environmental standpoint before exporting, and prior approval from the nation that will be receiving the import must be obtained. Furthermore, Switzerland has ratified the Basel Convention, which prohibits them from illegally disposing of their E-waste in developing nations, with the exception of OECD nations (El-Nakib, 2012). Approval from the Swiss Federal Government is required to export recycled products (SWICO, 2011)

Canada

E-waste continues to be a significant challenge for the world and Canada. In 2010, an estimated 224,000 tonnes of E-waste were produced in the country, whereas in 2014, 725,000 tonnes of E-waste were generated with 20.4 kg per resident (VanderPol, 2014). However, Kumar & Holuszko (2016) mentions that the E-waste generated in 2014 was still less than in any other developed nations such as Germany, the United States, or the United Kingdom. The reason was due to the possibility of lower population densities.

In Canada, E-wastes are disposed of by Canadian households either by throwing it away, returning it to the providers, dropping it off at a depot, or donating it. Various E-waste collection programs have been initiated to dispose of E-wastes efficiently and in an environmentally responsible manner, but the collection rate is far less than the total amount of E-waste produced yearly, which is approximately 20%.

There may be several causes for this, including the difficulty of recycling programs, the lack of drop-off places or depots that accept all sorts of E-waste, and a lack of public understanding of or willingness to follow the right E-waste disposal system (Statistics Canada, 2015).

The Ministry of the Environment is responsible for managing electronic waste in Canada. To control the management of E-waste, the Canadian Council of Ministers of the Environment (CCME) issued a set of guidelines for Electronics Product Stewardship. Electronics Product Stewardship Canada (EPSC), a non-profit industry-led group, was created in 2003 to create, promote, and execute sustainable approaches to recycling electronic waste (About Electronics Product Stewardship Canada, 2022).

Regarding the landfilling of E-waste, Canada banned this type of disposal. It increased the average land tipping fee, discouraging the residents from landfilling, and encouraging them to divert, recycle and reuse end-of-life electronics (Kerr & Bryant, 2013). By joining an international treaty that forbids the export of hazardous waste to developing nations, Canada has made an effort to tackle the illegal trade in E-waste on a global scale (The Basel Convention) and by putting into action a federal law that prohibits the export of hazardous waste (Behrens, 2013). Unlike the United States, Canada has ratified and signed the Basel Convention, which forbids the export of hazardous waste from developed to developing nations (Basel Secretariat, 2005).

Japan

East Asia has an environmentally sound E-waste management infrastructure where most E-waste imports are controlled, Japan being one of the nations (Baldé et al., 2022). Japan is one of the developed countries which has come up with two laws to deal with the E-waste situation. Law for the Promotion of Effective Utilisation of Resources (LPUR) and Law for Recycling Specified Kinds of Home Appliances (LRHA) are the two laws implemented in Japan. The former maintains small-sized batteries and personal computers, whereas the latter takes care of air conditioners, televisions, refrigerators, washing machines, and clothes dryers (Pariatamby & Victor, 2013). Furthermore, the current regulations of Japan require manufacturers or producers to recycle their products and retailers to take back their sold goods (Yoshida & Yoshida, 2014). In order to return the UEEE, the consumers have to pay a small amount to the traders when buying the product so that its recycling process becomes more effortless in the future. This rule is bound by law, so any non-compliance will result in strict penalties as the law specifies a target rate (Widmer et al., 2005).

In the Japanese E-waste management system, the e-wastes are collected and transferred to a 380 E-waste collection spot which is then shifted to the facilities with a proper distribution system (Li et al., 2006). This distribution system includes a primary disassembly procedure, which is one of the essential characteristics of the Japanese system. This procedure means filtering the extensive residues more adequately and accurately by reducing the number of plastics and then reusing those parts (Fishbein, 2002).

Above all, equivalent legislation is a significant force that maintains a proper collection and recycling of UEEEs; it has been prevalent since 2003. The overall success of Japanese E-waste recycling systems has been due to their environmental awareness, social responsibility, and the citizen's concern and responsibility toward their regulations (Gaidajis et al., 2010)

E-Waste Management System of Developing Countries

While considering the various systems in developed nations, it is also imperative to consider the methods of E-waste cycles in developing countries. Especially when developed countries' systems are still being developed into standardized and systematic processes, multiple areas must be worked upon. When considering the health of the ecosystems in countries like India, China, and other Southeast Asian countries along with Africa, hazardous chemicals and toxicants present in systems most prevalent in developing nations require much attention to be looked into.

China

Southeast Asia mainly imports uncontrolled used EEE and E-waste and has no capacity for treating it (Baldé et al., 2022). As of 2018, China is one of the world's highest producers of E-waste, contributed by its highly advanced fast-paced technological development and the E-waste dumped from various developed nations, which constitutes almost 70% of other nations' waste by weight (Fu et al., 2018; Zeng et al., 2017). Baldé et al. (2015) note that although the government has intervened and implemented E-waste laws and regulations, the informal sector is a significant issue affecting the improper disposal of such wastes. The disposal measures through open-ground incineration, open landfills, and illegal sale and recycling are significant issues when dumping and recycling such wastes. Another thing is, E-waste scattered in various parts of the country, especially near unorganized dumping grounds, has noticeable effects which can affect all age groups, especially children, who are more susceptible to such hazardous toxins. The reason is that the children's height closer to the soil level is relative to the toxins found in the body, possibly life-threatening (Chen et al., 2022). Moreover, the number of peer-reviewed research papers written for China is the highest compared to other nations, such as Japan and The US, which also indicates that China is facing an issue that is detrimental to its nation (Fu et al., 2018).

The Reverse Supply Chain (RSC) model, which the Chinese government introduced in 2012, was to distinguish between the formal and informal sectors by providing those companies licensed and registered with subsidies that would lure them into complying with the regulations (Wang et al., 2022). Wang et al. (2022) state that three criteria must be fulfilled for an RSC system to be implemented correctly. The first is that formal systems need to restructure the WEEE management systems. Secondly, formal enterprises should decide whether to collect E-waste from either independent collectors of waste or through the public's effort of collectively working together (which may not be effective because there is almost a 95% chance that citizens of the nation have no idea how to dispose of their degraded batteries). Lastly, there should be a decision on whether the disposal companies need to invest in green technology.

Baldé et al. (2022), in their constant research on E-waste, have clearly stated that the informal management systems in Central Asia cannot safely dispose of waste properly even with the system implemented by the Basel Convention. Along similar lines, even though much thought has been given to improving the E-waste channels in China, the economic benefit from the transboundary movement and the advancement of technology in the nation is on a level that drastically beats other nations worldwide.

India

Southeast Asia and East Asia are primary import and export hubs. As per the Global E-waste Monitor Report 2020, the rapid growth of information technology and electronics in India is also one of the primary reasons it has become the world's third-largest E-waste generating country (Bandyopadhyay, 2009; Sengupta et al., 2022; Forti et al., 2020). Further studies have noted that only a tiny percentage of individuals, about 30%, know the hazardous effects of E-waste. Moreover, an estimated 17 kilos of E-waste is generated by every Indian from the state of Chandigarh annually (Ravindra & Mor, 2019). Moreover, 90% of the E-waste generated by Indians is consumed by the informal sector (Halder & Roy, 2015). Southern Asia has strong informal local markets for E-waste and shows little transboundary movement (Baldé et al., 2022).

Previously before 1989, India had no legislative compliance to handle the E-waste generation and dumping in its country (Biswas & Singh, 2020). India had only begun implementing policy initiatives concerning handling E-waste since introducing the E-waste Hazardous Amendment Act in 2003. E-waste was then defined as WEEE under schedule 3 of the act, which was explicitly applicable to only a tiny niche of E-waste that excluded those containing batteries (Halder & Roy, 2015). Since 2012, due to the inefficiency of the Act, E-waste (Management and Handling) Rules have been introduced (Biswas & Singh, 2020). Later on, in 2016, the rules were amended, and E-waste (Management) Rules 2016 was introduced where the Central Pollution Control Board (CPCB), along with the State Pollution Control Board and Pollution Control Committees, joined forces to tackle the rising issue of sloppy and unsystematic handling of E-waste in the nation through the systematic implementation of EPR (Biswas & Singh, 2020; Sengupta et al., 2022). By implementing EPR, the nation's government also strictly ensured that producers and consumers abide by PRO regulations and agreements. With much need, in 2018, the government has also restructured its targets concerning the informal sectors in the E-waste management industry.

E-waste recycling and upcycling are significant issues in developing nations because of the strong presence of the informal sectors; only about 2% of India's total E-waste is adequately recycled according to the legislative compliances of the nation (Mohammed, 2022; Borthakur & Sinha, 2014). It is crucial to comprehend the effects of unorganized sectors in developing nations that threaten physical health in the long run, especially when the poor people and children working in such conditions are illiterate. This statistic is also a tiny estimate of what is happening in the nation because it lacks proper statistical records of its E-waste system (Arya & Kumar, 2020). The stakeholders involved in the immense economic growth and transboundary movement of E-waste from OECD nations are a critical factor that is leading to the steep rise in the E-waste accumulated in the nation, which contributes to the rapid growth of informal workers such as kawaris and ragpickers (Borthakur & Sinha, 2014).

Africa

Africa is a nation that has poor waste management systems enforced in its nation, much of which is because of the lack of documentation on the E-waste generated and dumped from developed countries (Maes & Preston-Whyte, 2022). Ghana has the largest landfill accumulated, whereas Agbogbloshie in Ghana is known for being the most extensive and prominent E-waste Ghanian market in sub-Saharan Africa (Daum et al., 2017; Grant & Oteng-Ababio, 2012; Lebbie et al., 2021). Elevated levels of toxins such as zinc, iron, copper, and other chemicals are found in the blood samples of those working in areas exposed to the E-waste accumulated for long periods that pose a higher risk to neonatal and infants than adults (Daum et al., 2017).

The rate of growth of E-waste being almost three times higher than the growth rate of solid municipal wastes indicates that there is a lack of organization of the systems in place in both developed and developing nations for a such excessive rise in the electronic matter that poses no use but a detriment to the environment and health of various organisms (Davis & Herat, 2008). Although the government has intervened and implemented controls and measures to curb transboundary movement and excessive accumulation of landfills in the nation, the issue persists on a large scale. The Bamako Convention was introduced to support the Basel Convention, Nairobi Convention, and Stockholm Convention, along with other conventions and the National Environmental Management Authority (NEMA). These conventions and authorities in place have taken courses of action to control and ban the importation of E-wastes from other nations but lack consistency in their framework to regulate the unethical practices of disposal and formation of unorganized sectors in their continent (Otieno & Omwenga, 2015; Maes & Preston-Whyte, 2022).

Bimir (2020) in their research mentions that E-waste is prominent across various African regions, including Kenya, Ghana, Nigeria, Egypt, and South Africa, where the E-waste imported to these areas is significantly higher. Moreover, Egypt and South Africa are regions in Africa that have much more organized systems in place compared to the other cities, as mentioned earlier. The former countries lack efficient trade block routes to reduce transboundary dumping with the open presence of underground corruption and informal sectors.

LEGISLATIVE COMPLIANCE

Different nations, regardless of how developed, developing, or undeveloped they are, have created laws, policies, and initiatives to address the massively expanding problem of e-waste. Based on the significant initiatives and efforts made by the various developed countries, such as Canada, USA and Unions, such as the European Union, and their legislation and regulatory mechanisms, it is evident that some initiatives are based on ad hoc measures on the one hand, while on the other hand, they have specific goals and are prepared but fail to anticipate the growing implications that the world will have to deal later. Thus, economic incentives, a lack of adequate financial support for the formal sectors, and other factors could make it easier to implement any E-waste regulations. These factors could also encourage non-compliance and favor illegal markets and informal sectors (Kumar & Singh, 2016).

Reality is at stake because Canadian and American laws manage E-waste use contingent, rather than essential, ontologies of E-waste (Law, 2010). After the publishing of European Commission papers on priority waste streams in the middle of the 1990s, legal evaluations of E-waste rules started to emerge in 2000 (Cooper, 2000). Since then, a plethora of literature has concentrated on international law, such as the Basel Convention and the development of E-waste regulations in Europe and the US. However, a few studies have examined legislation in Asia, particularly in China and India.

Legislation and compliance provide the largest obstacle to developed nations like the USA, Canada, and Australia. For instance, the US has the ability to develop special federal legislation for WEEE, but it is unlikely to happen soon. However, this is not a smaller problem than how specific their regulations are or will be (Ongondo et al., 2011). Regarding developing nations, China is one of those that impact the flow of WEEE throughout Asia and beyond. Although it has laws governing WEEE management, only 18% of the WEEE produced in 2016 (7.2 MT) was formally collected and recycled due to the activities of the informal recycling sector still dominating the management system (Baldé et al., 2017).

Therefore, developed nations have conventions, regulations, and legislation to control their disposal, most of which are based on extended producer responsibility. Manufacturers take back items collected by retailers and municipal authorities to secure material recovery or destruction. However, it can be challenging to ensure compliance as it usually conflicts with financial incentives (Sthiannopkao et al., 2013).

INITIATIVES FOR SUCCESSFUL E-WASTE MANAGEMENT SYSTEMS

According to Baldé et al. (2022), the world currently generates about 53.6 million tonnes of E-waste every year, which is expected to increase in the subsequent years. The estimated amount of E-waste generated will be about 110 million tonnes by 2050, which is over twice as much in less than 30 years. In order to tackle the issue requires the implementation of innovative initiatives that will reduce the contamination and excessive generation of waste. A few of the initiatives include the following given below.

Reverse Supply Chain

The Reverse Supply Chain (RSC) is an initiative that has played an important role in reducing waste generation but also has improved the quality of the waste generated. Although it is evident that companies in developed countries have implemented an RSC in their nation, from which they collect E-waste from consumers directly to dispose further and recycle, developing nations in recent years have also implemented this practice in their countries. The RSC is a mechanism where wastes collected are further processed and utilized again by other manufacturers to produce proper goods and services for sale. The waste collected also affects the quality, quantity, and supply of the waste collected for further processing and manufacturing, affecting the pricing decisions because of the demand in the market (Hosseini-Motlagh et al., 2020). This strains and puts financial pressure on the manufacturer when they cannot produce as much in a given period; this, in turn, affects the firm's long-term survival. Moreover, using RSC requires extensive mathematical and statistical equations that are difficult to understand and comprehend without extensive prior knowledge (Isernia et al., 2019; Koshta et al., 2022).

Extended Producer Responsibility

The OECD (2001) has already defined Extended Producer Responsibility (EPR). However, in simple terms, EPR ensures sustainable measures are taken to control waste generation by extending the producer's responsibility. This means that the producer will be held responsible for the waste generated from their products even after the post-consumer stage of a product's life cycle. This mechanism ensures that, for the producer to benefit and receive incentives for being environmentally friendly, they need to ensure that the materials used to manufacture are not severely detrimental to the health of living beings and their habitat (Esenduran et al., 2019).

Extended Producer Responsibility (EPR) can be of two types that are, Individual Producer Responsibility (IPR) and Collective Producer Responsibility (CPR) (Atasu & Subramanian, 2012). Individual Producer Responsibility puts the entire responsibility on the individual to ensure that from the point of manufacture till the EoL of the product, the wastes are correctly disposed of and recycled, for which they will receive an incentive if they ensure to follow the measures correctly. Collective Producer Responsibility would refer to where all the manufacturers involved in the production would receive an incentive if they collect and dispose of the wastes from their customers that begin from the production till the EoL of the product. Here, every manufacturer involved in the collaborative collection process receives a slightly lesser incentive (Atasu & Subramanian, 2012).

A pivotal point to note is that most developed countries, such as the US, Japan, Canada, and Switzerland, have already implemented EPR in their E-waste management systems (Leclerc & Badami, 2020). Cao et al. (2016) mention that although EPR is implemented in developing nations, the issue arises when EPR cannot apply to electronic products in developing countries such as China, India, and Thailand. Kojima et al. (2009) state that it is difficult for the government to track and collect funds from imitation, smuggled, or counterfeit electronics, which is highly prevalent in developing countries. Moreover, corruption and an unorganized sector led to an increase in the statistical data needing to be appropriately recorded, where collectors and recyclers can hike the amount of waste collected and its price.

THE IMPACT OF COVID-19 ON E-WASTE

With the coronavirus disease outbreak in 2019, the year 2020 will go down in pandemic history as a turning point after the Spanish flu in 1918–1920. (Goldstein, 2020). The virus initially surfaced in Wuhan, China, in December 2019, and within two to three months, it spread throughout the entire planet. More than 150 million cases and 3 million fatalities were recorded worldwide as of May 2, 2021. (WHO, 2021).

People started to depend more on EEE as they were restricted from going outside during the lockdown period. The transportation facilities also stopped, which hindered the collection and transportation of the e-wastes. Due to this, the workers were laid off for their safety and the unavailability of transportation (Kumar et al., 2020).

Additionally, organic, and inorganic compounds included in E-waste contaminate the environment when dumped in the environment. Since the coronavirus can survive on metals and plastics for over five days, it is pretty likely that E-waste had a role in the infection of a significant number of people (Khan et al., 2021).

Schools, hospitals, workplaces, and other businesses were compelled to close during the lockdown, and local business travel also decreased. Only a few essential services were running. For instance, students were at home when schools were closed, and strategies were made to keep them occupied. The closure prompted online learning, which raised both the amount of EEE used and the amount of WEEE generated (Adejumo & Oluduro, 2021).

Developing nations, such as India, must deal with their E-waste and the transportation of E-waste from developed nations to these nations (Arya & Kumar, 2020). Due to this rapid increase in EEE production, the potential increase in WEEE output and the associated managerial difficulties faced by developing countries have come to light (Cunningham, 2020; Skelton, 2020). Despite the lack of statistics, it is expected that household WEEE generation is expected to be higher because of the increase in EEE consumption during the lockdown (Tejedor et al., 2020).

CONCLUSION

Understanding the differences between the E-waste management systems in developed and developing nations has helped us dig deeper into the issue rather than just understanding the persisting issues on the surface. From such an extensive study conducted, we infer that although developed nations have systematic order in their management and regulations, they lack ethics and resort to corrupt means to ensure that their nation can control the generation of WEEE effortlessly.

While political leaders and big companies obstruct developing nations from implementing economic and environmentally sound regulations for their countries, there is a need for stricter regulations with penalties and charges that are difficult to forgo and ignore. Furthermore, one of the most significant issues in the management of E-waste by developed countries is the transboundary movement which can be controlled with treaties signed by various countries on a partner agreement mutually.

In addition, with the onset of the pandemic lockdown in 2020, there has been a steep rise in purchasing electronic products to keep up with the evolving means of communication. This has significantly increased to E-waste generated from obsolete electronic devices incompatible with the recent modes and applications.

All things considered, developing nations have taken learnings and measures to improve their WEEE management systems by considering what works well in systematic management systems in developed countries. There is a need to improve regions in Africa like Ghana, Nigeria, and Kenya, which are economically poor regions where a majority of the population is involved in collecting, recycling, and disposing of such wastes to meet their daily means.

REFERENCES

Abalansa, S., El Mahrad, B., Icely, J., & Newton, A. (2021). Electronic waste, an Environmental Problem Exported to Developing Countries: The GOOD, the BAD and the UGLY. *Sustainability*, *13*(9), 5302. doi:10.3390u13095302

About Electronics Product Stewardship Canada. (2022). *Electronics Product Stewardship Canada*. https://epsc.ca/about-epsc/

Adejumo, O., & Oluduro, O. F. (2021). *Ticking Time Bomb: Implications of the Covid-19 Lockdown on E-Waste Management in Developing Countries*. UCL Press., doi:10.14324/111.444/000065.v1

Agarwal, R. (2012). E-Waste Law: New Paradigm or Business as Usual? *Economic and Political Weekly*, *47*(25), 14–16.

Alavi, M., & Leidner, D. E. (2001). Review: Knowledge Management and Knowledge Management Systems: Conceptual Foundations and Research Issues. *Management Information Systems Quarterly*, *25*(1), 107. doi:10.2307/3250961

Alias, A. F., Ishak, M. B., Zulkifli, S. N. A. M., & Jalil, R. A. (2014). E-waste management: An emerging global crisis and the Malaysian scenario. *International Journal of Environmental Sciences*, *4*(4), 444–457. https://www.ipublishing.co.in/ijesarticles/fourteen/articles/volfour/EIJES41039.pdf

Andeobu, L., Wibowo, S., & Grandhi, S. (2021). A Systematic Review of E-Waste Generation and Environmental Management of Asia Pacific Countries. *International Journal of Environmental Research and Public Health, 18*(17), 9051. doi:10.3390/ijerph18179051 PMID:34501640

Arya, S., & Kumar, S. (2020). E-waste in India at a glance: Current trends, regulations, challenges and management strategies. *Journal of Cleaner Production, 271*, 122707. doi:10.1016/j.jclepro.2020.122707

Askari, A., & Ghadimzadeh, A. Chandima Gomes, & M.D. Bakri Ishak. (2014). E-Waste Management: Towards an Appropriate Policy. *European Journal of Business and Management, 6*(1), 37–46. http://psasir.upm.edu.my/id/eprint/37547/

Atasu, A., & Subramanian, R. (2012). Extended Producer Responsibility for E-Waste: Individual or Collective Producer Responsibility? *Production and Operations Management, 21*(6), 1042–1059. doi:10.1111/j.1937-5956.2012.01327.x

Atlason, R. S., Giacalone, D., & Parajuly, K. (2017). Product design in the circular economy: Users' perception of end-of-life scenarios for electrical and electronic appliances. *Journal of Cleaner Production, 168*, 1059–1069. doi:10.1016/j.jclepro.2017.09.082

Awasthi, A. K., Zeng, X., & Li, J. (2016). Comparative Examining and Analysis of E-waste Recycling in Typical Developing and Developed Countries. *Procedia Environmental Sciences, 35*, 676–680. doi:10.1016/j.proenv.2016.07.065

Baldé, C. P., D'Angelo, E., Luda, V., Deubzer, O., & Kuehr, R. (2022). *Global Transboundary E-waste Flows Monitor - 2022. United Nations Institute for Training and Research*. UNITAR.

Baldé, C. P., Forti, V., Gray, V., Kuehr, R., & Stegmann, P. (2017). *The Global E-waste Monitor 2017: Quantities, Flows and Resources. United Nations Institute for Training and Research*. UNITAR.

Baldé, C. P., Wang, F., Kuehr, R., & Huisman, J. (2015). The global e-waste monitor – 2014. United Nations University, IAS – SCYCLE, Bonn, Germany.

Bandyopadhyay, A. (2009). Electronics waste management: Indian practices and guidelines. *International Journal of Energy and Environment, 1*(5), 793–804. https://www.ijee.ieefoundation.org/

Basel Secretariat. (2005). *Basel Convention on the Control of Transboundary Movements of Hazardous Wastes and Their Disposal*. Basel. http://www.basel.int/text/con-erev.doc

Batoo, K. M., Pandiaraj, S., Muthuramamoorthy, M., Raslan, E. H., & Krishnamoorthy, S. (2021). Behavior-based swarm model using fuzzy controller for route planning and Ewaste collection. *Environmental Science and Pollution Research, 29*(14), 19940 19954. doi:10.1007/s11356-021-12824-x

Behrens, A. (2013). Electronic Waste Management in Canada (Unpublished manuscript). *University of Calgary*, Calgary, AB. doi:10.11575/PRISM/30091

Bhutta, M. K. S., Omar, A., & Yang, X. (2011). Electronic Waste: A Growing Concern in Today's Environment. *Economic Research International*, *2011*, 1–8. doi:10.1155/2011/474230

Bimir, M. N. (2020). Revisiting e-waste management practices in selected African countries. *Journal of the Air &Amp. Waste Management Association*, *70*(7), 659–669. doi:10.1080/10962247.2020.1769769 PMID:32401631

Biswas, A., & Singh, S. G. (2020). *E-waste Management in India: Challenges and Agenda*. Centre for Science and Environment.

Borthakur, A., & Sinha, K. (2014). Electronic Waste Management in India: A Stakeholder's Perspective. *Electronic Green Journal*, *1*(36). Advance online publication. doi:10.5070/G313618041

Cao, J., Lu, B., Chen, Y., Zhang, X., Zhai, G., Zhou, G., Jiang, B., & Schnoor, J. L. (2016). Extended producer responsibility system in China improves e-waste recycling: Government policies, enterprise, and public awareness. *Renewable & Sustainable Energy Reviews*, *62*, 882–894. doi:10.1016/j.rser.2016.04.078

Chen, H., Wang, L., Hu, B., Xu, J., & Liu, X. (2022). Potential driving forces and probabilistic health risks of heavy metal accumulation in the soils from an e-waste area, southeast China. *Chemosphere*, *289*, 133182. doi:10.1016/j.chemosphere.2021.133182 PMID:34883131

Cooper, T. (2000). WEEE, WEEE, WEEE, WEEE, all the way home? An evaluation of proposed electrical and electronic waste legislation. *European Environment*, *10*(3), 121–130. doi:10.1002/1099-0976(200005/06)10:3<121::AID-EET226>3.0.CO;2-N

Coronavirus, W. H. O. (COVID-19) Dashboard. (n.d.). With Vaccination Data. https://covid19.who.int/?gclid=Cj0KCQiAwMP9BRCzARIsAPWTJ_Hjh ep5qv-AvH0RG2ssFAeoKmTIsrq-_W1wjzQSyh0gBXo1v1aK4r4aAgNwEALw_wcB

Cunningham, P. B. K. (2020, November 17). New study highlights the rise in e-waste during global pandemic. *Recycling Today*. https://www.recyclingtoday.com/article/study-highlights-pandemic-drives-increase-e-waste/

Daum, K., Stoler, J., & Grant, R. (2017). Toward a More Sustainable Trajectory for E-Waste Policy: A Review of a Decade of E-Waste Research in Accra, Ghana. *International Journal of Environmental Research and Public Health*, *14*(2), 135. doi:10.3390/ijerph14020135 PMID:28146075

Davis, G., & Herat, S. (2008). Electronic waste: The local government perspective in Queensland, Australia. *Resources, Conservation and Recycling*, *52*(8–9), 1031–1039. doi:10.1016/j.resconrec.2008.04.001

Debnath, B., Roychoudhuri, R., & Ghosh, S. K. (2016). E-Waste Management – A Potential Route to Green Computing. *Procedia Environmental Sciences*, *35*, 669–675. doi:10.1016/j.proenv.2016.07.063

Doan, L. T. T., Amer, Y., Lee, S. H., & Phuc, P. N. K. (2019). Strategies for E-waste management: A literature review. *International Journal of Energy and Environmental Engineering*, *13*(3), 157–162.

El-NakibI. (2012). Reverse Logistics: A Comparison of Electronic Waste Recycling Between Switzerland and Egypt. *Social Science Research Network*. https://papers.ssrn.com/sol3/Delivery.cfm/SSRN_ID2053344_code1603574.pdf?abstractid=2053344&mirid=1&type=2

Esenduran, G., Atasu, A., & Van Wassenhove, L. N. (2019). Valuable e-waste: Implications for extended producer responsibility. *IISE Transactions*, *51*(4), 382–396. doi:10.1080/24725854.2018.1515515

European Commission. (2008). Directive 2008/98/EC of the European Parliament and of the Council of 19 November 2008 on Waste and Repealing Certain Directives. http://eurlex.europa.eu/legalcontent/EN/TXT/PDF/?uri=CELEX:32008L0098&fromEN

Fadaei, A. (in press). E-waste management status worldwide: Major challenges and solutions. *International Academy of Ecology and Environmental Sciences, 12(4)*(22208860). http://www.iaees.org/publications/journals/piaees/onlineversion.asp

Fishbein, B. K. (2002). End-of-life management of electronics abroad. Waste in the wireless world: the challenge of cell phones, *INFORM Inc*., New York, http://www.informinc.org

Forti, V., Baldé, C. P., Kuehr, R., & Bel, G. (2020). The Global E-waste Monitor 2020: Quantities, flows and the circular economy potential. *United Nations Institute for Training and Research (UNITAR)*. https://ewastemonitor.info/

Fu, J., Zhang, H., Zhang, A., & Jiang, G. (2018). E-waste Recycling in China: A Challenging Field. *Environmental Science & Amp. Technology*, *52*(12), 6727–6728. doi:10.1021/acs.est.8b02329 PMID:29862810

Gaidajis, G., Angelakoglou, K., & Aktsoglou, D. (2010). E-waste: Environmental Problems and Current Management. *Journal of Engineering Science and Technology Review*, *3*(1), 193–199. doi:10.25103/jestr.031.32

Goldstein, J. L. (2020). The Spanish 1918 Flu and the ÇOVID-19 Disease: The Art of Remembering and Foreshadowing Pandemics. *Cell*, *183*(2), 285–289. doi:10.1016/j.cell.2020.09.030 PMID:33064981

Grant, R., & Oteng-Ababio, M. (2012). Mapping the Invisible and Real "African" Economy: Urban E-Waste Circuitry. *Urban Geography*, *33*(1), 1–21. doi:10.2747/0272-3638.33.1.1

Greenpeace. (2009). Where does e-waste end up? https://media.greenpeace.org/archive/Where-Does-E-Waste-End-Up----English-Version-27MZIF2D3F8X.html

Halder, S., & Roy, S. (2015). Problem, Policies and Management of Electronic Waste in India: A Review. Coherence (Seminar Edition, Vol. 2). The Bhawanipur Education Society College.

Herat, S., & Agamuthu, P. (2012). E-waste: A problem or an opportunity? Review of issues, challenges, and solutions in Asian countries. Waste Management &Amp; Research. *The Journal for a Sustainable Circular Economy*, *30*(11), 1113–1129. doi:10.1177/0734242X12453378 PMID:22851536

Hopson, E., & Puckett, J. (2016). *Scam recycling: E-dumping on Asia by us recyclers*. The E-Trash Transparency Project.

Hosseini-Motlagh, S. M., Nematollahi, M., & Ebrahimi, S. (2020). Tri-party reverse supply chain coordination with competitive product acquisition process. *The Journal of the Operational Research Society*, *73*(2), 382–393. doi:10.1080/01605682.2020.1824550

Ilankoon, I., Ghorbani, Y., Chong, M. N., Herath, G., Moyo, T., & Petersen, J. (2018). E-waste in the international context – A review of trade flows, regulations, hazards, waste management strategies and technologies for value recovery. *Waste Management (New York, N.Y.)*, *82*, 258–275. doi:10.1016/j.wasman.2018.10.018 PMID:30509588

Isernia, R., Passaro, R., Quinto, I., & Thomas, A. (2019). The Reverse Supply Chain of the E Waste Management Processes in a Circular Economy Framework: Evidence from Italy. *Sustainability*, *11*(8), 2430. doi:10.3390u11082430

Islam, M. T., Dias, P., & Huda, N. (2018). Comparison of E-waste management in Switzerland and in Australia: A qualitative content analysis. *International Journal of Environmental and Ecological Engineering*, *12*(10), 610–616.

Kahhat, R., Kim, J., Xu, M., Allenby, B., Williams, E., & Zhang, P. (2008). Exploring e-waste management systems in the United States. *Resources, Conservation and Recycling*, *52*(7), 955–964. doi:10.1016/j.resconrec.2008.03.002

Kahhat, R., & Williams, E. (2012). Materials flow analysis of e-waste: Domestic flows and exports of used computers from the United States. *Resources, Conservation and Recycling*, *67*, 67–74. doi:10.1016/j.resconrec.2012.07.008

Kalmykova, Y., Sadagopan, M., & Rosado, L. (2018). Circular economy – From review of theories and practices to development of implementation tools. *Resources, Conservation and Recycling*, *135*, 190–201. doi:10.1016/j.resconrec.2017.10.034

Kerr, S., & Bryant, L. (2013). Design for Environment Report. *Electronic Product Stewardship Canada*. https://epsc.ca/wpcontent/uploads/2022/03/2013_EPSC_Design_Env_Report.pdf

Khan, A. H., Tirth, V., Fawzy, M., Mahmoud, A. E. D., Khan, N. A., Ahmed, S., Ali, S. S., Akram, M., Hameed, L., Islam, S., Das, G., Roy, S., & Dehghani, M. H. (2021). COVID-19 transmission, vulnerability, persistence and nanotherapy: A review. *Environmental Chemistry Letters*, *19*(4), 2773–2787. doi:10.100710311-021-01229-4 PMID:33846683

Kirchherr, J., Reike, D., & Hekkert, M. (2017). Conceptualizing the circular economy: An analysis of 114 definitions. *Resources, Conservation and Recycling*, *127*, 221–232. doi:10.1016/j.resconrec.2017.09.005

Kojima, M., Yoshida, A., & Sasaki, S. (2009). Difficulties in applying extended producer responsibility policies in developing countries: case studies in e-waste recycling in China and Thailand. *Journal of Material Cycles and Waste Management, 11*(3), 263 269. doi:10.1007/s10163-009-0240-x

Koshta, N., Patra, S., & Singh, S. P. (2022). Sharing economic responsibility: Assessing end user's willingness to support E-waste reverse logistics for circular economy. *Journal of Cleaner Production*, *332*, 130057. doi:10.1016/j.jclepro.2021.130057

Kumar, A., & Holuszko, M. (2016). Electronic Waste and Existing Processing Routes: A Canadian Perspective. *Resources*, *5*(4), 35. doi:10.3390/resources5040035

Kumar, A., Luthra, S., Mangla, S. K., & Kazançoğlu, Y. (2020). COVID-19 impact on sustainable production and operations management. *Sustainable Operations and Computers*, *1*, 1–7. doi:10.1016/j.susoc.2020.06.001

Kumar, U., & Singh, D. (2016). E–Waste Management through Regulations. International *Journal of Engineering Inventions*, 6–14(3(2)). https://www.researchgate.net/file.PostFileLoader.html?id=579a0881cbd5c2324e13a611&assetKey=AS%3A388818185277444%401469712887190

Law, J. (2010). Reality Failures. FESRC Centre for Research on Socio-Cultural Change. http://heterogeneities.net/publications/Law2010RealityFailures5.pdf

Lebbie, T. S., Moyebi, O. D., Asante, K. A., Fobil, J., Brune-Drisse, M. N., Suk, W. A., Sly, P. D., Gorman, J., & Carpenter, D. O. (2021). E-Waste in Africa: A Serious Threat to the Health of Children. *International Journal of Environmental Research and Public Health*, *18*(16), 8488. doi:10.3390/ijerph18168488 PMID:34444234

Leclerc, S. H., & Badami, M. G. (2020). Extended producer responsibility for E-waste management: Policy drivers and challenges. *Journal of Cleaner Production*, *251*, 119657. doi:10.1016/j.jclepro.2019.119657

Lee, D., Offenhuber, D., Duarte, F., Biderman, A., & Ratti, C. (2018). Monitor: Tracking global routes of electronic waste. *Waste Management (New York, N.Y.)*, *72*, 362–370. doi:10.1016/j.wasman.2017.11.014 PMID:29137831

Li, J., Tian, B., Liu, T., Liu, H., Wen, X., & Honda, S. (2006). Status quo of e-waste management in mainland China. *Journal of Material Cycles and Waste Management*, *8*(1), 13–20. doi:10.100710163-005-0144-3

Maes, T., & Preston-Whyte, F. (2022). E-waste it wisely: Lessons from Africa. *SN Applied Sciences*, *4*(3), 72. Advance online publication. doi:10.100742452-022-04962-9 PMID:35155992

Manomaivibool, P. (2009). Extended producer responsibility in a non-OECD context: The management of waste electrical and electronic equipment in India. *Resources, Conservation and Recycling*, *53*(3), 136–144. doi:10.1016/j.resconrec.2008.10.003

Mihai, F. C., Gnoni, M. G., Meidiana, C., Ezeah, C., & Elia, V. (2019). Waste Electrical and Electronic Equipment (WEEE): Flows, Quantities, and Management—A Global Scenario. *Electronic Waste Management and Treatment Technology*, 1–34. doi:10.1016/B978-0-12-816190-6.00001-7

Mmereki, D., Li, B., Baldwin, A., & Hong, L. (2016). *The Generation, Composition, Collection, Treatment and Disposal System, and Impact of E-Waste*. E-Waste in Transition - From Pollution to Resource., doi:10.5772/61332

Mohammed, S. I. (2022). E-Waste Management in Different Countries: Strategies, Impacts, and Determinants. In (Ed.), Advances in Green Electronics Technologies. IntechOpen. doi:10.5772/intechopen.106644

Morf, L. S., Tremp, J., Gloor, R., Schuppisser, F., Stengele, M., & Taverna, R. (2007). Metals, non-metals and PCB in electrical and electronic waste – Actual levels in Switzerland. *Waste Management (New York, N.Y.)*, *27*(10), 1306–1316. doi:10.1016/j.wasman.2006.06.014 PMID:17008085

Moyo, T., Sadan, Z., Lötter, A., & Petersen, J. (2022). Barriers to recycling e-waste within a changing legal environment in South Africa. *South African Journal of Science*, *118*. Advance online publication. doi:10.17159ajs.2022/12564

Murthy, V., & Ramakrishna, S. (2022). A Review on Global E-Waste Management: Urban Mining towards a Sustainable Future and Circular Economy. *Sustainability*, *14*(2), 647. doi:10.3390u14020647

Olsen, S. (2006). High Tech: Anything But Green. *NEWS.COM*. http://news.com.com/High+tech+Anything+put+green/2008-1041_3-6090899.html

Ongondo, F., Williams, I., & Cherrett, T. (2011). How are WEEE doing? A global review of the management of electrical and electronic wastes. *Waste Management*, *31*(4), 714 730. doi:10.1016/j.wasman.2010.10.023

Organization for Economic Cooperation and Development (OECD). (2001). Extended producer responsibility: A guidance manual for governments.

Osibanjo, O., & Nnorom, I. (2007). The challenge of electronic waste (e-waste) management in developing countries. Waste Management &Amp; Research. *The Journal for a Sustainable Circular Economy*, *25*(6), 489–501. doi:10.1177/0734242X07082028 PMID:18229743

Otieno, I., & Omwenga, E. (2015). E-Waste Management in Kenya: Challenges and Opportunities. *Journal of Emerging Trends in Computing and Information Sciences*, *6*(12), ISSN 2079-8407.

Pak, P. (2008). Haste makes e-waste: A comparative analysis of how the United State should approach the growing e-waste threat. *Cardozo J. Int'l & Comp. L.*, *16*, 241.

Pariatamby, A., & Victor, D. (2013). Policy trends of e-waste management in Asia. *Journal of Material Cycles and Waste Management*, *15*(4), 411–419. doi:10.100710163-013-0136-7

Patil, R. A., & Ramakrishna, S. (2020). A comprehensive analysis of e-waste legislation worldwide. *Environmental Science and Pollution Research*, 27(13), 14412 14431. 1 doi:10.1007/s11356-020-07992-1

Policy Practices for E-waste Management. (n.d.). *ITU*. https://www.itu.int/en/ITUD/Environment/Pages/Toolbox/WEEE-A
frica-Toolkit.aspx

Raharjo, S., & Utomo, A. H. (2021). Comparative Study of Electronic Waste Management in Developed Countries and Indonesia. *Andalasian International Journal of Applied Science. Engineering and Technology*, *1*(1), 21–32. doi:10.25077/aijaset.v1i1.4

Ravindra, K., & Mor, S. (2019). E-waste generation and management practices in Chandigarh, India, and economic evaluation for sustainable recycling. *Journal of Cleaner Production*, *221*, 286–294. doi:10.1016/j.jclepro.2019.02.158

Seeberger, J., Grandhi, R., Kim, S. S., Mase, W. A., Reponen, T., Ho, S. M., & Chen, A. (2016). Special Report: E-Waste Management in the United States and Public Health Implications. *Journal of Environmental Health*, *79*(3), 8–16. PMID:29120130

Sengupta, D., Ilankoon, I., Dean Kang, K., & Nan Chong, M. (2022). Circular economy and household e-waste management in India: Integration of formal and informal sectors. *Minerals Engineering*, *184*, 107661. doi:10.1016/j.mineng.2022.107661

Shinkuma, T. (2003). On the Second-best Policy of Household's Waste Recycling. *Environmental and Resource Economics*, *24*(1), 77–95. doi:10.1023/A:1022842617469

Sinha-Khetriwal, D., Kraeuchi, P., & Schwaninger, M. (2005). A comparison of electronic waste recycling in Switzerland and in India. *Environmental Impact Assessment Review*, *25*(5), 492–504. doi:10.1016/j.eiar.2005.04.006

Skelton, S. K. (2020, November 19). Covid-19 shift to remote working adds to Earth's growing e-waste problem. ComputerWeekly.com. https://www.computerweekly.com/news/252492372/Covid-19-shift-to-remote-working-adds-to-Earths-growing-e-waste-problem

Stahel, W. R. (2013). Policy for material efficiency sustainable taxation as a departure from the throwaway society. *Philosophical Transactions of the Royal Society A: Mathematical, Physical and Engineering Sciences, 371*(1986), 20110567. doi:10.1098/rsta.2011.0567

Statistics Canada. (2015). Manufacturing Sales, by Subsector, by Province and Territory, *Statistics Canada* http://www.statcan.gc.ca/tables-tableaux/sum-som/l01/cst01/manuf33aeng.htm

StEP solving the E-waste Problem - StEP Initiative. (n.d.). https://www.step-initiative.org/

Sthiannopkao, S. (2012). Managing E-Waste in Developed and Developing Countries. The Handbook of Environmental Chemistry, 263–278. doi:10.1007/698_2012_195

Sthiannopkao, S., & Wong, M. H. (2013). Handling e-waste in developed and developing countries: Initiatives, practices, and consequences. *The Science of the Total Environment*, *463–464*, 1147–1153. doi:10.1016/j.scitotenv.2012.06.088 PMID:22858354

SWICO. (2011). *SWICO Recycling*. https://www.swico.ch/media/filer_public/01/0c/010c5855-4dcc-41d7-9807-34901c386158/activity_rapport_2011.pdf

Tejedor, S., Cervi, L., Pérez-Escoda, A., & Tusa, F. (2020). Smartphone usage among students during COVID-19 pandemic in Spain, Italy, and Ecuador. *Eighth International Conference on Technological Ecosystems for Enhancing Multiculturality*. 10.1145/3434780.3436587

Thakur, P., & Kumar, S. (2021). Evaluation of e-waste status, management strategies, and legislations. *International Journal of Environmental Science and Technology*, *19*(7), 6957–6966. doi:10.100713762-021-03383-2

Trivedi, V., Pandey, K. K., & Trivedi, A. (2022). Analyzing the challenges of e-waste management practices in India during COVID-19. *Management of Environmental Quality*, *33*(6), 1611–1628. Advance online publication. doi:10.1108/MEQ-12-2021-0273

UNEP 2009 annual report. (n.d.). *UNEP - UN Environment Programme.* https://www.unep.org/resources/annual-report/unep-2009-annual-report

UNEP Annual Evaluation Report 2007. (n.d.). *UNEP - UN Environment Programme.* https://www.unep.org/resources/synthesis-reports/unep-annual-evaluation-report-2007

VanderPol, M. (2014). Overview of E-waste Management in Canada. *International E-waste Management Network (IEMN).* https://19january2017snapshot.epa.gov/sites/production/files/2014-08/documents/canada_country_presentation.pdf

Walls, M. (2011). Deposit-Refund Systems in Practice and Theory. SSRN *Electronic Journal.* doi:10.2139/ssrn.1980142

Wang, J., Li, W., Mishima, N., & Adachi, T. (2022). Exploring the optimal reverse supply chain for e-waste treatment under Chinese government subsidy. *Waste Management (New York, N.Y.)*, *137*, 128–138. doi:10.1016/j.wasman.2021.10.031 PMID:34752946

Wath, S. B., Vaidya, A. N., Dutt, P., & Chakrabarti, T. (2010). A roadmap for development of sustainable E-waste management system in India. *The Science of the Total Environment*, *409*(1), 19–32. doi:10.1016/j.scitotenv.2010.09.030 PMID:20951410

Wei, L., & Liu, Y. (2012). Present Status of e-waste Disposal and Recycling in China. *Procedia Environmental Sciences*, *16*, 506–514. doi:10.1016/j.proenv.2012.10.070

Widmer, R., Oswald-Krapf, H., Sinha-Khetriwal, D., Schnellmann, M., & Böni, H. (2005). Global perspectives on e-waste. *Environmental Impact Assessment Review*, *25*(5), 436–458. doi:10.1016/j.eiar.2005.04.001

Williams, E., Kahhat, R., Allenby, B., Kavazanjian, E., Kim, J., & Xu, M. (2008). Environmental, social, and economic implications of global reuse and recycling of personal computers. *Environmental Science & Technology*, *42*(17), 6446–6454. doi:10.1021/es702255z PMID:18800513

Wong, M., Wu, S., Deng, W., Yu, X., Luo, Q., Leung, A., Wong, C., Luksemburg, W., & Wong, A. (2007). Export of toxic chemicals – A review of the case of uncontrolled electronic-waste recycling. *Environmental Pollution*, *149*(2), 131–140. doi:10.1016/j.envpol.2007.01.044 PMID:17412468

Yoshida, F., & Yoshida, H. (2014). E-Waste Management in Japan: A Focus on Appliance Recycling. *Advanced Materials Research*, *878*, 420–423. . doi:10.4028/www.scientific.net/AMR.878.420

Zeng, X., Yang, C., Chiang, J. F., & Li, J. (2017). Innovating e-waste management: From macroscopic to microscopic scales. *The Science of the Total Environment*, *575*, 1–5. doi:10.1016/j.scitotenv.2016.09.078 PMID:27723459

Zhang, K., Schnoor, J. L., & Zeng, E. Y. (2012). E-Waste Recycling: Where Does It Go from Here? *Environmental Science &Amp. Technology*, *46*(20), 10861–10867. doi:10.1021/es303166s PMID:22998401

Zhong, H., & Schiller, S. (2011). Design of the expense allocation mechanism in e-waste recycling deposit system under EPR framework. *ICSSSM11*. doi:10.1109/ICSSSM.2011.5959352

Chapter 7
E-Waste Management System (EMS):
An Online Platform to Sell Crafts Made From Tech Waste

Soumya George
https://orcid.org/0000-0002-7256-7677
St. George's College, Aruvithura, India

Anjala Michael
St. George's College, Aruvithura, India

ABSTRACT

The usage of electronic devices increased rapidly. The over-usage of electronic devices always led to a huge amount of waste in the form of electronic products like monitors, wires, discs, etc. e-waste is hazardous, involving things containing toxic components that are dangerous to human health, such as mercury, lead, cadmium, polybrominated flame retardants, barium, and lithium. The negative health effects of these toxins on humans include brain, heart, liver, kidney, and skeletal system damage. Every organization follows a throwback policy of dumping all these electronic wastes into the environment. This chapter proposes some tips and tricks on how to recycle hazardous e-wastes into beautiful reusable products, and methods for the distribution of these objects by creating an e-waste management system.

INTRODUCTION

E-waste is the abbreviation term for the various electrical and electronic types of

DOI: 10.4018/978-1-6684-7573-7.ch007

equipment or components that are reaching the end of their working life. Computers, TV sets, cables or wires, batteries, thermometers, toys, etc. are some common products included in this category. These are also called “e-scrap” or “end-of-life electronics”. Nowadays, the usage of electrical and electronic devices increased rapidly. This over usage of electronic devices always led to a huge amount of waste in the form of monitors, wires, discs, etc. (Misra, 2021).

E-waste is hazardous containing toxic components that are dangerous to human health, such as mercury, lead, cadmium, polybrominated flame retardants, barium, and lithium (Zeng, 2022); (Frazzoli, 2022); (Chakraborty, 2022); (Roy, 2022); (Lin, 2022). The negative health effects of these toxins on humans include brain, heart, liver, kidney, and skeletal system damage. Every organization follows a throwback policy of dumping all these electronic wastes into the environment. Figure 1 shows a dump of computer wastes accumulated on the roadside. To reduce the scale of this great horrible pollution route, we must turn to the recycling and reusing habits of these products.

The main mantra for e-waste management is “Reduce, Recycle, Reuse (RRR)”. This chapter proposes some beautiful tips and tricks to recycle computer waste, one of the main contributors of e-waste in this digital era, and then sell them through online platforms by developing an e-waste management system (EMS).

The main objective of the system is to efficiently control the throwback policy of electronic waste to the environment. Even though there are many government-approved well-established centers to collect e-waste, these systems don’t give any provision to mention what they are doing with the collected electronic waste. The proposed work, an e-waste management system is mainly focused to collect electronic waste from nearby areas and convert or recycle it into a useful one. This system will help an individual to sell his/her electronic waste at a few prices and can purchase new or recycled computer parts or crafts that are made from electronic waste.

BACKGROUND

The amount of e-waste is increasing day by day and studies reported that Asia ranked top in the generation of e-waste at about 24.9 Mt, followed by America, Europe, etc. (Forti V., 2020). Even though e-waste contains various toxic and hazardous components, it also contains precious platinum metals and non-metals (Murthy, 2022). e-waste management sectors include formal sectors like government agencies with licenses and informal sectors that collect e-waste from the doorstep (Sengupta, 2022).

According to the statistics of leading countries based on the generation of electronic waste worldwide in 2019, China is the largest producer of e-waste followed by the United States and India (Statista, 2020). A reverse logistics system is implemented

Figure 1. e-waste dumped on the roadside

in Brazil, the largest producer of e-waste in Latin America, to manage the e-waste where the main responsibility lies with the different stakeholders including producers, manufacturers, consumers, etc. (Santos, 2022). In India, the e-waste (Management and handling) rules come into effect on 1st May 2012 (Meity, 2011). The rules got strengthened in 2016 which introduces new restrictions to every producer of electrical and electronic equipment that they should ensure no samples or parts contain lead, mercury, or other pollutants above a maximum concentration value (Vakkalanka, 2018).

A behavior-based swarm model for e-waste collection and its transportation by optimizing the vehicle routes and thereby reducing the time and cost of e-waste collection from households is proposed in (Batoo, 2022) by integrating the fuzzy controller logic which helps to collect the e-waste on time. An integration of formal and informal e-waste recyclers to collect e-waste is proposed in (Sengupta, 2022) where a portion of e-waste is recycled by the formal sectors and the rest is processed by informal sectors to improve the e-waste management in India.

Sources of e-waste

Electrical or electronic products that are no longer useful or cease functioning in their associated task can be considered e-waste. The main sources of e-wastes are houses, hospitals, government. and other private sectors like industries, hotels, etc. (Mohd, 2018). Fans, lights, Personal computers (PCs), TVs, cell phones, Air coolers (ACs), etc. are common forms of e-waste from all sources. Household appliances include electric cookers, heaters, microwaves, iron boxes, CD or DVD players, etc. e-waste from hospitals includes different medical equipment like microscopes, ECG devices, incubators, thermometers, etc. E-scraps from government and industries include different office types of equipment like printers, scanners, fax machines, Photostat machines, incubators, etc. Boilers, mixers, electric gas stoves, induction cookers, etc. are some forms of e-waste from hotels.

Not all e-waste can be considered toxic or hazardous. But most e-wastes contain toxic materials when not properly disposed of or managed get released into the air and water that in turn will harm the people and animals. Fluorescent lamps, older batteries, thermometers, toys, etc. contain mercury wastes (DTSC, n.d.). Exposure to mercury in a small amount can cause severe health problems and can cause harmful effects on the digestive and immune systems, nervous systems, lungs, kidneys, etc. It is also corrosive to skin, eyes, etc. due to the presence of inorganic salts in it. In computer wastes, Cathode Ray Tubes (CRTs) from Computer monitors and televisions are considered hazardous because it contains several toxic materials like barium, lead, fluorescent powders, cadmium, etc. (Lecler, 2015).

Just like a coin has two sides, e-waste is a rich source of different types of precious metals. e-waste contains precious metals like gold, platinum, silver, palladium, etc., and useful metals like aluminium, iron, copper, etc. All these are so costlier items that when properly extracted will help to gain a high turnover or profit. Recently researchers used graphene to extract gold from e-waste by just dissolving grounded e-waste into a solution containing graphene oxide (Irving, 2022); (Tipre, 2021).

Computer wastes are the main sources of precious metals. In addition to the above-mentioned metals, other valuable metals from tech waste include nickel, tantalum, cobalt, zinc, tin, and neodymium. All these metals can be extracted from different computer parts like printed circuit boards, hard drives, computer chips, CPU heat sinks, capacitors, keyboard membranes, etc. (McFadden, 2020).

e-waste Management Rules in India

To reduce the amount of e-waste, different countries adopted different rules to reduce the pollution of water and soil and to save human and animal life. In India, the first rule on e-waste (Management and Handling) Rules, 2011 come into effect on 1st May 2012 which applies to every producer, manufacturer, consumer of electrical and electronic equipment or components and collection centre, dismantler, and recycler of e-waste (Meity, 2011). It sets responsibilities for each type of user involved. Both producer and consumer should ensure that e-waste is properly collected and channelized to a registered e-waste recycler or dismantler. Also, producers, collection centres, recyclers, or dismantlers should obtain authorization and registration from the state pollution control board.

These rules got strengthened by Environment Protection Act, 1986, and e-waste (Management) Rules, 2016 which should be followed by the parties involved in e-waste like consumers, producers, collection centres, dismantlers, recyclers, etc. and they should be responsible to properly channelize the e-waste generated by them (Vakkalanka, 2018); (Regtech, 2016). According to this rule, consumers or bulk consumers should ensure that e-waste not containing or not mixed with radio-active material as governed by the Atomic Energy Act, 1962. The manufacturer should keep a record of the details of e-waste handled by him. The producer is responsible for the proper disposal of e-waste generated by him or for properly channelizing it to recyclers, called extended producer responsibility (EPR). Collection centres, dismantlers, and recyclers should ensure that no harm or damage is caused to the environment while dismantling and recycling e-waste and should obtain authorization from the state pollution control board for the dismantling or recycling process. The violation of rules or acts shall be punished with "imprisonment" which may extend to 5 years or with "fine" which may extend to Rs. 1 lakh, or both.

The e-waste (Management) Rules, 2016 again revised by the e-waste (Management) Rules, 2018 with the mission of properly channelizing e-waste to authorized dismantlers and recyclers to formalize the e-waste recycling sectors and new rules have been introduced for new producers (Press Information Bureau, 2018).

Existing System

There are many government-approved well-established centres to collect e-waste. But these systems don't give any provision to mention what they are doing with the collected electronic waste. The current system does not provide an option to generate details about the product.

PROPOSED SYSTEM

The main aim of the proposed system is the collection of electronic waste materials, mainly computer waste from nearby areas to reduce the rate of e-waste to the environment and thereby reduce pollution to keep the surroundings neat and clean. People can inform the availability of electronic waste through E collections.

It consists of two steps as shown in Figure 2.

- Recycling of e-waste into Stunning Crafts – The first step is to convert the computer waste items collected from nearby areas into beautiful craft products like pen stands, tree models, chains, etc.
- Distribution of recycled objects through an online e-waste Management System, EMS – Next step is to sell these crafted products through EMS by fixing a selling price so that customers from any place can buy these products by Cash on Delivery (COD) method.

Figure 2. The proposed system model of recycling e-waste into stunning crafts and its distribution through the e-waste Management system

Recycling of e-waste into Stunning Crafts

e-waste in the form of monitors, keyboard keys, wires, etc. can be recycled into many useful products. The first step is to collect e-waste and craft beautiful reusable products from it as shown in Figure 3. Figure 4 shows some beautiful recycled and reusable products from motherboard wires, keyboard keys, etc. These products can be easily crafted as an initiative to recycle computer e-waste by collecting different unusable computer wastes. Figure 5 shows the various recycled beautiful products from e-wastes shared via pinterest.com by Rike Technologies (RikeTechnologies, n.d.).

Housewives with crafting skills can turn their e-wastes into beautiful crafts as given above in the form of trees, pen stands, hand chains, clocks, etc. Once crafted, the next step is to find suitable customers to purchase the product. These steps can be made easy by creating an e-waste management system where users can both sell and purchase recycled craftworks.

Distribution of Recycled Objects – e-waste Management System

Even though there are many government-approved well-established centers to collect e-wastes, these systems don't give any provision to mention what they are doing with the collected electronic waste. The e-waste Management system deals with this situation effectively by creating an online website where sellers can sell their beautiful crafts and customers can purchase them. This system can be created

Figure 3. Conversion of computer waste into beautiful crafts

E-waste

Crafts out of e-waste

easily where a central admin control adding products for sale with details including price and purchase of products from the website.

System Design

- e-waste management system has 3 modules – Admin, Sales, and Purchase.
- The admin has superior power over the system.
- Admin can input new products to the purchase module named ECG cart.
- Admin can view the products and details of individuals who are registered to E collections.
- Individuals can input details about their e-waste on E-collections and purchase products from ECG carts.

The system is developed using PHP as the front end with MySQL as the backend. MySQL is a fast, easy-to-use relational database. It is currently the most popular open-source database. It is very commonly used in conjunction with PHP scripts to create powerful and dynamic server-side applications. MySQL is used for many small and big businesses. It is developed, marketed, and supported by MySQL AB, a Swedish company. It is written in C and C++. PHP stands for Hypertext Pre-processor. PHP is an interpreted language, i.e., there is no need for compilation.

Some features of PHP include:

- PHP is a server-side scripting language.

Figure 4. Beautiful Recycled and reusable products from motherboard wires, keyboard keys, etc.

Tree

Pen Stand

Hand Chain

Photo Frame

Key Chain

Pen Stand

Clock

Figure 5. Stunning recycled computer crafts by Rike Technologies (Adapted from: pinterest.com by Rike Technologies)

- Performance: Script written in PHP executes much faster than those scripts written in other languages such as JSP & ASP.
- Open-Source Software: PHP source code is freely available on the web; you can develop all the versions of PHP according to your requirement without paying any cost.
- Platform Independent: PHP is available for WINDOWS, MAC, and LINUX & UNIX operating systems. A PHP application developed in one OS can be easily executed in another OS also.
- Compatibility: PHP is compatible with almost all local servers used today like Apache, IIS, etc.
- Embedded: PHP code can be easily embedded within HTML tags and scripts.

Database Structure

A database includes the complete details about a particular system arranged into separate tables by grouping related data together. Data in tables are arranged in rows and Columns. Rows are called RECORDS, and the Columns of each row are called FIELDS. Each field has different properties including the type of field like Varchar (when character type data are stored), Int (for integer type of data or numbers), Longblob (for images), etc., different constraints defined for each field like a primary key (unique field), not, null, etc. Data can be inserted into the database using front-end forms or by bulk uploading from the backend.

e-waste Management System (EMS) consists of the following tables.

1. Login
2. Update
3. Product

Login

The login table is used to store user credentials including username and password, denoted by 'Username' and 'Password' respectively. Users will be verified at the time of login by checking the entered username and password with the entries in the login and only authorized users are allowed to login into the e-waste management system. The Login table is shown in Table 1.

Update

It consists of details about various products added for sale in EMS. It includes the name of the item denoted by 'Item Name', the item price denoted by 'Price', the description of the item 'Description', and an image of the item 'Image'. Using this,

Table 1. Login Table

#	Name	Type
1	Username	Varchar(25)
2	Password	Varchar(8)

the added details of the products in the ECG cart can be updated. The Update table is shown in Table 2.

Table 2. Update Table

#	Name	Type
1	Item_Name	Varchar(25)
2	Price	Int(11)
3	Description	Varchar(100)
4	Image	Longblob

Product

It consists of details of products added to the ECG cart which include details of customers like a unique id for the customer, 'CustomerID', name of the customer, 'CustomerName', address details ('HouseName', 'PostOffice', 'Place', 'NativeTown', 'PinCode', 'District', 'Phone', and 'Email') and product details like name of the product, 'ProductName', description, 'Description', Image of the product to be added, 'Image' and an expecting price, 'ExpectingPrice'. Each customer can view his/her added products in the ECG cart after logging into the system. The Product table is shown in Table 3.

There is only one login for the admin who has complete access to each module. The admin can view the updated products to E-collections, and from the same section, he can download a pdf for the dealer to collect the product. That document contains only the necessary information about the product and the customer. The admin can also update products to the purchase module named ECG cart. Here the products updated by the admin are reusable computer parts or recycled crafts from the collected electronic waste. Also, the customer can purchase products from the ECG cart through telephone contact.

A sample layout of the e-waste management system is given in Figure 6. A more detailed layout is given in Figure 7. Figure 8 shows the status of E-collections

Table 3. Product Table

#	Name	Type
1	CustomerID (Primary key)	Int(11)
2	ProductName	Varchar(30)
3	Description	Varchar(250)
4	Image	Longblob
5	ExpectingPrice	Float
6	CustomerName	Varchar(30)
7	HouseName	Varchar(30)
8	PostOffice	Varchar(30)
9	Place	Varchar(30)
10	NativeTown	Varchar(30)
11	PinCode	Varchar(30)
12	District	Varchar(30)
13	Phone	Bigint(20)
14	Email	Varchar(30)

including item name, image, price, customer details, etc., and its generated report is shown in Figure 9. Figure 10 shows the form to update a new product to the ECG cart. The complete working model of the system is shown in Figure 11.

System Environment

- Hardware specification:
  - Processor: Intel(R) Core (TM) i3-7100U CPU @ 2.40GHz 2.42 GHz
  - RAM: 4 GB
  - Hard Disk: 128 GB
  - Input Device: Standard Keyboard and Mouse
- Software specification:
  - Operating System: Windows 10
  - Front End: PHP
  - Back End: MySQL

Figure 6. A sample website layout of the e-waste Management System

PURPOSE

The main objective of the system is to efficiently control the throwback policy of electronic waste to the environment. This gives the facility to give up electronic waste from society. Admin can generate details of electronic waste that are registered in the E- collections in a single click, it saves a lot of time and reduces the chance of error.

Figure 7. A detailed layout for distribution, sales, and purchase of recycled useful products from e-waste

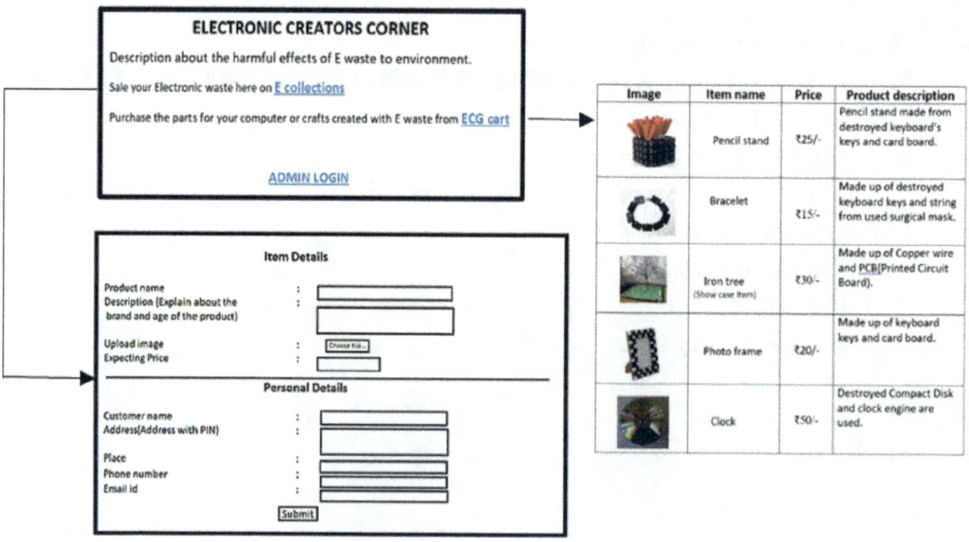

Figure 8. Status of E collections

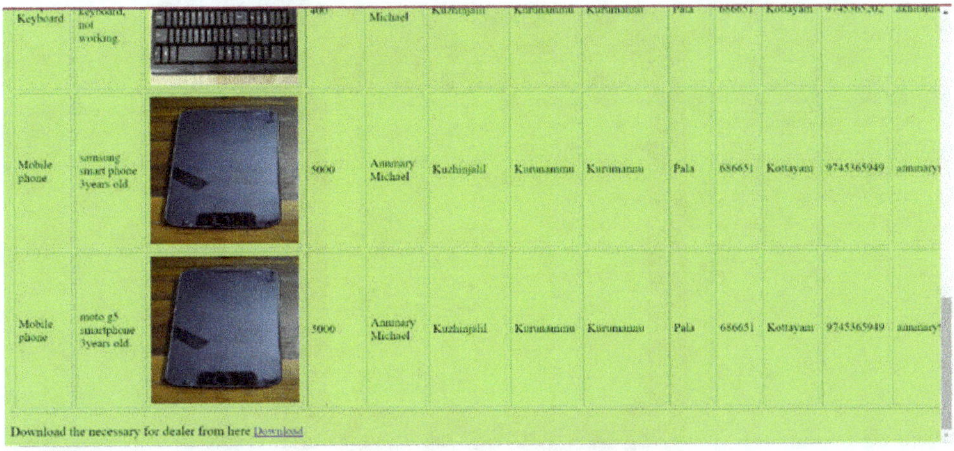

SCOPE

The system will be helpful to common people. Any number of individuals can update products to E-collections which will reduce the accumulation of e-waste. Also, customers can purchase useful things and crafts at a cheap rate from the ECG cart.

Figure 9. Report generated from status of E collections

E Collections report

Product	Expected price	Customer	Phone	Email
Mobile phone	3000	Georgia Michael	9400921598	georgiamichaelk@gmail.com
Television	5000	Baby John	9875624391	johnbaby@gmail.com
Fan	3000	Biju K G	9874563210	bijukg123@gmail.com
Mouse	100	Michael K M	9447568616	devkmichael@gmail.com
Computer Keyboard	500	Aleena Michael	9846438298	aleenamichael916@gmail.com
Computer Keyboard	400	Ahkila Michael	9745365202	akhilamichaelz@gmail.com
Mobile phone	5000	Annmary Michael	9745365949	annmarymichaelgmai.com
Mobile phone	5000	Annmary Michael	9745365949	annmary92@gmail.com

Figure 10. Updating a new product to the ECG cart

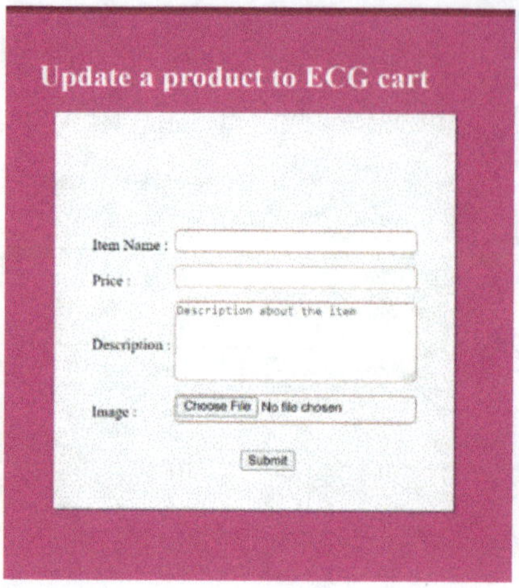

CONCLUSION

e-waste is hazardous containing toxic components that are dangerous to human health, such as mercury, lead, etc. The amount of e-waste is also increasing day by day and get accumulated in the environment causing various health issues to living organisms and also polluting the environment. To reduce the scale of this great

Figure 11. Complete working model of the system

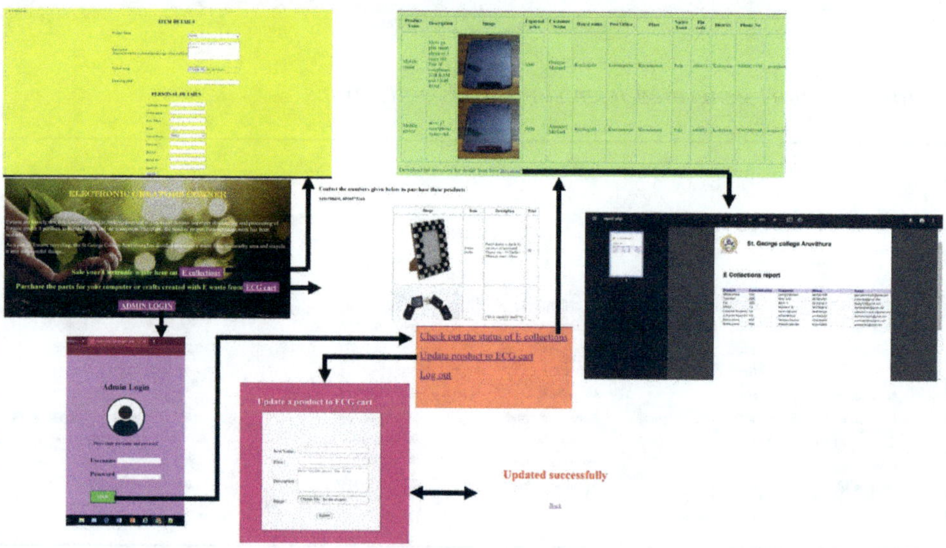

horrible pollution route, we must turn to the recycling and reusing habits of these products. This chapter proposes some beautiful tips and tricks to recycle computer waste, one of the main contributors of e-waste in this digital era, and then sell them through online platforms by developing an e-waste Management System (EMS). The main strategy is to first recycle e-waste into beautiful stunning craft products and then use the proposed e-waste management system to sell and purchase these products. The main objective of the system is to efficiently control the throwback policy of electronic waste to the environment. This helps to reduce the accumulation of e-waste.

REFERENCES

Batoo, K. M., Pandiaraj, S., Muthuramamoorthy, M., Raslan, E. H., & Krishnamoorthy, S. (2022). Behavior-based swarm model using fuzzy controller for route planning and e-waste collection. *Environmental Science and Pollution Research International*, *29*(14), 19940–19954. doi:10.100711356-021-12824-x PMID:33743154

Chakraborty, S. C., Qamruzzaman, M., Zaman, M. W. U., Alam, M. M., Hossain, M. D., Pramanik, B. K., Nguyen, L. N., Nghiem, L. D., Ahmed, M. F., Zhou, J. L., Mondal, M. I. H., Hossain, M. A., Johir, M. A. H., Ahmed, M. B., Sithi, J. A., Zargar, M., & Moni, M. A. (2022). Metals in e-waste: Occurrence, fate, impacts and remediation technologies. *Process Safety and Environmental Protection*, *162*, 230–252. doi:10.1016/j.psep.2022.04.011

DTSC. (n.d.). *Universal Waste*. DTSC. https://dtsc.ca.gov/universalwaste/

Forti, V. B. C. (2020). The Global e-waste Monitor 2020: Quantities,flows and the circular economy potential. United Nations University (UNU)/United Nations Institute for Training and Research (UNITAR) – co-hosted SCYCLE Programme,International Telecommunication Union (ITU) & International Solid Waste Association (ISWA).

Frazzoli, C. R., Ruggieri, F., Battistini, B., Orisakwe, O. E., Igbo, J. K., & Bocca, B. (2022). E-WASTE threatens health: The scientific solution adopts the one health strategy. *Environmental Research*, *212*, 113227. doi:10.1016/j.envres.2022.113227 PMID:35378120

Harshini Vakkalanka. (2018, June 19). The A-Z of e-waste management. *The Hindu*. https://www.thehindu.com/sci-tech/energy-and-environment/what-about-e-waste/article24193081.ece

Irving, M. (2022, August 21). Graphene efficiently recovers gold from waste electronics. *New Atlas*. https://newatlas.com/technology/graphene-recovers-gold-waste-electronics/

Lecler, M. T., Zimmermann, F., Silvente, E., Clerc, F., Chollot, A., & Grosjean, J. (2015). Exposure to hazardous substances in Cathode Ray Tube (CRT) recycling sites in France. *Waste Management (New York, N.Y.)*, *39*, 226–235. doi:10.1016/j.wasman.2015.02.027 PMID:25776743

Lin, S. A., Ali, M. U., Zheng, C., Cai, Z., & Wong, M. H. (2022). Toxic chemicals from uncontrolled e-waste recycling: Exposure, body burden, health impact. *Journal of Hazardous Materials*, *426*, 127792. doi:10.1016/j.jhazmat.2021.127792 PMID:34802823

McFadden, C. (2020, September 20). *How Much Gold Is in Your Computer and How Efficient It Is to Reclaim It*. Interesting Engineering: https://interestingengineering.com/innovation/how-much-gold-is-in-your-computer-and-how-efficient-it-is-to-reclaim-it

Meit, Y. (2011, May 12). *Ministry of Electronics & Information Technology*. Meity. https://www.meity.gov.in/writereaddata/files/1035e_eng.pdf

Misra, N. R., S. K. (2021). A Review on e-waste: Fostering the Need for Green Electronics. *2021 International Conference on Computing, Communication, and Intelligent Systems (ICCCIS)*, (pp. 1032-1036). IEEE. 10.1109/ICCCIS51004.2021.9397191

Mohd, S. &. (2018). e-waste management in India: current practices and challenges. *ResearchGate A Comprehensive Study on Electrical*, 355.

Murthy, V., & Ramakrishna, S. (2022). A review on global e-waste management: Urban mining towards a sustainable future and circular economy. *Sustainability*, *14*(2), 647. doi:10.3390u14020647

Press Information Bureau. G. o. (2018, March 23). *e-waste Management Rules amended for effective management of e-waste in the country*. Union Environment Minister. https://pib.gov.in/newsite/PrintRelease.aspx?relid=177949

Regtech, T. (2016). *Environment Protection Act, 1986 and e-waste (Management) Rules, 2016*. Team Leaser G Tech. https://www.teamleaseregtech.com/resources/acts/article/29/environment-protection-act-1986-and-e-waste-management-rules-2016/

RikeTechnologies. (n.d.). *Recycled Computer Projects*. Pinterest. https://in.pinterest.com/riketech/recycled-computer-projects
/

Roy, H. R.-M., Rahman, T. U., Suhan, M. B. K., Al-Mamun, M. R., Haque, S., & Islam, M. S. (2022). A comprehensive review on hazardous aspects and management strategies of electronic waste: Bangladesh perspectives. *Heliyon*, *8*(7), 09802. doi:10.1016/j.heliyon.2022.e09802 PMID:35815143

Santos, S. M. (2022). e-waste management in Brazil: Challenges and opportunities of a reverse logistics model. . *Environmental Technology & Innovation, 102671.*

Sengupta, D. I., Ilankoon, I. M. S. K., Dean Kang, K., & Nan Chong, M. (2022). Circular economy and household e-waste management in India: Integration of formal and informal sectors. *Minerals Engineering*, *184*, 107661. doi:10.1016/j.mineng.2022.107661

Statista. (2020, July). Retrieved from Leading countries based on generation of electronic waste worldwide in 2019: https://www.statista.com/statistics/499952/ewaste-generation
-worldwide-by-major-country/

Tipre, D. R., Khatri, B. R., Thacker, S. C., & Dave, S. R. (2021). The brighter side of e-waste—A rich secondary source of metal. *Environmental Science and Pollution Research International*, *28*(9), 10503–10518. doi:10.100711356-020-12022-1 PMID:33438127

Zeng, X. L., Liu, D., & Wu, W. (2022). PM2. 5 exposure and pediatric health in e-waste dismantling areas. *Environmental Toxicology and Pharmacology*, *89*, 103774. doi:10.1016/j.etap.2021.103774 PMID:34800720

Chapter 8
Strategies for Waste Recycling, Reuse, and Reduction

Arun Shrivastava
Maulana Azad National Institute of Technology, India

Vikash Kumar
Maulana Azad National Institute of Technology, India

ABSTRACT

Electronic waste (e-waste) represents a fast-growing disposal scenario worldwide. The ever-increase in e-waste volume from electronic industries and households has been creating health and environmental concerns since this waste contains some toxic substances. Additionally, the shipment of toxic wastes has generated a huge public outcry. E-waste is the discarded electrical or electronic devices or equipment. Used electronics which are destined for refurbishment, reuse, resale, salvage recycling through material recovery, or disposal are also considered e-waste. Electronic scrap components, such as CPUs, contain potentially harmful materials such as lead, cadmium, beryllium, or brominates flame retardants. Recycling and disposal of e-waste involve significant risk to health and environment. E-waste is electronic products nearing the end of their useful life. Computers, televisions, VCRs, stereos, copiers, printers, cell phones and fax machines are common electronic products. Many of these products can be reused, refurbished, or recycled.

INTRODUCTION

The waste generated due to the discarded, obsolete, end of life electrical and electronics equipment is known as E-waste world over. It is well known by acronym WEEE

DOI: 10.4018/978-1-6684-7573-7.ch008

(Waste from Electrical and Electronics Equipment). However, various definitions have been presented by different agencies. The waste management hierarchy of prevention, reuse, recycle and disposal in a landfill is accepted as a universal guideline for waste management. However, when it is desirable to shift from one stage of hierarchy to another, would depend on several factors such as cost, impact to the environment and risk perceived by the public. A schematic diagram showing the influence of these factors on the different stages of waste management hierarchy. The waste management hierarchy can typically be depicted by an inverted triangle with reuse at the top which has maximum width, signifying maximum preference to this management option. For minimization of environmental impact, the ideal scenario would be maximum possible reuse and disposal in a landfill only when it cannot be reused or recycled. Typically, that would mean maximum possible time span of life cycle of waste. The case would be the same while minimizing perceived risk, as it has been observed that, people perceive minimum risk for reuse and maximum for disposal in a landfill. As recycling is a preferred option than disposal in a landfill for the objectives of minimization of environmental impact and perceived risk, recycling of the waste would be preferred even after it is no longer economically attractive than disposal. This would mean a delay in shift from recycling stage of hierarchy to disposal, as compared to the scenario of priority to minimization of cost. However, when the objective shifts to minimization of cost, reuse will be preferred only till it is economically more attractive than recycle and recycling would be continued only till it is economically more attractive than disposal in a landfill. This would mean a shorter reuse span/life cycle than under the priority of minimization of perceived risk or environmental impact.

The sustainable use of natural resources depends on the better management of secondary resources, which are the residual materials". The general goal of this Policy is to divert at least 65% of "waste" and to reuse and recover these residual materials. The Policy wisely postulates that we can only attain "these objective if all segments of society contribute to the effort". Because these residues can be reused, recovered, or recycled, this is not a question of waste but one of resources. Considering our "waste products" as resources means having an awareness that will lead to the adoption of new Diverting, reusing and recovering this waste also helps to reduce the risks to our health and to the environment. In fact, waste products that decompose in a landfill form a toxic mix that, through leachate (liquid created when rainwater passes through wastes), contaminates the groundwater, surface water and soil. Moreover, the decomposition of organic matter generates biogases: methane, $CO2$, nitrogen, carbon monoxide and other volatile organic compounds such as toluene and benzene, both of which are carcinogenic. Burning waste does not solve the problem: toxic heavy metals (lead, mercury, arsenic, cadmium, etc.) are, in fact, sent back into the atmosphere as dangerous organic compounds (dioxins, furans, phenolic compounds, etc.). The resulting ash, which sometimes represents up to a third of the waste products' mass, also contains heavy metals and dangerous organic compounds

The goals established in the Quebec Policy for the Management of Residual Materials, which replaces the government's Policy on Integrated Solid Waste Management of 1989, are:

- The application of the 4Rs principles;
- The increased accountability of producers;
- Citizens participation;
- Partnerships between those active in the domain;
- The regionalization of responsibilities and powers.

To reduce wastes that are sent to the dump or incinerated by 65%, the government set objectives for each sector of society and for each residual material.

In the Municipalities (and Communities)

- 60% of glass, plastic, metal, fibers (paper and cardboard), bulky residues and putrescible matter (food waste and green waste);
- 80% of non-refillable beer and soft drink containers;
- 50% of textiles; • 20% non-returnable metals;
- 75% of oils, paints and pesticides (dangerous household materials);
- 60% of all other dangerous household materials.

In Industries, Businesses, and Institutions

- 70% of plastic and fibers (paper and cardboard), wood and textiles;
- 95% of metals and glass;
- 60% putrescible matter (food waste and green waste);
- 85% of tires (consumers', industries', businesses' and institutions' out-of-service tires).

In the Construction, Renovation and Demolition Industry

- 60% of all recoverable materials.

Aboriginal communities are sensitive to the effects of their actions on their environment, but their sensitization, training, and human and financial resources in matters concerning waste reduction and the reuse, recycling and reclamation of residual materials remain limited. Members of the First Nations undoubtedly want to participate in the general effort to develop at least 65% of residual materials that can be systematically recovered.

The Life Cycle

The food we buy at the grocery store, the electricity we use to operate our made-in-China electronic devices: all consumer goods require raw materials, water and energy to exist and to be disposed of.

"Like a living thing, a waste product has its own life, and we can talk about its life cycle. Here are the major stages:

Figure 1. Life cycle

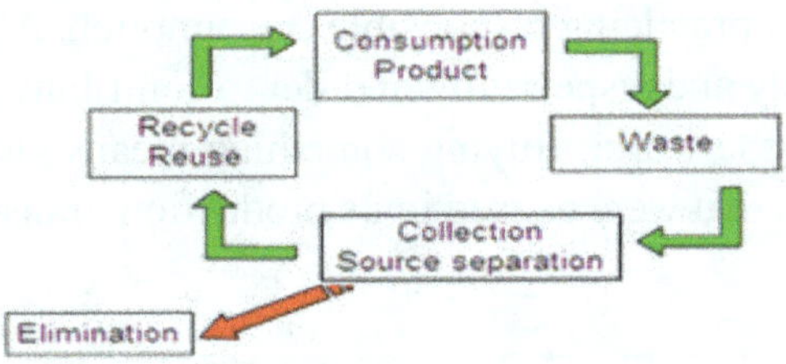

Man intervenes at each stage in the cycle. We therefore have a great responsibility in the management of a waste product.

A new kind of analysis is being developed in universities around the world: life-cycle analysis. This is "a method for evaluating the environmental impact of a product, process, or service, from the extraction of natural resources to the elimination of waste, including fabrication, delivery, product use and dismantlement of the product at the end of its life cycle.

Life-cycle management has several advantages for industries, companies and service providers, advantages that benefit the environment and its resources, health and us, the clients! Considering the life cycle helps:

- To significantly reduce production costs by reducing the use of energy and the consumption of raw materials.
- To reduce polluting emissions.

Reduce

We live in an era of over-consumption. Everything can be bought. Everything is monetized. Durable products are rare. And this consumption goods production, transportation, packaging and waste management is all too often at the expense of the environment. At the expense of life. At our children's and grandchildren's expense.

On average, a six-month-old Canadian baby will have already consumed as many resources as a person living in a developing country will in all his or her lifetime. In other words, if the 6.5 billion humans on Earth consumed as much as we do, we would need three to five planets to meet our needs. The waste an individual produces in his or her lifetime is 750 times his or her adult weight.

Reducing one's consumption means going after the source of waste production, rather than managing the consequences of this production: This means choosing a box in which muffins are not individually wrapped.

We can also reduce our consumption, our production of waste and our impact on the environment by practicing responsible consumption. All it takes is using our buying power carefully and respectfully in order to contribute positively to society, the environment and the world. Buying something means guaranteeing the waste and greenhouse gases that were caused in its production, transportation, packaging and future removal.

Reuse

Reuse is thus about extending the life or giving a second life to something that we previously considered as "garbage". "You are reusing each time you go to: libraries, rental stores (video, tools), repair shops (shoe repair shop, electronics help desk), garage sales, used sport equipment stores, used-clothing counters and stores, antique dealers, furniture refinishers, used-book and cd stores, used construction material stores, stores that sell product in bulk and reuse containers, and computer and home appliances stores."

Recycle

Like reuse, recycle means recovering an object and giving it a second life. While reusing an object means using it without really modifying it or favoring multi-purpose objects and products over single-use ones, recycling means bringing an object back to a state of raw material: paper goes back to pulp, plastics are melted and molded into new products, etc. What was once considered waste becomes a resource, thus breaking with the linear extraction-production-consumption-destruction logic. With this new mindset, an environmentally friendly cycle and sustainable development are established, and that reduces consumption and its negative impact.

After collection, recuperated materials head to a recycling center where paper, cardboard, plastics, glass and metals are just sorted. Each material is then compressed in a cubic tone and sold to companies that do the actual recycling of these resources.

Most of the materials thrown in the garbage can be used and processed in ways other than being destroyed. This is what is called recovering. Reusing, recycling is the most frequently used methods for recovering waste. When it's not possible to reuse or recycle objects such as dead batteries, ink cartridges or cell phones, which all contain toxic elements labeled hazardous household waste there is one last option before throwing them away: scrap dealers, recycling workers and recoveries.

Another alternative is recovering the energy stored in residual material. That means turning waste into a fuel for manufacturing processes or equipment designed to produce energy. Various mechanical, biological and caloric systems and technologies can convert, reprocess or break up wastes into new materials or energy. For example, the methane caused by rotting materials in dump sites can be recycled. This gas is converted into power, and thus eliminates its harmful effects on the environment (methane is a greenhouse gas 20 times more powerful than CO2).

Electronic Devices

Electronic devices were sent to a dumping site or incinerator. More precisely, we're talking about:

- 1,8 million computers; 552 000 laptops;
- 2,4 million screens; 2,4 million cell phones;
- 3 million computers, scanners, and faxes.

"These mountains of wastes will continue to grow as the lifetime of devices shortens: The life cycle of the personal computer went from 6 to 2 years in one decade!" We buy over 3.5 million IT and communications devices each year: computers, printers, televisions, cell phones, sound systems, small appliances, scanners, etc. These devices have an average lifespan of 2 to 3 years, after which individuals, companies and institutions replace them with models that have better performance. This generates 30,000 tons of waste per year (140,000 tones in Canada) of which only 6% is recycled! Yet, 82% of a computer's weight is recyclable. The recovery industry that collects this waste and gets rid of the hazardous materials they contain exist already.

Non-rechargeable batteries are expensive and have a short lifespan. That means they soon become hazardous household waste. Apply the first R—Reduce—by purchasing products that do not require batteries. Or use adapters. Or buy rechargeable batteries.

When used with products with high energy consumption, rechargeable batteries can replace from 50 to 300 alkaline batteries. Not only does this represent a major savings, it also effectively applies the 2nd R—Reuse!

However, only 4% of the 48 million batteries bought each year in Québec are rechargeable.6 Not only are alkaline batteries not rechargeable, but they also contain heavy metals (including mercury, cadmium and lead) and very high collection costs means that few retailers recover them. Rechargeable batteries also contain heavy metals, but they can be reused between 500 to 1,000 times and can be recycled. Their original casing and circuits are kept, only the cells are changed. The refurbishing process is less expensive in most cases than buying new batteries, and these batteries perform better.

Over 50,000 retailers, businesses and city collection centers participate in the Rechargeable Battery Recycling Corporation's (RBRC) battery recycling program. This industry-funded non-profit organization collects, transports, and recycles rechargeable batteries and cellular phones. The collection containers are too often behind the retailer's counter or back store, and you simply can't see it. Ask for it! This rechargeable battery and cellular phone recycling program is free for all participants: consumers, retailers, communities, public organizations, and companies. RBRC pays the costs of shipping and recycling rechargeable batteries and cellular phones, and provides free collection containers, including prepaid and preaddressed shipping labels, safety instructions and small plastic bags to hold each used battery and cellular phone. It's simple and free.

LITERATURE REVIEW

Life cycle analysis (LCA) forms the core theme of this paper. Hence, relevant literature pertaining to LCA is being reviewed and cited. There is now a widespread agreement amongst industry, government and other stakeholders, that environmental issues and impacts must be considered from a life cycle perspective. Several researchers have adopted LCA based methodology to characterize environmental considerations with respect to an array of pollutants (Powell et al., 1996). LCA is currently being used in several countries to evaluate different strategies for integrated solid waste management (ISWM) and to evaluate treatment options for specific waste fractions. The first solid waste management models were optimization models viewing specific aspects of the problem in the life cycle perspective. More recent models based on the life cycle approach are focused on integrated waste management. For computer

Figure 2. Recycling Program Works

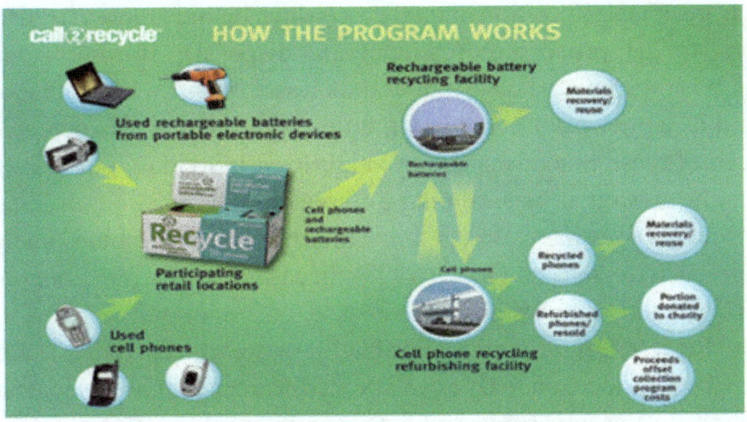

waste management, it is important to find out optimum waste flow from source to reuse, recycle and disposal facilities, over the life cycle/span of waste based on the governing socio-economic scenario. Although, several studies have been done related to solid waste management utilizing life cycle approach, it is to be noted that the optimum material flow from one management option to other has not been covered so far. Also the approach to estimate environmental impact and estimation of risk perceived by the public, from management of computer waste is not being stated in any of the cited studies. The objective of this paper is to present a model which gives the best possible configuration of computer waste management facilities and allocates wastes to these facilities, to achieve the required objective (minimization of cost, minimization of perceived risk (PR) minimization of environmental impact (EI) or a compromise between cost, PR and EI) in the life cycle perspective.

PROPOSED MODEL FORMULATION

The proposed model for the management of computer waste is based on life cycle approach and is multi-time step and multi-objective. Since printers and computer peripherals are integral to a personal computer and are quite substantial in volume, computer waste would henceforth refer to a mixed waste of personal computers, printers, and other computer peripherals. Herein, we would like to define 'time step' as that period of time, for which the waste generation and associated costs per unit quantity of weight, for a management activity (transportation, storage, disposal, etc.) remain constant. The time step could be 1 year, 6 months or any other unit of time for which the waste generation and associated management costs are assumed constant. The objectives addressed in the present mathematical formulation are: (i) minimization of total cost (which includes cost of transportation, segregation, storage, processing/treatment, and disposal), (ii) minimization of EI (which has been formulated as a function of waste quantity being disposed of in landfills and its composition), and (iii) minimization of total risk perceived by the people. Each of these objectives can be minimized individually or a compromise solution can be arrived at by assigning different weightings to each objective. As each of these objectives has different units, they have been combined using a utility function approach (Nema and Gupta, 1999). In case of utility function, objective = minimize (U), where U = weighting to cost × (cost/minimum achievable cost) + weighting to PR × (PR/minimum achievable PR) + weighting to EI × (EI/minimum achievable EI). The decision maker can assign different weightings to cost, PR and EI, depending on the governing socio-economic scenario. Various scenarios of different weightings to each of the objectives can also be analyzed to arrive at the various tradeoffs between these objectives.

The problem is subjected to the following constraints:

- Mass balance of wastes at each node (i.e., all source cum storage nodes, processing, and disposal nodes/facilities).
- Allowable capacities at various facilities.
- Logical constraints at processing and disposal sites.

Estimation of Cost

Costs considered are the cost of segregation and storage at source nodes, cost of transportation of waste from source nodes to processing facilities, cost of processing waste at processing facilities, transportation cost of reusable waste types to reuse facilities, transportation cost of non-reusable, non-recyclable portion of waste from source nodes to disposal facilities, transportation cost of non-recyclable residue of waste from processing facilities to disposal facilities, capital cost for locating facilities (processing and disposal), cost of disposal, cost recovered from the sale of recyclable portion of generated waste and cost recovered from the sale of reusable portion of generated waste.

Estimation of PR

Risk managers and risk-management institutions are faced with an ever-increasing set of challenges to foster good relationships with the public, as illustrated by the conflicts that exist in sitting the new facilities. There is a distinction between scientifically assessed risk and risk perceived by the people. The public's beliefs about environmental risk are often very different from the experts stated that the public perception of health risk in close proximity to a hazardous waste site is higher than the assessments of experts. PR is often assumed to be a central factor in social and political dilemmas. It is related to the "acceptance" of a technology and a lifestyle and hence is an important concept. The PR associated with various activities associated with computer waste management (storage, segregation, transportation, reuse, recycling/processing, and disposal) has been estimated as a function of the waste quantity at each activity multiplied by the PR per unit quantity of waste at that activity. PR for various management options for each waste type was estimated using expert opinion. Information regarding the hazardous constituents of e-waste and the potential hazards and relative possibility of accident was provided to various experts. The experts were asked to relatively rank their PR associated with a particular management activity at a certain facility, as compared to certain similar management activities (SMA).

Estimation of EI

The need to encourage the minimization of EI associated with electronic products across their total life cycle has already been emphasized. This includes upstream impacts arising from the choice of materials and from the manufacturing process as well as the downstream impacts, i.e., from the use and disposal of products. The EI and human health concerns associated with computer hardware are many and varied, particularly when measured over their whole life cycle. The most immediate of these concerns is those surrounding disposal and the high toxic content of computer hardware. The list of toxic materials in computer components includes lead and cadmium in circuit boards, lead oxide in computer monitors' cathode ray tubes, mercury in switches and flat screen monitors, cadmium in computer batteries, polychlorinated biphenyls (PCBs) in older capacitors and transformers, and brominates flame retardants on printed circuit boards, cables and plastic casing.

There are three main environmental concerns surrounding computer hardware disposal:

- The high volumes of hardware reaching end of life introduces the immediate concern of landfill space provision for these bulky items.
- Presence of hazardous materials, such as; chlorinated and brominates substances, toxic metals, photoactive materials, plastics and plastic additives.
- Disposal of computer hardware is the loss of potentially valuable resources. Recovery of these materials may reduce raw materials extraction and the environmental impacts of computer hardware production. For example, the primary production of metals accounts for 10% of global CO2 emissions.

Figure 3. E-waste processing chart

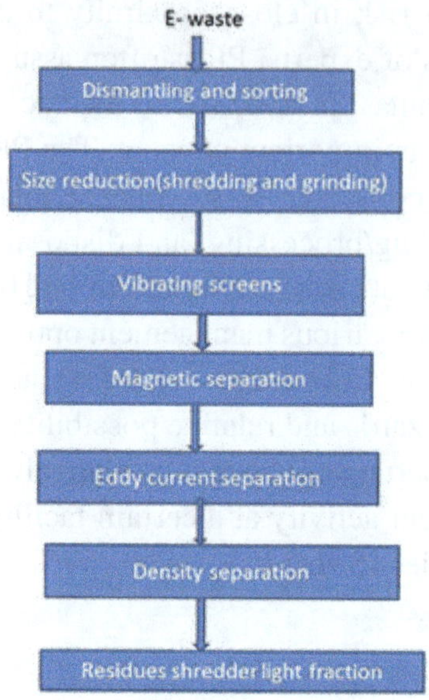

Recycling of these metals will save 70–95% of the energy required for raw materials.

E-Waste Treatment Technologies

Scientific dismantling and recycling are of prime importance when we talk about managing the e-waste in an appropriate manner. For environmentally sound treatment, following the steps could be implemented to ensure an efficient processing of e-waste.

Environmentally sound E-waste treatment technologies are used at three levels as described below:

- 1st level treatment
- 2nd level treatment
- 3rd level treatment

All the three levels of e-waste treatment are based on material flow. The material flows from 1st level to 3rd level treatment. Each level treatment consists of unit operations, where e-waste is treated and output of 1st level treatment serves as input to 2nd level treatment. After the third level treatment, the residues are disposed of either in TSDF (Treatment, Storage, and Disposal Facility) or incinerated. The

Figure 4. Treatment steps

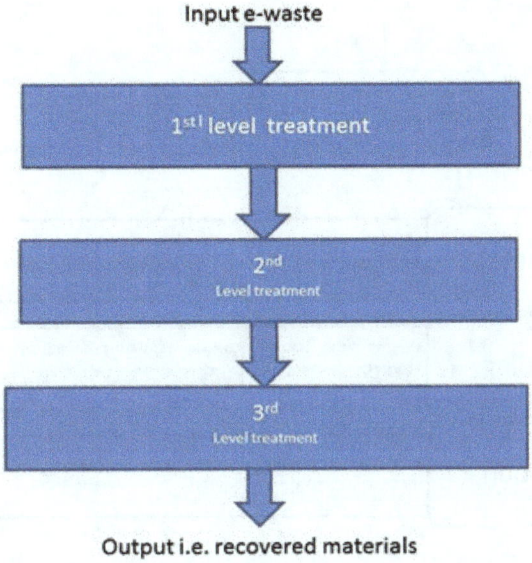

efficiency of operations at first and second level determines the quantity of residues going to TSDF. The simplified version of all the three treatments is shown below.

For non-CRT E-waste, the major e-waste treatment facilities in India use the following technologies.

1. Dismantling
2. Pulverization/ Hammering
3. Shredding
4. Density separation using water.

First Level of Treatment

There are three-unit operations at first level of e-waste treatment as shown below;

The following terms explain the individual process steps in more detail:

- **Decontamination** is the first step in the treatment of e-waste to render it non-hazardous. This means removal of all types hazardous substances present in e-waste. Decontamination is carried out prior to processing e-waste for recovery so that it is safe for recycling. Decontamination also involves removal of all types of liquids and gases under negative pressure which are then stored safely. Some equipment may not require decontamination.

Figure 5. First level of treatment

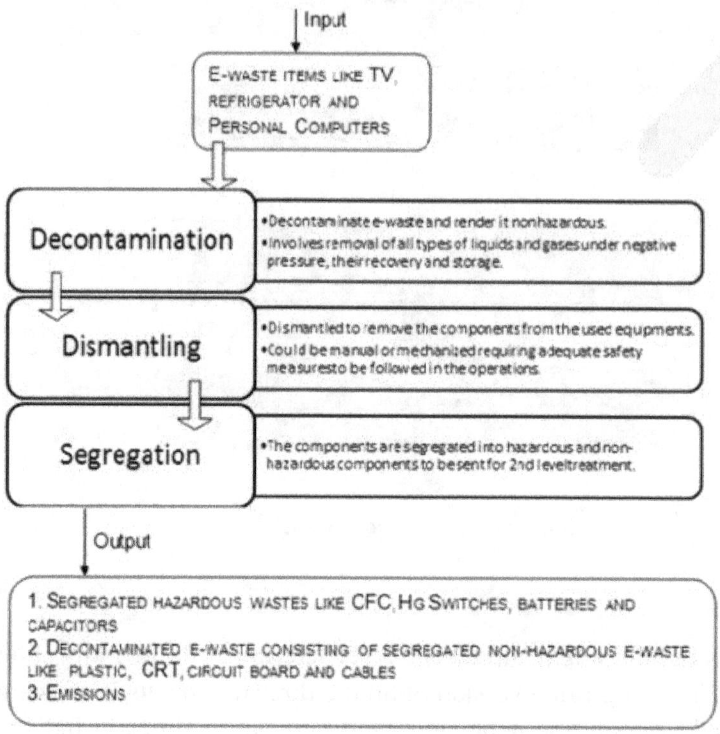

- **Dismantling** is the process by which the components are removed from the used equipment. This is the first operational stage in e-waste recycling and is essential for all used equipment. Dismantling could be manual or mechanized. Dismantling operations require adequate safety measures.
- **Segregation** of the dismantled e-waste fractions is an important step in separating the hazardous and non-hazardous fraction and for channelization of these fractions for the 3rd level Treatment.

Second Level of Treatment

The two major unit operations in the second level of treatment are hammering and shredding. The proposed technology for sorting, treatment, including recycling and disposal of e-waste is fully based on dry process using mechanical operations. The pre-comminuting stage includes separation of Plastic, CRT and remaining non CRT based e-waste. Equipment like hammer mill and shear shredder will be used at comminuting stage to cut and pulverize e-waste and prepare it as a feedstock to magnetic and eddy current separation. A heavy-duty hammer mill grinds the material to achieve separation of inert materials and metals. After separation of metals from inert material, metal fraction consisting of Ferrous and Non-Ferrous metals are subjected to magnetic current separation. After separation of Ferrous containing fraction, Non-ferrous fraction is classified into different non-metal fractions, electrostatic separation and pulverization.

For non-CRT e-waste the following steps can be applied for the second treatment step of e-waste:

1. The proposed technology for sorting, treatment, including recycling and disposal of E-waste is fully based on dry process using mechanical operations.
2. The pre-comminuting stage includes separation of Plastic, CRT and remaining non CRT based E-waste. Equipment like hammer mill and shear shredder will be used at comminuting stage to cut and pulverize e-waste and prepare it as a feedstock to magnetic and eddy current separation.
3. A heavy-duty hammer mill grinds the material to achieve separation of inert materials and metals.
4. After separation of metals from inert material, metal fraction consisting of Ferrous and Non-Ferrous metals are subjected to magnetic current separation. After separation of Ferrous containing fraction, Non-ferrous fraction is classified into different non-metal fractions, electrostatic separation and pulverization.
5. The ground material is then screened and de dusted subsequently followed by separation of valuable metal fraction using electrostatic, gravimetric separation and eddy current separation technologies to recover fractions of Copper (Cu), Aluminum (Al), residual fractions containing Gold (Au), Silver (Au) and other precious metals. This results in recovery of clean metallic concentrates, which are sold for further refining to smelters. Sometimes water may be used for separation at last stage.
6. Electric conductivity-based separation separates materials of different electric conductivity (or resistivity) mainly different fractions of non-ferrous metals from E-waste. Eddy current separation technique has been used based on electrical conductivity for non-ferrous metal separation from e-waste. Its operability is based on the use of rare earth permanent magnets. When a conductive particle is exposed to an alternating magnetic field, eddy currents will be induced in that object, generating a magnetic field to oppose the magnetic field. The interactions between the magnetic field and the induced eddy currents lead to the appearance of electro dynamic actions upon conductive non-ferrous particles and are responsible for the separation process.
7. The effiency of the recycling system is dependent on the expected yields/ output of the recycling system. The expected yields/ output from the recycling system

are dependent on the optimization of separation parameters like particle size, particle shape, feeding rate (RPM) and optimum operations. Size properties are essential for choosing an effective separation technique. Therefore, eddy current separator is best for granular nonferrous materials having size greater than 5mm. The eddy current separation will ensure better separation of Al fraction in comparison to fraction containing Cu, Ag and Au.

8. Particle shape is dependent on comminuting and separation. Since hammer mills and screens will be used in the proposed technology, the variations are expected to be the same as that of Best Available Technology (BAT).
9. The feeding rate can be optimized based on the speed and width of the conveyor.

For CRTs, the following steps can be implemented:

1. CRT is manually removed from plastic/ wooden casing.
2. Picture tube is split and the funnel section is then lifted off the screen section and the internal metal mask can be lifted to facilitate internal phosphor coating.
3. Internal phosphor coating is removed by using an abrasive wire brush and a strong vacuum system to clean the inside and recover the coating. The extracted air is cleaned through an air filter system to collect the phosphor dust.

Different types of splitting technology used are Ni-Chrome hot wire cutting, Thermal Shock, Laser Cutting, Diamond Wire Method, Diamond Saw Separation, Water Jet separation.

Below a depiction of the second e-waste treatment step is given:

Third Level of Treatment

The 3rd level E-waste treatment is carried out mainly to recover ferrous, nonferrous metals, plastics, and other items of economic value. The major recovery operations are focused on ferrous and nonferrous metal recovery, which is either geographically carried out at different places or at one place in an integrated facility. The input, output and unit operations at 3rd level treatment are described in the table below.

RESULTS AND DISCUSSION

The proposed model was solved using LINGO version 9.0 (LINDO Systems Inc.), an integer linear programming solver. The example problem has been solved for the following sets of joint functions of cost, PR and EI: (i) minimization of cost (cost weighting = 1, EI and PR weighting = 0); (ii) equal weightings to cost, EI

Figure 6. Second level of treatment

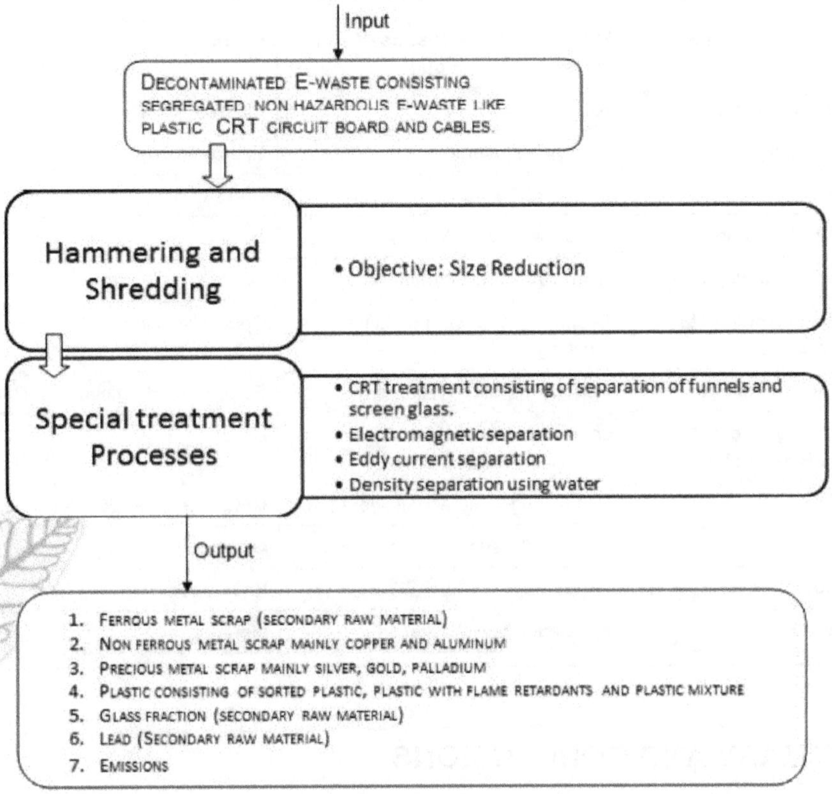

and PR (cost, EI and PR weighting = 1/3); (iii) minimization of PR (PR weighting = 1, EI and cost weighting = 0); (iv) minimization of EI (EI weighting = 1, PR and cost weighting = 0). The constraints were checked for each time step and the results were summarized over all the time steps to give total cost and total risk over all the thirty time steps. The results are summarized in Fig. 6a–c. Time step (in years) at which various facilities were sited under different scenarios of waste generation and different objectives. Cost for various factors was taken from the actual and secondary data and has been converted to US$ (Rs 45 = 1 US$). The results for scenario 1 are discussed in detail. The trend of results for the other two waste generation scenarios is similar.

Table 1. Third level of e-waste treatment

Input/ WEEE residues	Unit Operations/ Disposal/ Recycling Technique	Output
Sorted Plastic	Recycling	Plastic Product
Plastic Mixture	Energy Recovery/ Incineration	Energy Recovery
CRT	Breaking/ Recycling	Glass Cullet
Lead Smelting	Secondary Lead Smelter	Lead
Ferrous metal scrap	Secondary steel/ iron recycling	Iron
Non Ferrous metal scrap	Secondary copper and aluminum smelting	Copper/aluminum
Precious metals	Au/Ag separation (refining)	Gold/ Silver/ Platinum and Palladium
Batteries (Lead Acid/ NiMH and Li ion)	Lead recovery and smelting remelting and separation	Lead
CFC	Recovery/ Reuse and Incineration	CFC/ Energy Recovery
Oil	Recovery/ Reuse and Incineration	Oil recovery/ Energy
Capacitors	Incineration	Energy recovery
Mercury	Separation and Distillation	Mercury

SUMMARY AND CONCLUSIONS

A life cycle based multi-time step; multi-objective decision support model has been presented in the study. The proposed model can assist the decision makers to select the optimum configuration of waste management facilities and transportation routes, allocate waste to the waste management facilities and, to arrive at the optimum reuse time span of a particular waste which, as already stated, can be interpreted as the life cycle of a waste. This inference could guide the authorities to protect infiltration of computers coming in the name of donations and charity, by restricting their import after their optimum life span. Moreover, analysis of tradeoffs between conflicting objectives, such as minimization of cost, minimization of PR and minimization of EI significantly facilitates a compromise between conflicting interests, e.g., it was observed in the case study that a slight increase in the cost incurred on computer waste management, though decreased the PR by the same degree, significantly reduced the impact to the environment. Delhi is presently not only generating substantial quantity of e-waste itself, but also attracting computer waste from other metropolitan cities in India like Bombay and Chennai. Computer waste management under the present scenario is only limited to rudimentary methods of segregation and recycling, at small scale centers scattered all over Delhi. Presence of such centers located in the heart of densely populated areas is posing grave hazards to the surrounding population and the environment. Integrated approach of computer waste management as analyzed by the proposed model is expected to greatly resolve environmental and other concerns such as associated health and PR. Moreover, addressing the risk perceived by the public may significantly ensure public support for such management activities.

REFERENCES

Ahluwalia, P. K., & Nema, A. K. (2006). Optimal planning for computer waste. In: *Proceedings of workshop on efficient e-waste management jointly organized by T E R I*. The Institute of Waste Management and Contaminated Site Treatment, DUT.

Arena, U., Mastellone, M. L., & Perugini, F. (2003). The environmental performance of alternative solid waste management options: A life cycle assessment study. *Chemical Engineering Journal*, *96*(3), 207–222. doi:10.1016/j.cej.2003.08.019

Barton, J. R., Dalley, D., & Patel, V. S. (1996). Life cycle assessment for waste management. *J Waste Manage*, *161*(3), 35–50. doi:10.1016/S0956-053X(96)00057-8

Berkhout, F., & Howes, R. (1997). The adoption of life cycle approaches by industry: Patterns and impacts. *Resources, Conservation and Recycling*, *20*(2), 71–94. doi:10.1016/S0921-3449(97)01199-3

Bevc, C. A., Marshall, B. K., & Picou, J. S. (2007). Environmental justice and toxic exposure: Toward a spatial model of physical health and psychological well-being. *Social Science Research*, *36*(1), 48–67. doi:10.1016/j.ssresearch.2005.11.001

Craighill, A. L., & Powell, J. C. (1996). Lifecycle assessment and economic evaluation of recycling: A case study. *Resources, Conservation and Recycling*, *17*(2), 75–96. doi:10.1016/0921-3449(96)01105-6

Denison, R. A. (1996). Environmental life-cycle comparisons of recycling, landfilling, and incineration: A review of recent studies. *Annual Review of Energy and the Environment*, *21*(1), 191–237. doi:10.1146/annurev.energy.21.1.191

Denison, R. A. (1996). Environmental life-cycle comparisons of recycling, landfilling, and incineration: A review of recent studies. *Annual Review of Energy and the Environment*, *21*(1), 191–237. doi:10.1146/annurev.energy.21.1.191

Ekvall, T. (1999). Key methodological issues for life cycle inventory analysis of paper recycling. *Journal of Cleaner Production, 74*(4), 281–294. doi:10.1016/S0959-6526(99)00149-3

Environment Australia. (2001). *Developing a product stewardship strategy: for electrical and electronic appliances in Australia – Discussion Paper.* Commonwealth of Australia.

Environment Australia. (2001). *Developing a product stewardship strategy: for electrical and electronic appliances in Australia – Discussion Paper.* Commonwealth of Australia.

Finnveden, G. (1999). Methodological aspects of life cycle assessment of integrated solid waste management systems. *Resources, Conservation and Recycling, 26*(3-4), 173–187. doi:10.1016/S0921-3449(99)00005-1

Moy P., (2005). *A health risk comparison of landfill disposal and waste-to-energy treatment of municipal solid wastes in New York city.* Research submitted for partial fulfillment of an M.P.H. at the Mailman School of Public Health at Columbia University.

Nema, A. K., & Gupta, S. K. (1999). Optimization of regional hazardous waste management systems: An improved formulation. *Waste Management (New York, N.Y.), 19*(7-8), 441–451. doi:10.1016/S0956-053X(99)00241-X

Powell, J., Craighill, A., & Brisson, I. (1996). The life cycle assessment and valuation of waste management options: a U.K. study. *Proceedings of Air and waste management association's 89th annual meeting and exhibition.* IEEE..

Solano, E., Dumas, R. D., Harrison, K. W., Ranjithan, S., Barlaz, M. A., & Brill, E. D. (2002). Life cycle-based solid waste management. II: Illustrative applications. *Journal of Environmental Engineering, 128*(10), 993–1005. doi:10.1061/(ASCE)0733-9372(2002)128:10(993)

Chapter 9
A Study of the Printed Circuit Board (PCB) E-Waste Recycling Process

S Selvakumar
https://orcid.org/0000-0001-8056-097X
Aarupadi Veedu Institute of Technology, Vinayaka Mission Research Foundation, India

Sarma Adithe
BVC Institute of Technology and Science, India

J. Samson Isaac
https://orcid.org/0000-0001-9860-2816
Karunya Institute of Technology and Sciences, India

Raghuram Pradhan
https://orcid.org/0000-0002-5008-210X
PACE Institute of Technology and Sciences, India

V. Venkatesh
Rajalakshmi Engineering College, India

Sampath B.
Mythayammal Engineering College (Autonomous), India

ABSTRACT

One of the waste sources that is growing the quickest globally is electronic garbage. Metal deoxidation and the loss of carbon in the gaseous fraction were inhibited by inert atmospheres. The composition, hazardous and toxic components of various e-wastes are explained. The concentration of precious metals in printed circuit boards (PCBs) makes them a significant source of metals. The metallic phases were discovered to include notable concentrations of Ag, Au, Pd, and Pt. In very small amounts, precious metals were found in the carbons and oxides of non-metallic fraction. Copper-based alloys and carbon resources were successfully recovered through high-temperature pyrolysis. Additionally, this method reduced the amount

DOI: 10.4018/978-1-6684-7573-7.ch009

of material that needed to be managed and processed further. Recycling e-waste should improve metal recovery, preserve natural resources, and offer a solution that is environmentally sustainable.

INTRODUCTION

One of the fastest-growing solid waste sources in the world is electronic garbage (or "e-waste"). Because of the high value and small mineral deposits, urban mining from waste PCBs is particularly appealing. Waste PCBs were subjected to high temperature pyrolysis in an argon-filled horizontal tube furnace at temperatures between 750°C and 1350°C. To prevent thermal shock, samples were placed in the furnace's cool zone for ten minutes before being subjected to high temperatures for up to twenty(Khalid et al., 2021). The global production of outdated equipment was estimated at 41.8 million metric tonnes. The worldwide production of E-waste is largely a result of technological advancements and the planned obsolescence of EEE. The majority of e-waste is either dumped or disposed of in landfills; only a small portion is processed or repurposed. E-waste contains a vast variety of materials in the form of a highly complicated amalgamation of polymers. EEE can have up to 60% non-metal components, such as polymers, resins, and fibres, and up to 40% metals. There is currently no recognised standard procedure for treating e-waste, which causes wide variances in the recyclable materials (Kumar et al., 2023; Morf et al., 2007). The majority of e-waste also includes harmful and dangerous substances. Ten percent to twenty percent of discarded PCBs may include copper. Open burning, manual dismantling and disassembly, and pyrometallurgy are a few of the techniques used. The operation of these processes may result in the production of extra dangerous chemicals if they are insufficient or unsafe. The creation of these dangerous and poisonous pollutants has the potential to pollute at high levels and seriously harm both the environment and human health (Boopathi, 2022a). The disposal of these residues and related emissions discharge can contaminate nearby recycling sites' water, soil, and air. As the particles move through the air or water, they might also spread over larger areas. It is vital to create environmentally sound methods for recycling e-waste and to lessen the amount of waste that is thrown out or dumped. Increasing the lifespan of precious materials and reducing the quantity of natural resources needed to produce EEE are both important goals (Hanafi et al., 2012).

A significant investigation on the behaviour of discarded polycyclic boards (PCBs) at high temperatures will be carried out by the University of Bristol. The study's objectives are to maximise the amount of metal recovered from used PCBs and minimise secondary by-product production and greenhouse gas emissions. Effect of operating circumstances on the recovery of precious metals from waste PCBs, including gold, silver, platinum, palladium, etc. the behaviour of phase separation and metal loss in the polymeric or ceramic residue. Temperature and time of pyrolysis have an impact on how well copper and other metals are extracted from e-waste(Xing et al., 2009). Compared to virgin mining, urban mining is a more affordable method of recovering metals from obsolete devices. Copper, silver, gold, and palladium are among the rare and precious metals produced by recycling abandoned electrical gadgets. Businesses are investing in cutting-edge industrial infrastructure to deal with e-waste in a sustainable way(Boopathi, 2021a; Gunasekaran et al., 2022). End-of-life recycling, remanufacturing, and reuse are prioritised by Design for Environment (DFE) and Eco-Design Concepts to minimise environmental impacts. Regulations in the EU limit the use of hazardous materials in EEE and encourage the collection and recycling of WEEE-contained hazardous materials. ARF was created to compensate for costs at the end-of-life stage because the EPR law has not resulted in effective e-waste collection or recycling. International corporations provide WEEE take-back on a voluntary basis; however, recycling and collection are hampered by ignorance. E-waste processing has a lower environmental impact when recycled in a regulated setting. The 3R (reduce, reuse, and recycle) method is used by e-waste management solutions to minimise e-waste scrap and increase its use. A significant portion of electronic waste ends up in the informal sector since it is a self-reliant network that is exempt from government norms and regulations. E-waste chemicals have been discovered to have major negative effects on health, contributing to global warming. Global e-waste management can be improved via consumer actions. Through awareness campaigns, conferences, and events, raise knowledge of WEEE hazards and appropriate disposal techniques. To ensure the proper disposal of e-waste, consumer education is crucial. E-waste must be understood by living communities in order to enable efficient collection and recycling.

COMPOSITIONS OF E-WASTE

E-waste is a heterogeneous collection of materials whose composition is constantly changing as a result of the quick advancements in electronic devices' features and architecture. Plastics, resins, glass fibres, and ceramics are a few examples of non-metals. Up to 1,000 distinct chemicals could be present.

Plastics

Due to qualities like electrical insulation and mechanical robustness, plastics are one of the primary materials in electronics. Thermosets and thermoplastics are the two principal categories of plastics used in EEE. In most cases, polymeric materials with strong mechanical and thermal capabilities are referred to as "thermoplastic."

Thermoplastics are more easily recycled than thermosets, which cannot be melted and moulded into new goods. E-waste typically contains 8 to 12 different kinds of polymers. High-impact polystyrene (HIPS) accounts for 56 percent of the total polymers used in EEE(Abdelbasir et al., 2018; Boopathi et al., 2023; Kumara et al., 2023; Vanitha et al., 2023). For impact prevention, CRT displays and TVs employ indigenous plastic (ABS). It possesses high melt viscosity, flame retardant qualities, and chemical and heat resistance. ABS is made up of a continuous phase of styrene-acrylonitrile and scattered polybutadiene. Excellent mechanical qualities, high stability, and flame resistance are all attributes of polycarbonate (PC). PC has a high melt viscosity in addition to traditional halogen and/or non-halogen flame retardants. PPO (polyphenylene oxide) is extremely strong and temperature resistant. Electricity is isolated by polyethylene (PE) and polyvinyl chloride (PVC)(Ramesh Kumar et al., 2020; Sampath et al., 2022; Trojovský et al., 2023). Other polymers utilised in the electrical sector include epoxy, polyurethane (PU), polyamide (PA), polypropylene (PP), polystyrene (PS), and polyurethane (POM).

Copper and Precious Metals

One of the most commonly used metals for a variety of purposes is copper. 18 million metric tonnes of refined copper were used worldwide, with Asia seeing the greatest growth. Urban garbage has the potential to be a significant source of several resources, including materials and energy. Because of their great chemical stability, corrosion resistance, and electrical conductivity, copper and precious metals are extensively used in the electronics sector. The primary metal used in PCBs is copper, which is used as a sheet bonded to the substrate. Gold is used as a bonding wire, in contacts, and in integrated circuits; silver is used in contacts, switches, and solders(Kamberovic et al., 2011).

The majority of the metal used in the creation of EEE comes from primary industry. Around 0.5 to 1 percent of copper is often present in different ores. Mining for metals has a number of drawbacks, including the use of resources like land, water, and energy. There is a strong correlation between high levels of greenhouse gas emissions and the primary production of precious metals like gold and platinum (Zhao et al., 2009). Palladium and platinum each release 10,000 and 14,000 metric tonnes of CO_2, but 1 tonne of gold produced by mining produces 17,000 metric tonnes of CO_2.

Printed Circuit Boards

Electronic circuits, called printed circuit boards (PCBs), are created by attaching electronic parts to a nonconductive board and creating connections between them.

Epoxy resin, fibreglass, and copper are the three basic components of PCBs. The most common substrate for PCBs is FR4, which is made of brominated flame retardant and epoxy resin bonded with fibreglass. There is a risk that a significant portion of the outdated printed circuit boards still contain dangerous lead. Most vintage electronics use Pb-Sn solder. Tin content in PCBs ranges between 1 and 5.62 weight percent. Printed circuit boards (PCBs) fall into one of three categories. PCBs can also be divided into cheap and rich boards based on their composition. Large cylindrical capacitors, a lot of polymers, few or no processors, and thickness are characteristics of poor boards. Rich boards are slimmer than poor boards and include many more processors and components. They also have fewer or no capacitors(Zhao et al., 2008).

Rich boards, like PCBs from smartphones and laptops, contain significant amounts of metal. PCBs are typically made of polymers, ceramics, and metals. The PCBs present in trash from various appliances, manufacturers, and/or eras vary greatly in composition. For instance, when batteries and capacitors are removed, PCBs from TVs and computers may contain up to 70% organic material. Waste PCBs are a significant material resource, containing millions of metric tonnes. Most valuable metals, like copper, can make up between 40% and 70% of the overall value in various forms of waste. The section that follows provides a general review of copper and precious metals. This article is the first in a series on the physical and chemical characteristics of used polychlorinated biphenyls (PCBs), with a focus on the gaseous phase and the metallic and metallic fractions anticipated during various heat treatments(Arshadi et al., 2018). At Aberystwyth University in the UK, experiments will be conducted to better understand the generation, characterization, and structural changes of residual carbons that are anticipated following the high temperature treatment of waste PCBs. In this section, comprehensive descriptions of the experimental tools and methods are offered.

HAZARDOUS AND TOXIC COMPONENTS ASSOCIATED WITH E-WASTE

There are various concentrations of toxic, hazardous, and non-toxic pollutants in various types of e-waste. Hazardous substances found in e-waste make up primary components; by-products or leftovers produced during inefficient recycling processes are considered secondary products. Tertiary emissions are substances produced during recycling that, if managed improperly, could have a serious negative impact on the environment and human health.

Primary Contaminants

E-waste contains significant amounts of heavy metals. If the glass is broken, lead may leak out into the ground. Lead exposure could have disastrous effects on the human body. Pb, Sn, Ni, Sb, and Zn are present in electronic trash and are released after burning or disintegrating. Cadmium and other dangerous and harmful substances can be found in toners and cartridges as well. Kidney CD accumulation has been linked to cancer (Abdelbasir et al., 2018; Domakonda et al., 2022; Samikannu et al., 2022; Vennila et al., 2022).

Flame retardants are present in nearly 12% of the polymers used in electrical and electronic equipment. Thyroid hormone levels may be impacted by PBDE exposure, which can also have genotoxic effects. The endocrine disruptor TBBPA is known to interfere with thyroid homeostasis. Fluorescent lighting, switches, thermostats, and sensors are just a few items that contain mercury. In landfills, mercury can evaporate or leach out. The harm that Hg can do to the central, peripheral, genitourinary, and fish nervous systems makes its consequences for health irreversible. Polychlorinated biphenyl (PCB), which is used in capacitors, condensers, and transformers, raises the risk of cancer in people due to oxidative stress. Chlorofluorocarbons (CFCs), which harm the ozone layer and contribute to global warming and climate change, are found in refrigerators, air conditioners, and freezers. Americium, chromium, gallium, selenium, and beryllium are further elements that are found in e-waste and are used in smoke detectors, data cassettes, semiconductors, and rectifiers. Toxic hexavalent chromium compounds can be ingested or inhaled and can result in allergies, ulcers, gastroenteritis, renal and hepatocellular deficiencies, and lung cancer.

Secondary Contaminants

Volatile chlorine and nitrogen compounds, used acids, and sludges that remain after the process can all be discharged during hydrometallurgical processes. Following leaching, waste acids are produced that include heavy metals, PBDEs, PCBs, and polycyclic aromatic hydrocarbons (PAHs). Plastics that include flame retardants may release harmful by-products during combustion, including PAHs and PHAHs. When burned openly, insulated copper wires can produce 100 times as much dioxin as burning household garbage. These contaminants are dispersed through the air and are present around recycling facilities in the air, dust, and water. Long-term exposure to PAHs may result in cancer as well as lung, skin, and bladder disorders. Human health problems related to gastrointestinal, hepatic, cutaneous, and neurologic toxicity can result from exposure to PCDDs and Fs. Animals exposed to PBDD/F may have mortality, immunotoxicity, and reproductive problems. Around PCB workshops in Guiyu, China, metals and other potentially harmful compounds have been detected in the air, soil, waters, and rivers(Arshadi et al., 2018; Zhao et al., 2009). The amount of PCDDs and Fs in the air was discovered to be the highest ever recorded anywhere in the world, as well as the amount of PBDDs and Fs. Workers were exposed to significant quantities of Sn, Pb, Cd, Cu, Ni, Ag, and Zn through dust samples from PCB factories. The health of individuals suffers when dangerous compounds are dispersed through water, soil, and air. Along with illnesses of the skin, stomach, and breath, high amounts of lead and copper have been discovered in both personnel and residents. Leukaemia cases have been reported, and umbilical cord chromium levels have also increased. (Krishna, 2009). Metal, plastic, ceramic, and silica dusts generated during the pre-processing stage of the shredding process can be inhaled or come into contact with the skin and persistent in the environment are HBCD, PBDEs, and PBDDs. These can alter thyroid hormone levels and impair a new-born's ability to learn and remember things.

Tertiary Contaminants

Pyrometallurgical processes may result in tertiary emissions, such as the release of some metals in the gas fraction. Particular concerns have been expressed about high temperature procedures for recycling e-waste due to PCDDs/Fs mixed chlorinated-brominated dibenzodioxins and furans. High concentrations of sulfuric acid are used in the leaching and solvent extraction stages of hydrometallurgical processes to recover mostly copper and precious metals found in e-waste. After the process, the leftover waste acids pose major risks to the environment and human health. Environmental challenges are also posed by the treatment's high secondary waste generation rates. When leaching PCBs with aqua regia or nitric acid, it is possible for nitrogen oxide and chlorine vapours to be formed, hence the gas emission must be effectively controlled (Chakraborty et al., 2018; Ren et al., 2012). Hydrometallurgical techniques have been used in a number of investigations to recover metals from e-waste. As lead and mercury have been seen to precipitate, these sludges may contain heavy metals. After processing, additional particles, including plastic, other metals, and activated carbon, are still present. Ineffective e-waste management can harm the environment, wildlife, and human health severely and permanently. If proper recycling procedures are implemented, there is a potential material resource of about 40 million metric tonnes per year in electronic waste that may be used repeatedly.

STAGES OF E-WASTE RECYCLING PROCESSES

The recycling chain, shown in Figure 1, is a four-step procedure that is typically used to recycle e-waste. gathering, disassembling or disassembling, pre-processing or upgrading, and material recovery, including refining and by-product disposal (Ren et al., 2012).

Figure 1. Various stages of PCB recycling process

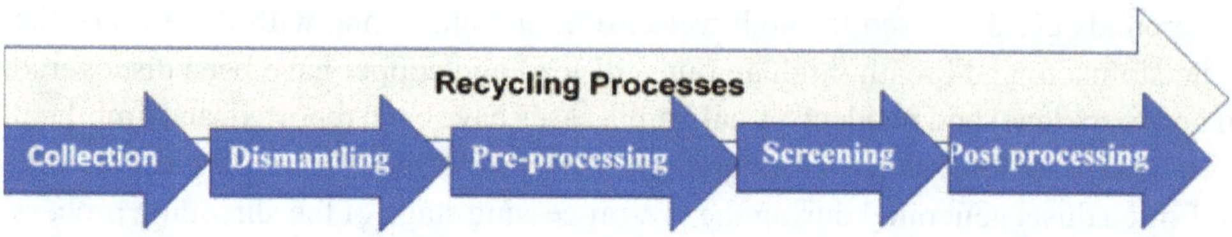

1. **Collection**

The first phase in the recycling process is collection, which establishes the amount of material that may be recovered. Social and governmental factors, for example, have an impact on the collection system. The effectiveness of each phase in the recycling process determines its overall efficacy.

2. **Dismantling**

The material is released during the dismantling or disassembly process, and some parts are taken out. Batteries, glass, and metallic components are separated during the process for later processing.

3. **Pre-Processing technologies**

Physical and mechanical procedures are used to accurately separate the materials. Typically, mechanical sorting based on the properties of the material comes after shredding. Some materials can be processed directly in the materials recovery stage without needing to go through the pre-processing stage.

4. **Screening and Shape Separation**

Metals, polymers, and ceramics all have a variety of particle sizes and shapes. Screening is used to uniformize material size and boost the amount of metal in the material.

Magnetic Separation

Magnetic separation techniques separate ferromagnetic metals from non-ferrous metals and other non-magnetic materials using a permanent or electric magnet. The development of permanent magnets constructed of rare earth alloys has led to an increase in the use of high-intensity magnetic separators, which provide strong field strengths and gradients. Cupro-nickel alloys are typically separated from waste streams using high-intensity separators. A strong magnetic field is used to recover copper from materials with high mass susceptibility, like aluminium and multi-compound bronze.

Electric Conductivity-Based Separation

The use of triboelectric separators allows for the separation of materials with varying electric conductivity or resistivity. Glass, metal, and plastic are just a few of the materials that can be used with this approach. The following description of three techniques includes triboelectricity, corona electrostatics, and eddy currents. Separation of eddy currents: Using rare-earth permanent magnets, eddy current separators remove non-ferrous metals like copper and aluminium. These magnets produce eddy currents and a magnetic field around the component as they rotate quickly(Chen et al., 2014). Since the polarity of the metal and the revolving magnet are the same, these metals repel one another. Copper and precious metals can be extracted from used PCBs using this method as well. Triboelectric separation is a technique for classifying plastics based on their electric characteristics. For instance, when ABS and HIPS are combined, ABS charges positively and HIPS charges negatively. Separation based on density: Materials can be divided depending on their specific gravities. These impede the travel of a fluid, such as water or air, which relies on the particle's size, shape, density, and other factors.

Optical Sorting

By using sensors to record data, optical sorting processes work better and can be used in a number of commercial and consumer products. Recycling of **plastics**: mechanical, chemical, and thermal methods can be used to recycle plastics from e-waste. Identification and separation of plastics are essential for mechanical recycling. a process that could encounter issues like the inclusion of additives or contaminants.

EQUIPMENTS INVOLVED

High Temperature Horizontal Tube Furnace

In a high temperature horizontal tube furnace, experimental studies on recycling used PCBs were conducted. The furnace is built of highly pure alumina and was supplied by Laboratory Equipment (Sydney, Australia). The set-point temperature is displayed on one digital panel, and the current furnace temperature is displayed on the other. To maintain an inert atmosphere inside the furnace throughout the procedure, argon gas was continually fed through it both before and during. The furnace has a gas intake so that during test runs, gas from a pipeline may be connected to travel through it.

Off-Gas Analysis

At the University of Aberdeen in Scotland, a gas analyzer has been installed to examine carbon monoxide and carbon dioxide produced during heat treatments. A piece of software that took input every 10 seconds automatically saved the data.

Thermogravimetric Analysis (TGA) & Derivative Thermogravimetry (DTG)

A well-known approach to estimating a sample's mass loss as a function of temperature and/or time is thermogravimetric analysis (TGA). Under an inert atmosphere of nitrogen, pulverised waste PCBs were heated continuously at a rate of 20 °C per minute at temperatures as high as 1350 °C. Mass loss happens as a result of polymer breakdown or heat decomposition and volatile release; this loss is measured and noted.

Fourier transform infrared spectroscopy (FTIR)

The analysis and characterization of volatile chemicals released during the pyrolysis process can be done using the Fourier transform infrared spectroscopy (FTIR) analytical technique. A temperature of 600 °C was reached by heating pulverised waste PCB samples in a nitrogen environment at a rate of 20 °C per minute.

Inductively Coupled Plasma (ICP) – Optical Emission Spectrometry (OES)

The components were dissolved in aqua regia, a solution of hydrochloric and nitric acids in a ratio of 3:1. The ICP-OES technology was used to quantitatively analyse the solution that contained elements in their atomic state. This technique uses a spectrum of rays and wavelengths to identify specific elements in order to detect trace metals using plasma energy. By ionising argon gas with high-frequency electric current, plasma is produced.

Scanning Electron Microscopy (SEM) – Energy Dispersive Spectroscopy (EDS)

The analysis of leftover products formed after heat treatments was done using a scanning electron microscope (SEM) of type Hitachi S3400X. A carbon tape was used to affix the solid items, which were then wrapped in plastic and attached to a stainless-steel stub. Before the samples were transported for X-ray diffraction microscopy investigation, they were dried for 24 hours in an oven kept at 35 °C.

X-Ray Diffraction (XRD)

Inspections were performed in steps of 0.02° to 0.05° over the 2-angle range of 10° to 80°. MPD is a goniometer system made up of programmable slits, a curved monochromator, and an automatic sample changer. The carbonaceous and slag fractions produced during the recycling process for usage in the construction industry were analysed. Before doing diffraction experiments on the samples, they were finely powdered. High-score software was used to carry out phase identification.

Carbon Analysis

Using the LECO CN TruSpec Analyzer, the carbon content of the carbonaceous residue is determined. The ballast volume, gas lines, and atmospheric gases that entered during sample loading were all purged. A crucible is used to burn the sample at 950 °C while being flushed with pure oxygen to speed up combustion. The evolved gases are measured by a number of infrared detectors to determine the carbon concentration. The after-burner furnace, furnace filter, pre-cooler, and thermoelectric cooler all travel through the combustion products before they are collected in the ballast volume. Results are presented as weight percentages for carbon.

Raman Spectroscopy

Raman uses an argon ion laser with a 514 nm wavelength. Data collected between 1000 cm^{-1} and 2000 cm^{-1} are represented by spectra.

Surface Area Analysis

Scientists at the University of Bristol in the UK have used a number of approaches to explore the pores produced by escaping gases during polymer degradation. The nitrogen physisorption technique was employed to analyse surface area and pore distribution while samples were dried for three hours at 150 °C under vacuum. The BJH model (Condon, 2006) and the BET model were used to calculate the pore size and distribution of the adsorption and desorption isotherms, respectively. Analyzing samples from various organic and inorganic materials allowed researchers to identify the pore distribution.

Initial Characterization of Waste PCBs

Waste PCBs are heterogeneous mixes of materials that include silicon, refractories, alkalis, and alkaline-earth oxides. Some boards also contained other phases like barium titanite, potassium, magnesium, and aluminium silicates. Alkali nitride, silicon oxide, aluminium oxide, carbon monoxide, and others were also discovered in the substance. It has been established that the carbon concentration of these samples is consistent with a PCB that is rich in polymers, paper, and ceramics.

HIGH TEMPERATURE PYROLYSIS METHOD

Sample Selection

For the study of the carbonaceous/slag product, various types of computer waste PCB components, such as ceramic-rich and polymer-rich ones, were employed (Figure 2). Steinert also provided sporadic bits of waste polymer-rich PCBs for studies aimed at characterising carbonaceous slag. To determine the weight loss of each sample over time, weigh the samples both before and after heat treatments. In the experimental investigations, crushed discarded PCBs were used.

Experimental Procedure

Organic samples underwent high temperature pyrolysis, or heat treatments, in a furnace that was constructed and kept at temperatures between 750 °C and 1550 °C (Figure 3). The samples were kept in an alumina crucible with a graphite rod for 10 minutes before being pushed into the hot zone and kept there for an additional 10 minutes. An infrared (IR) analyzer has been installed in the heat treatment furnace monitor the CO, CO_2, and CH_4 gases produced. To capture particles carried by the gas

Figure 2. PCB Sample for the pyrolysis process

flow, two filters were placed in the line leading from the furnace to the IR analyzer. When operating at temperatures of 800 °C, 900 °C, and 1000 °C, "red" sheets were produced. Copper made up the majority of this percentage (Asamoah et al., 2018).

The carbonaceous/slag fraction contained both "white" and "red" metallic particles. Both the metallic fraction and the carbonaceous/slag phase obtained at temperatures above 1150 °C are displayed. When operating at temperatures of 800 °C, 900 °C, and 1000 °C, some wires were also seen in the "white" metallic portion, which was primarily made up of droplets. This phase produced a tin-lead alloy that was rich in lead. Copper's carbonaceous and metallic phases were examined using SEM-EDS, Raman spectroscopy, and x-ray diffraction methods. ICP-OES analysis showed that the carbonaceous residue still contained a modest amount of metals. Slag phases and carbons were difficult to separate, but an approximate estimate of the relative proportion of carbon in the residue was obtained. During SEM studies, x-ray microanalysis was used to evaluate the chemical composition of carbon foams and fibres (Hosoda et al., 2014).

PHASE SEPARATION PROCESS

The stages of PCB-Waste to Metal Extraction is shown in Figure 4. The characterization of the reaction products was produced by various heat treatments. A variety of analytical methods were used to analyse the products. Measurements of weight loss, infrared analysis, and thermogravimetric analysis were used to analyse the gaseous fraction. The LECO carbon analyzer, BET surface area analysis, and ICP-OES were used to analyse the carbonaceous/slag phases. In later chapters, SEM research

Figure 3. High Temperature Pyrolysis Apparatus

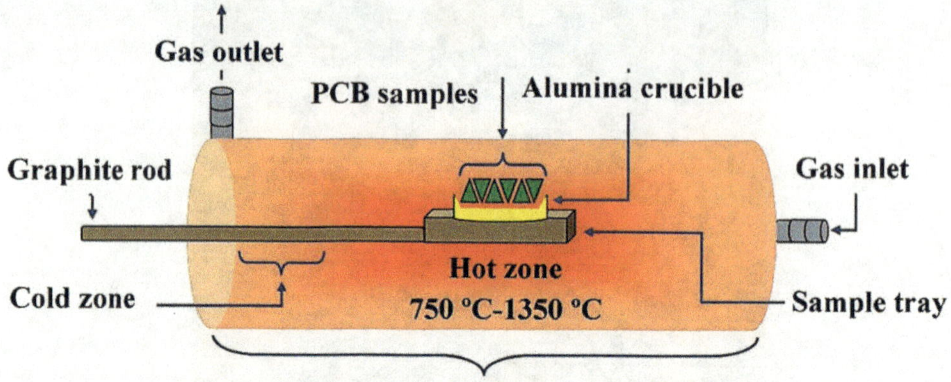

on these goods is presented. Due to the extremely small sizes/amounts of metallic products, X-ray diffraction could not be used.

Figure 4. PCB-Waste to Metal Extraction Stages

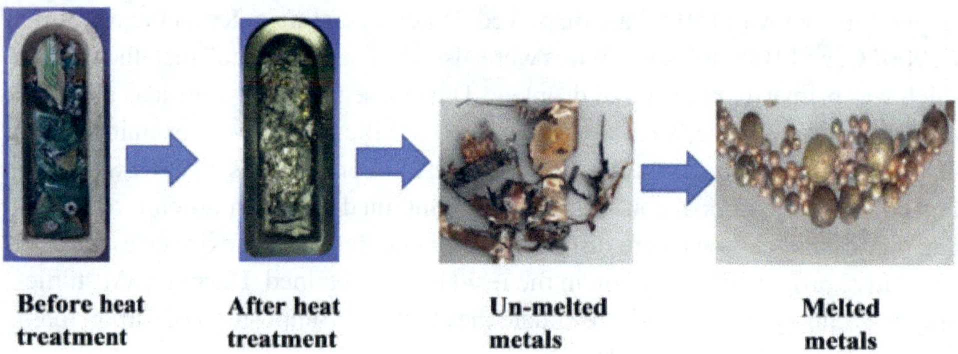

Gaseous Products

Gases are produced when discarded PCBs are heated in a horizontal tube furnace between 750 and 1350 degrees Celsius. The majority of the gases produced were CO, with a minor amount of CO2 as well. The creation of methane was often found to be fairly low. Compared to CO2 and CH4, the amount of CO emitted was found to be substantially higher. During the heating process, CO was still being released, but by the time the PCBs were breaking down, it was very little. Waste PCBs showed two stages of weight loss after thermal degradation at temperatures between 320 °C and 460 °C. It is anticipated that PCBs will weigh substantially less when they decompose at higher temperatures than pure polymers and mixtures of coke and polymers. This is due to the complex mixture of ceramics, metals, and polymers they contain, which decompose at significantly higher temperatures (Ren et al., 2012).

At different durations and temperatures between 800 °C and 1350 °C, the weight loss of waste PCB samples was measured. The main causes of this weight loss are gas release, volatile release, and polymer degradation. The weight loss was generally consistent at lower temperatures, but it increased to 60% at higher temperatures. Reduction reactions that might occur in this temperature range could be the cause of this rise. In contrast to pyrolysis experiments, where temperatures grew exclusively at higher temperatures, in TGA temperatures climbed constantly at a rate of 20 °C per minute. The heat produced at 1150 °C and above is an indication of further reduction reactions/degradation of different waste PCB components. These results differ slightly from TGA data, which demonstrated a gradual but consistent weight reduction under these conditions. Spectra derived from gases produced during the thermal decomposition (TGA) of polymer-rich waste PCBs in a nitrogen atmosphere. The FTIR technique is frequently employed to describe the gaseous components that emerge during pyrolysis. It was anticipated that the generation of volatile products would take place mostly between 310 °C and 510 °C. After 310 °C, it was discovered that volatile release increased and that additional absorption bands were clearly visible. Only trace levels of CO and CO_2 were found after 470 °C, suggesting that carbonization and char production may have occurred. These substances were recognised by the distinctive bonds in them: HBr for HBr, and 2300–2400 cm-1 for CO_2.

Metallic Products

After 10, 20, 30, and 60 minutes of heat treatment, red foils were produced, while the "white" fraction was created by metallic substance droplets and wires. Metallic concentrations change based on the process parameters and the type of metal used. While some metals tend to form oxides, which can be extremely stable in these circumstances, others have a high vapour pressure. After 10 minutes of heat treatment, copper predominated as the main metal in the red fraction, whereas tin predominated as the main metal in the white droplets. As a result of the ceramic component in the raw waste PCBs, metal ions such as Ni, Zn, and Fe were found in the slag/carbonaceous phase. Base metals and slag-forming metals were divided into two classes(Arshadi et al., 2018). Levels of the three metals are typically high in both metallic fractions at temperatures above 1100 °C, indicating a better affinity and stability of these metals. At all temperatures, Mg and Al levels are quite low, notably in the Sn-Pb fraction. A higher proportion was only seen in the copper-rich fraction at 1000 °C. This might be because a particular sample has higher levels of aluminium than would typically be expected.

Slag phase and Carbonaceous Residue

The non-metallic fraction (NMF), which was formed by heat treating e-waste, was shown to form at all temperatures. Silica, magnesium, and alumina made up the majority of the slag that was produced, with trace amounts of other elements.

Carbon Content

Using a LECO analyzer, the carbon content of heat-treated PCB residues was calculated; the results are presented. The carbon content was determined to average around 60 wt. percent, despite some data dispersion caused by variances in the compositions of the PCB specimens under analysis. The carbon content of the residue, which is defined by the nature/relative proportions of thermoplastic/thermosetting polymers, has been compared to the weight loss data in the gaseous phase of waste PCBs. The temperature range that was investigated was a little different, with experiments beginning at 750 °C and concluding at 1,000 °C.

Impurity Levels

The principal metals included in waste Cu, Pb, and Sn in the carbonaceous residue varied from 0.1 to 0.7 weight percent at temperatures lower than 1000 °C, it was discovered. The concentrations of these metals increased to 5.7 wt. percent, 2.3 wt. percent, and 5.1 wt. percent, respectively, at temperatures exceeding 1150 °C. The carbonaceous residue left behind when a reactor is quenched contains higher concentrations of the primary metals than the gaseous or vaporized states of the system(Abdelbasir et al., 2018; Ramesh Kumar et al., 2020).

BEHAVIOUR OF PRECIOUS METALS FROM WASTE PCB

This chapter presents an extensive examination of the extraction and concentration of Ag, Au, Pd, and Pt during the recycling of used PCBs. In the temperature range of 800 °C to 1350 °C, it was examined how the concentration of precious metals was affected by copper, the primary component of PCBs. It is anticipated that the interaction of different metals with copper would greatly influence how those metals are extracted. 800 °C was chosen as the lowest temperature because, according to

Figure 5. Copper & Tin rich fractions and Carbon contents versus weight percentage of PCB Wastes

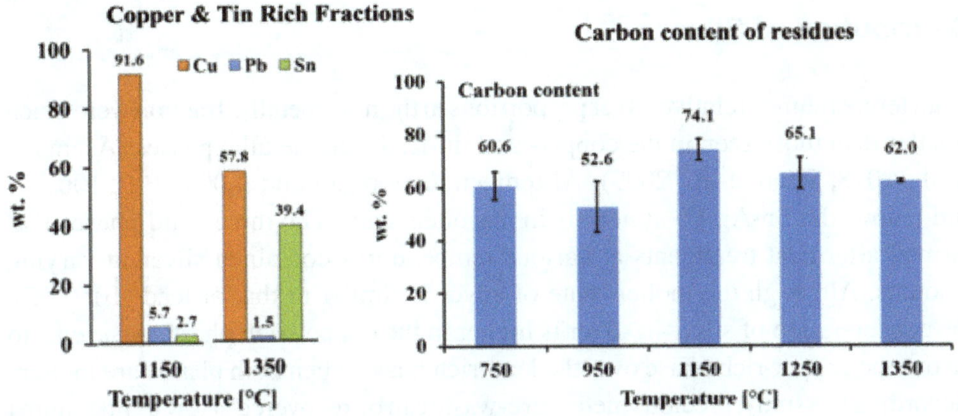

earlier research, it is high enough to avoid the development of dioxin and furan. Figure 5 illustrates Copper & Tin rich fractions and Carbon contents versus weight percentage of PCB Wastes.

Concentration of Precious Metals in Different Phases

The copper rich proportion was mostly discovered as foils at temperatures below the melting point of copper (800 °C). As a carbonaceous/slag residue, about 22 to 25 percent of the waste PCBs had been phase separated. The metallic phase that is rich in copper was found to weigh between 4 and 8 percent, and the metallic phase that is rich in tin ranged from 7 to 17 percent. Copper was discovered to be abundant in the red fractions, but Sn and Pb concentrations were incredibly low. All three stages of decommissioning contained precious metals, but they tended to concentrate mostly in the two metallic portions.

Since a variety of motherboards were utilised in this experiment, the values shown in this table do not represent absolute values. The data does not provide an accurate comparison of the metal contents of different waste PCBs since the distribution of different precious metals may not be consistent across different trash PCBs.

A heterogeneous blend of metals, polymers, and ceramics makes up PCB. The different phases were separated by the process conditions used in this experiment. Assuming a negligible loss of precious metals, the weight of silver, gold, palladium, and platinum in a gaseous form was roughly 20% of the initial sample (Ren et al.,

2012; Zhao et al., 2008). The relative weight proportion of these elements was calculated after they were divided into three solid fractions.

Distribution of Silver

At all temperatures, relative silver proportions in the non-metallic fraction were much smaller than those seen in the copper- and tin-lead-rich metallic phases (Asamoah et al., 2018; Zhao et al., 2009). At temperatures of around 4,000 °F (2,100 °C) and below, the Sn-Ag-Pb system is in a molten state. The three solid phases that formed after heat treatments at various temperatures contained silver in varying amounts. Although the molten state of silver is similar to that of lead (1050 °C), the concentration of silver in silver is higher in the copper-rich phase. Ag seems to favour the copper-rich phase over the lead-rich phase when both phases are molten. According to a study, precious metals in e-waste can be recovered at low temperatures (800 °C). The NMFs had just a very small amount of silver, which suggests that silver has little affinity for carbon or ceramics. This can result in a large decrease in the amount of waste that needs to be treated for further refinement.

Distribution Behaviour of Gold

Following heat treatment at temperatures ranging from 800 °C to 1350 °C, the gold concentration was discovered to be extremely low. At these temperatures, ICP-OES analysis was unable to detect the concentration of gold with any degree of accuracy. The distribution of gold in the three solid fractions produced by thermal treatments is shown. The majority of the gold was discovered at 800 °C in the copper-rich fraction. Gold was fairly evenly distributed across the two metallic portions at 900 °C. Gold displayed a greater propensity to concentrate in the tin-lead fraction at 1000 °C. The greater copper concentration in this fraction might potentially have an impact on this factor. The related phase diagrams show that both silver and gold have a strong affinity for copper. Impurities such as polymers and ceramics in used PCBs don't seem to have any impact on the initial diffusion and concentration of gold and silver in either of the metallic fractions.

Distribution Behaviour of Palladium

At all heat treatment temperatures, palladium was 3:2 more predominant. Pd was found at very trace levels in the NMFs. In pure systems, Pd and Cu completely combine.

Distribution Behaviour of Platinum

The majority of the platinum was discovered in both the copper-rich and the tin-lead-rich metallic fractions, but at 800 °C, somewhat higher platinum proportions were seen in the NMFs than in the other three metals (Ag, Au, and Pd). Platinum exhibited no preference for either of the two metallic portions. In comparison to the copper-rich product at 1000 °C, 56.8% of the platinum was present in the Sn-Pb fraction at 900 °C. Pt contents were virtually evenly split between the two metallic phases at 1150 °C. A comparable outcome, however, was attained at 1350°C, when 69 percent of the Pt was discovered concentrated in this product. Platinum does not favour either of the two investigated metallic fractions. Due to the molten state of Sn and Pb in the system, Pt can also dissolve in these metals at high Sn and Pb concentrations. This percentage contains some Cu, which would aid in the absorption of platinum. Similar in characteristics to Au and Ag, Pt and Pd are miscible in both the liquid and solid forms. It is commonly known that copper and precious metals have an affinity for pure systems. Due to the heterogeneous nature of the components, such as ceramics and polymers, Precious metals typically form solid or liquid solutions with the metallic phase when phase separation occurs.

DISCUSSIONS

PCBs from e-waste were pyrolyzed at high temperatures to produce metallic fractions rich in copper and lead, carbonaceous ceramic residue, and a gaseous fraction. As a result, the volume of metallic fractions was significantly reduced, while the precious metals found in waste PCBs were effectively concentrated. Slags contain concentrated amounts of valuable metals, making them easier to remove and appropriate for further refining. Copper production methods serve as a suitable illustration for refining these metals because precious metals are still present in copper during the final phases of production. Low vapour pressures exist for platinum group metals, including Pd and Pt, between 900 and 1450 °C. Under vacuum or inert atmospheres, the production of these chemicals is greatly diminished(Arshadi et al., 2018; Krishna, 2009). The melting points of silver and gold are both low (982 °C and 1085 °C, respectively). In the temperature range that was examined, they existed in both the solid and molten phases. The loss of silver could occur if the experimental temperature was too high due to the volatilization of lead. The primary metals found in waste PCBs have a strong attraction for Au, Ag, Pd, and Pt. These metals served as solvents and collectors, drawing valuable metals into their liquid or molten states. At 1400 °C, Nanda et al. found no gold particles in the gaseous fraction.

FUTURE WORK EXTENSIONS

This work has been extended to optimize the various parameters of the metal extraction process to obtain the base and high value metals from E-waste sources. The design of experiments has been performed before conducting experiments using Taguchi(Boopathi, 2021b; Boopathi et al., 2012, 2022; Sampath et al., 2021; Saravanan et al., 2022) and RSM Methods(Boopathi & Sivakumar, 2013, 2016; Haribalaji et al., 2021; Kannan et al., 2022; Myilsamy & Sampath, 2021; Sampath & Myilsamy, 2021). The global optimum process parameters have been obtained through Genetic algorithm, evolutionary technique(Boopathi, 2022b; Boopathi et al., 2021; Yupapin et al., 2022), NSGA, particle swarm optimization, and various machine learning methods. The multi criteria optimization techniques like TOPSIS(Kannan et al., 2022; Sampath et al., 2023), VIKOR(Kannan et al., 2022; Sampath et al., 2023), CRITIC integrated methods have been applied for the improving the processes.

SUMMARY

In this chapter, there is an explanation of the composition, harmful, and toxic components of various E-wastes. With quantities of copper, tin, lead, and precious metals that are substantially higher than those found in their ore, printed circuit boards (PCBs) constitute an important source of metals. It was found that the metallic phases contained significant amounts of important metals. The carbons and oxides in the non-metallic fraction contain trace amounts of precious metals. Through the use of high-temperature pyrolysis, copper-based alloys and carbon resources were successfully recovered. The quantity of material that needed to be controlled and processed further was also decreased by this technique. Recycling e-waste ought to increase metal recovery, protect natural resources, and provide an environmentally sound alternative.

REFERENCES

Abdelbasir, S. M., Hassan, S. S. M., Kamel, A. H., & El-Nasr, R. S. (2018). Status of electronic waste recycling techniques: A review. *Environmental Science and Pollution Research International*, *25*(17), 16533–16547. doi:10.100711356-018-2136-6 PMID:29737485

Arshadi, M., Yaghmaei, S., & Mousavi, S. M. (2018). Content evaluation of different waste PCBs to enhance basic metals recycling. *Resources, Conservation and Recycling*, *139*, 298–306. doi:10.1016/j.resconrec.2018.08.013

Asamoah, A., Essumang, D. K., Muff, J., Kucheryavskiy, S. V., & Søgaard, E. G. (2018). Assessment of PCBs and exposure risk to infants in breast milk of primiparae and multiparae mothers in an electronic waste hot spot and non-hot spot areas in Ghana. *The Science of the Total Environment*, *612*, 1473–1479. doi:10.1016/j.scitotenv.2017.08.177 PMID:28903176

Boopathi, S. (2021a). An investigation on gas emission concentration and relative emission rate of the near-dry wire-cut electrical discharge machining process. *Environmental Science and Pollution Research International*, *29*(57), 86237–86246. doi:10.100711356-021-17658-1 PMID:34837614

Boopathi, S. (2021b). Improving of Green Sand-Mould Quality using Taguchi Technique. *Journal of Engineering Research*. doi:10.36909/jer.14079

Boopathi, S. (2022a). An Extensive Review on Sustainable Developments of Dry and Near-Dry Electrical Discharge Machining Processes. *Journal of Manufacturing Science and Engineering*, *144*(5), 1–37. doi:10.1115/1.4052527

Boopathi, S. (2022b). Experimental investigation and multi-objective optimization of cryogenic Friction-stir-welding of AA2014 and AZ31B alloys using MOORA technique. *Materials Today. Communications*, *33*, 104937. doi:10.1016/j.mtcomm.2022.104937

Boopathi, S., Jeyakumar, M., Singh, G. R., King, F. L., Pandian, M., Subbiah, R., & Haribalaji, V. (2022). An experimental study on friction stir processing of aluminium alloy (AA-2024) and boron nitride (BNp) surface composite. *Materials Today: Proceedings*, *59*, 1094–1099. doi:10.1016/j.matpr.2022.02.435

Boopathi, S., Kumar, P. K. S., Meena, R. S., Sudhakar, M., & Associates. (2023). Sustainable Developments of Modern Soil-Less Agro-Cultivation Systems: Aquaponic Culture. In Human Agro-Energy Optimization for Business and Industry (pp. 69–87). IGI Global.

Boopathi, S., Myilsamy, S., & Sukkasamy, S. (2021). Experimental Investigation and Multi-Objective Optimization of Cryogenically Cooled Near-Dry Wire-Cut EDM Using TOPSIS Technique. *Ijamt Preprint*. doi:10.21203/rs.3.rs-254117/v1

Boopathi, S., & Sivakumar, K. (2013). Experimental investigation and parameter optimization of near-dry wire-cut electrical discharge machining using multi-objective evolutionary algorithm. *International Journal of Advanced Manufacturing Technology*, *67*(9–12), 2639–2655. doi:10.100700170-012-4680-4

Boopathi, S., & Sivakumar, K. (2016). Optimal parameter prediction of oxygen-mist near-dry Wire-cut EDM. *International Journal of Manufacturing Technology and Management*, *30*(3–4), 164–178. doi:10.1504/IJMTM.2016.077812

Boopathi, S., Sivakumar, K., & Kalidas, R. (2012). Parametric Study of Dry WEDM Using Taguchi Method. *International Journal of Engineering Research and Development*, *2*(4), 1–6.

Chakraborty, P., Selvaraj, S., Nakamura, M., Prithiviraj, B., Cincinelli, A., & Bang, J. J. (2018). PCBs and PCDD/Fs in soil from informal e-waste recycling sites and open dumpsites in India: Levels, congener profiles and health risk assessment. *The Science of the Total Environment*, *621*, 930–938. doi:10.1016/j.scitotenv.2017.11.083 PMID:29223911

Chen, S. J., Tian, M., Zheng, J., Zhu, Z. C., Luo, Y., Luo, X. J., & Mai, B. X. (2014). Elevated levels of polychlorinated biphenyls in plants, air, and soils at an e-waste site in southern china and enantioselective biotransformation of chiral PCBs in plants. *Environmental Science & Technology*, *48*(7), 3847–3855. doi:10.1021/es405632v PMID:24628413

Domakonda, V. K., Farooq, S., Chinthamreddy, S., Puviarasi, R., Sudhakar, M., & Boopathi, S. (2022). Sustainable Developments of Hybrid Floating Solar Power Plants: Photovoltaic System. In Human Agro-Energy Optimization for Business and Industry (pp. 148–167). IGI Global.

Gunasekaran, K., Boopathi, S., & Sureshkumar, M. (2022). Analysis of a Cryogenically Cooled Near-Dry Wedm Process Using Different Dielectrics. *Materiali in Tehnologije*, *56*(2), 179–186. doi:10.17222/mit.2022.397

Hanafi, J., Jobiliong, E., Christiani, A., Soenarta, D. C., Kurniawan, J., & Irawan, J. (2012). Material Recovery and Characterization of PCB from Electronic Waste. *Procedia: Social and Behavioral Sciences*, *57*, 331–338. doi:10.1016/j.sbspro.2012.09.1194

Haribalaji, V., Boopathi, S., & Asif, M. M. (2021). Optimization of friction stir welding process to join dissimilar AA2014 and AA7075 aluminum alloys. *Materials Today: Proceedings, 50*(xxxx), 2227–2234. doi:10.1016/j.matpr.2021.09.499

Hosoda, J., Ofosu-Anim, J., Sabi, E. B., Akita, L. G., Onwona-Agyeman, S., Yamashita, R., & Takada, H. (2014). Monitoring of organic micropollutants in Ghana by combination of pellet watch with sediment analysis: E-waste as a source of PCBs. *Marine Pollution Bulletin*, *86*(1–2), 575–581. doi:10.1016/j.marpolbul.2014.06.008 PMID:24997873

Kamberovic, Z., Korac, M., & Ranitovic, M. (2011). Hydrometallurgical Process for Extraction of Metals from Eectronic Waste-Part II: Development of the Processes for the Recovery of Copper from Printed Circuit Boards (PCB). *Metalurgija-MJoM*, *17*(3), 139–149. http://www.metalurgija.org.rs/mjom/vol17/No3/4_Kamberovic_MJoM_1703.pdf

Kannan, E., Trabelsi, Y., Boopathi, S., & Alagesan, S. (2022). Influences of cryogenically treated work material on near-dry wire-cut electrical discharge machining process. *Surface Topography: Metrology and Properties*, *10*(1), 015027. Advance online publication. doi:10.1088/2051-672X/ac53e1

Khalid, F., Hashmi, M. Z., Jamil, N., Qadir, A., & Ali, M. I. (2021). Microbial and enzymatic degradation of PCBs from e-waste-contaminated sites: A review. *Environmental Science and Pollution Research International*, *28*(9), 10474–10487. doi:10.100711356-020-11996-2 PMID:33411303

Krishna, K. (2009). Effective electronic waste management and recycling process involving formal and non-formal sectors. *International Journal of Physical Sciences*, *4*(13), 893–905.

Kumar, P., Sampath, B., Kumar, S., Babu, B. H., & Ahalya, N. (2023). Hydroponics, Aeroponics, and Aquaponics Technologies in Modern Agricultural Cultivation. In Trends, Paradigms, and Advances in Mechatronics Engineering (pp. 223–241). IGI Global.

Kumara, V., Mohanaprakash, T. A., Fairooz, S., Jamal, K., Babu, T., & Sampath, B. (2023). Experimental Study on a Reliable Smart Hydroponics System. In *Human Agro-Energy Optimization for Business and Industry* (pp. 27–45). IGI Global. doi:10.4018/978-1-6684-4118-3.ch002

Morf, L. S., Tremp, J., Gloor, R., Schuppisser, F., Stengele, M., & Taverna, R. (2007). Metals, non-metals and PCB in electrical and electronic waste - Actual levels in Switzerland. *Waste Management (New York, N.Y.)*, *27*(10), 1306–1316. doi:10.1016/j.wasman.2006.06.014 PMID:17008085

Myilsamy, S., & Sampath, B. (2021). Experimental comparison of near-dry and cryogenically cooled near-dry machining in wire-cut electrical discharge machining processes. *Surface Topography: Metrology and Properties*, *9*(3), 035015. doi:10.1088/2051-672X/ac15e0

Ramesh Kumar, G., Santhosh, K. S., & Bharani, S. (2020). Influence of E-waste on properties of bituminous mixes. *Materials Today: Proceedings*, *37*(Part 2), 2719–2724. doi:10.1016/j.matpr.2020.08.539

Ren, Z. F., Huang, B., Liu, M., Bi, X. H., Sheng, G. Y., & Fu, J. M. (2012). The major components and characteristics of particles emitted during recycling of waste printed circuit boards in a typical e-waste workshop of South China. *Zhongguo Huanjing Kexue*. *Zhongguo Huanjing Kexue*, *32*(8), 1447–1451.

Samikannu, R., Koshariya, A. K., Poornima, E., Ramesh, S., Kumar, A., & Boopathi, S. (2022). Sustainable Development in Modern Aquaponics Cultivation Systems Using IoT Technologies. In *Human Agro-Energy Optimization for Business and Industry* (pp. 105–127). IGI Global.

Sampath, B., & Myilsamy, S. (2021). Experimental investigation of a cryogenically cooled oxygen-mist near-dry wire-cut electrical discharge machining process. *Strojniski Vestnik. Jixie Gongcheng Xuebao*, *67*(6), 322–330. doi:10.5545v-jme.2021.7161

Sampath, B., Naveenkumar, N., Sampathkumar, P., Silambarasan, P., Venkadesh, A., & Sakthivel, M. (2021). Experimental comparative study of banana fiber composite with glass fiber composite material using Taguchi method. *Materials Today: Proceedings*, *49*(5), 1475–1480. doi:10.1016/j.matpr.2021.07.232

Sampath, B., Sasikumar, C., & Myilsamy, S. (2023). Application of TOPSIS Optimization Technique in the Micro-Machining Process. In Trends, Paradigms, and Advances in Mechatronics Engineering (pp. 162–187). IGI Global.

Sampath, B., Yuvaraj, D., & Velmurugan, D. (2022). Parametric analysis of mould sand properties for flange coupling casting. *AIP Conference Proceedings*, *2460*(1), 070002. doi:10.1063/5.0095599

Saravanan, M., Vasanth, M., Boopathi, S., Sureshkumar, M., & Haribalaji, V. (2022). Optimization of Quench Polish Quench (QPQ) Coating Process Using Taguchi Method. *Key Engineering Materials*, *935*, 83–91. doi:10.4028/p-z569vy

Trojovský, P., Dhasarathan, V., & Boopathi, S. (2023). Experimental investigations on cryogenic friction-stir welding of similar ZE42 magnesium alloys. *Alexandria Engineering Journal*, *66*(1), 1–14. doi:10.1016/j.aej.2022.12.007

Vanitha, S., Radhika, K., & Boopathi, S. (2023). Artificial Intelligence Techniques in Water Purification and Utilization. In *Human Agro-Energy Optimization for Business and Industry* (pp. 202–218). IGI Global. doi:10.4018/978-1-6684-4118-3.ch010

Vennila, T., Karuna, M. S., Srivastava, B. K., Venugopal, J., Surakasi, R., & Sampath, B. (2022). New Strategies in Treatment and Enzymatic Processes: Ethanol Production From Sugarcane Bagasse. In Human Agro-Energy Optimization for Business and Industry (pp. 219–240). IGI Global.

Xing, G. H., Chan, J. K. Y., Leung, A. O. W., Wu, S. C., & Wong, M. H. (2009). Environmental impact and human exposure to PCBs in Guiyu, an electronic waste recycling site in China. *Environment International*, *35*(1), 76–82. doi:10.1016/j.envint.2008.07.025 PMID:18804865

Yupapin, P., Trabelsi, Y., Nattappan, A., & Boopathi, S. (2022). Performance Improvement of Wire-Cut Electrical Discharge Machining Process Using Cryogenically Treated Super-Conductive State of Monel-K500 Alloy. *Iranian Journal of Science and Technology - Transactions of Mechanical Engineering*, 1–17. doi:10.1007/s40997-022-00513-0

Zhao, G., Wang, Z., Dong, M. H., Rao, K., Luo, J., Wang, D., Zha, J., Huang, S., Xu, Y., & Ma, M. (2008). PBBs, PBDEs, and PCBs levels in hair of residents around e-waste disassembly sites in Zhejiang Province, China, and their potential sources. *The Science of the Total Environment*, *397*(1–3), 46–57. doi:10.1016/j.scitotenv.2008.03.010 PMID:18439655

Zhao, G., Wang, Z., Zhou, H., & Zhao, Q. (2009). Burdens of PBBs, PBDEs, and PCBs in tissues of the cancer patients in the e-waste disassembly sites in Zhejiang, China. *The Science of the Total Environment*, *407*(17), 4831–4837. doi:10.1016/j.scitotenv.2009.05.031 PMID:19539352

KEY TERMS AND DEFINITIONS

3R: This stands for Reduce, Reuse, Recycle.
ABS: This stands for Acrylonitrile Butadiene Styrene.
CFCs: This stands for Chlorofluorocarbons.
EDS: This stands for Energy Dispersive Spectroscopy.
FTIR: This stands for Fourier transform infrared spectroscopy.
HIPS: This stands for High Impact Polystyrene sheets.
ICP: This stands for Inductively Coupled Plasma.
IR: This stands for Infrared.

NMF: This stands for Non-Metallic Fraction.
OES: This stands for Optical Emission Spectrometry.
PA: This stands for Polyamide.
PAHs: This stands for Polycyclic Aromatic Hydrocarbons.
PC: This stands for Polycarbonate.
PCBs: This stands for Printed Circuit Boards.
PE: This stands for Polyethylene.
POM: This stands for Polyurethane.
PP: This stands for Polypropylene.
PPO: This stands for Polyphenylene Oxide.
PS: This stands for Polystyrene.
PU: This stands for Polyurethane.
PVC: This stands for Polyvinyl Chloride.
SEM: This stands for Scanning Electron Microscopy.
TGA: This stands for Thermogravimetric Analysis.
WEEE: This stands for Waste Electrical and Electronic Equipment.

Chapter 10
Integration of Sustainable Supply Chain Flexibility (SSCF) and the Circular Economy (CE): Waste Minimization Technique

Animesh Agrawal
https://orcid.org/0000-0001-8175-5410
National Institute of Technology, Raipur, India

Suraj Kumar Mukti
National Institute of Technology, Raipur, India

ABSTRACT

The Circular Economy (CE) is a backbone for sustainable economic growth and development. Companies now operate in a market that is more complicated, unpredictable, and fast-paced than ever before. The material closed-loop fluxes essential to a fully functional CE are the source of these difficulties. Addressing CE operational difficulties requires incorporating flexibility capabilities into sustainable supply networks. The development of this operational competence must not diminish performance, raise expenses, or lower product quality. Designers of supply chains now understand they must prioritize Sustainable Supply Chain Flexibility (SSCF) initiatives if they want to achieve CE-targeted performance. Material and energy recycling, a closed-loop supply chain, and minimal environmental impact all contribute to a product's performance aimed at achieving CE status. In this study, Decision Making Trial and Evaluation Laboratory (DEMATEL) technique has been used to integrate the most influencing factors of sustainable supply chain flexibility and circular economy.

DOI: 10.4018/978-1-6684-7573-7.ch010

INTRODUCTION

A new policy and commercial approach known as the "Circular Economy" (CE). Sustainability in production, distribution, and consumption is a primary goal of CE. (Cooper, 1999), (Sarkis and Zhu, 2018). CE has garnered a lot of interest because of the fact that it has the ability to minimize the effect that humans have on the environment while also improving economic development (Zhu, Geng, & Lai, 2010), (Geng et al., 2013) (Geng, Sarkis, & Bleischwitz, 2019), (Geng, Sarkis, & Ulgiati, 2016). One of the most important parts of CE's mission to promote sustainable supply chain management is supply chain management that takes the environment into account (Pan et al. 2015; Genovese et al. 2017).

Companies and their supply chains must adapt to a more volatile, unpredictable, and complicated market because of CE. Changes in the competitive landscape could be caused by a number of things, such as changing customer demand for greener products, stricter environmental rules, or the entry of more green competitors. The term "Supply Chain Flexibility" (SCF) refers to a supply chain's capacity to respond to market dynamics (such as demand and opportunity) with little time, cost, or performance consequences (Blome, Schoenherr, & Eckstein 2014) (Agrawal & Mukti, 2020). While working toward CE objectives, SSCF aids the focal organisation in a supply chain in adapting to changing greening standards and environmental legislation (Swafford, Ghosh, & Murthy, 2006; Bai and Sarkis, 2019). These businesses and their supplier networks aim to strike a balance between rapid response and environmental responsibility.

SSCF initiatives, which are especially useful in light of volatile business environments and changing market dynamics, can be used to protect focal companies from environmental risks such as those related to energy resource requirements, waste management, and greenhouse gas (GHG) emissions. This can be accomplished by protecting focal companies from environmental risks such as those related to energy resource requirements. These changes might make it easier to add a CE framework to a supply chain that is more environmentally friendly.

Due to the lack of certainty that the advantages of flexibility would outweigh the costs, many leading organisations are hesitant to undertake SSCF initiatives (Montabon, Sroufe, & Narasimhan 2007). So, if you want to combine SSCF and CE goals and targets well, you need to look at how they are related.

Research Gaps

By going through a rigorous literature review, authors have formulated some research gaps as follows:

RG 1: Rare literature have discussed the criteria affecting Sustainable Supply Chain Flexibility (SSCF).

RG 2: Rare literature have discussed the relationships between the degree of Sustainable Supply Chain Flexibility (SSCF) measures and Circular Economy (CE) target performance.

Research Questions

Based on the research gaps some research questions are as follows:

RQ 1: Is there any criteria affecting Sustainable Supply Chain Flexibility (SSCF)?

RQ 2: Is there any relationships between the degree of Sustainable Supply Chain Flexibility (SSCF) measures and Circular Economy (CE) target performance?

Specifically, this research focuses on three CE-targets and goals: decreasing pollution and waste output; reusing raw materials and energy; and achieving a closed-loop system for material circulation. These goals for Continuing Education (CE) are grounded in the '3Rs' of CE (Lieder and Rashid 2016). SSCF procedures can vary from company to company, but some examples include increasing the number of environmentally friendly vendors, shortening the time it takes to ship ecologically friendly goods, and reacting rapidly to pollution emergencies. As a result of these SSCF practises and the increased diversity of CE-targets, the management and decision-making processes involved in establishing and maintaining a sustainable supply chain have become more complicated and difficult. When investing in and adopting SSCF practises with the intent of improving CE-targets, a methodology that can detect correlations between SSCF practises and performance is a beneficial organisational tool.

Research Objectives

Two major objectives have been covered in this research; the objectives of this study are listed below:

1. To identify the criteria affecting Sustainable Supply Chain Flexibility (SSCF).
2. To analyse the relationships between the degree of Sustainable Supply Chain Flexibility (SSCF) measures and Circular Economy (CE)-target performance.

Traditional flexibility measures have only applied to a single company in a supply chain, but supply chain-level flexibility measures are needed by the industry (Mukti, Ahirwar and Agrawal, 2021). The supply lead time reported by a single provider is not always indicative of the lead time reported by the whole supply chain. At the supply chain level, a theoretical SSCF framework is established by conducting a literature study to develop a set of multi-item SSCF measurements. By constructing this preliminary framework, researchers took a step toward a middle theory (Carter and Rogers 2008). This research also contributes to the second aim by developing a practical SSCF decision support approach for taking into account a number of elements and their connections (Dou, Zhu, & Sarkis 2014). As a structured analytic tool, the DEMATEL approach (Gabus & Fontela 1972) can be used to establish connections between variables and produce causal diagrams that adequately characterise these associations. In order to extract potentially relevant knowledge data from Rough Set Theory (RST), data mining is a powerful method (Pawlak 1982). By using objective relational values instead of the subjective relational methods used in traditional DEMATEL, RST will make people less likely to trust the expert opinion of DEMATEL. The SSCF framework and a multi-stage evaluation technique are used in an exploratory empirical field research among 23 manufacturing firms in India. The results of this empirical research support the concept and methods. The study produced an SSCF framework that businesses can use to guide their efforts to improve their performance in relation to CE-targets. The remaining sections of this work will be structured as follows: In Section 2, the literature and highlights of SSCF measures have been evaluated, as well as the existing research needs for adopting and implementing SSCF initiatives. Section 3 applies the DEMATEL technique to collect the data from a field study of Indian manufacturing firms. The findings are discussed in Section 4, with contributions to theory and practise offered, before the study concludes in Section 5 by discussing its limits and suggesting directions for further study.

BACKGROUND OF THE STUDY

Sustainable Supply Chain Adaptability

The supply chain environment has become more unpredictable and difficult as a result of increasing market dynamic trends such as globalization, technological progress, outsourcing, shorter product life cycles, and more demanding customers. Fayezi, Zutshi, and O'Loughlin (2017) say that to keep a competitive edge in the market, firms and their supplier networks need to make their supply chains more flexible so they can deal with the market's complexity and uncertainty.

Greater emphasis has been placed on the flexibility of supply chains as a means of controlling the risks and uncertainties associated with supply chains (Gosling, Purvis, and Naim 2010; Esmaeilikia et al. 2016) (Alshurideh et al., 2022). The absence of environmental, ecological, green, or sustainable language in a recent systematic literature review on supply chain flexibility (Fayezi, Zutshi, & O'Loughlin 2017) is evidence that researchers have paid less attention to sustainable supply chain flexibility (Bai and Sarkis 2013a). The fact that researchers haven't given sustainable supply chains as much attention is evidence of this.

As part of a green supply chain, Bai and Sarkis (2013) presented a reverse logistics flexibility architecture that "closes the loop." Including multiple forms of economic and environmental flexibility (Bhide & Akarte, 2022), Bai and Sarkis (2017) created effective green flexibility metrics for manufacturing enterprises, incorporating economic and environmental flexibility types. Bai and Sarkis (2018) study the use of information systems to create green supply chain flexibility (Fu et al., 2022). Bai, Dhavale, and Sarkis (2016) say that problems with environmental sustainability require a green supply chain that is flexible enough to deal with complexity and unpredictability, which affect competitive advantage and long-term growth.

The concept of sustainable flexibility can be challenging and puzzling for academics as well as for professionals working in the field. In addition to the manufacturing process flexibility of green products, various environmental protection process flexibilities such as pollutant emissions, energy consumption, and other manufacturing process flexibilities are taken into account. More investigation and development should be put into the SSCF idea because of its significance on several fronts, including the economy, environment, and society (Makeel et al., 2022). The relationship between this matter and CE is one that cannot be avoided and is essential (Lehtokunnas et al., 2022).

The evaluation of supply chain flexibility efforts and the improvement of supply chain flexibility performance may both be made easier with the assistance of a measuring system for supply chain flexibility. There is a lot of nuance involved in the concepts of supply chain flexibility and green supply chain flexibility (Blome, Schoenherr, and Eckstein 2014; Bai and Sarkis 2017). There are many different perspectives on supply chain flexibility, with the bulk of these perspectives concentrating on core supply chain activities, such as the flexibility of procurement and sourcing, the flexibility of production, and the flexibility of distribution and logistics (Swafford, Ghosh, & Murthy 2006) (Richey et al., 2022). Although these investigations are primarily focused on the internal viewpoint of a particular company operating inside a supply chain, their goal is to increase the flexibility of the investigative border so that it extends all the way up to the level of the supply chain (Stevenson and Spring 2007). Despite this, it is difficult to characterise the adaptability of the whole system in terms of the adaptability of each supply chain function. For example, the lead time of a single supplier isn't always the same as the lead time of the whole supply chain. This is because a number of other things can affect the lead time (Pushpamali et al., 2021).

Research on the flexibility of supply chains at different levels is still in its infancy, particularly in terms of their impact on the environment and their capacity to be sustained. The results of this investigation will hopefully add to the research that has already been done. The SCOR model is an extensively utilised framework as a supply chain performance system, both in research and in practise (Bai et al. 2012). Shepherd and Günter (2006) discussed flexibility is not a single metric, and it is closely intertwined with other metrics like cost, time, quality, and innovation. Consequently, SCOR performance measurements will serve as the basis for the SSCF measures utilised in this study (Lukinskiy et al., 2023).

Few studies have examined the performance of flexibility in sustainable supply chain management. Using the SCOR model, researchers reviewed the pertinent literature and provided a framework to explain flexibility metrics in the cost, time, quality, and innovation dimensions. Given resource consumption, pollutant generation, environmental policies, and technology within sustainable supply chains, the absence of this study is particularly significant. Other supply chain and green flexibility studies also influenced the SSCF framework's development (Edwin Cheng et al., 2022).

Various authors have discussed about how SCOR performance measurements are categorised by cost, time, quality, flexibility, and innovation aspects but There have been a lot of sites across the world that have attempted to conserve the environment while still making money. To fulfil this sustainable supply chain adaptability and performance to CE-standards has been discussed further in the next section.

Sustainable Supply Chain Adaptability and Performance to CE-Standards

CE policies and ideas have assisted countries in preventing their economic progress and well-being from being directly linked to damage to the environment (Ghisellini, Cialani, & Ulgiati 2016). Because China bases its economic policy on CE, it is consistently ranked among the top performers in this sector (Yuan, Bi, & Moriguichi 2006). The Ellen MacArthur Foundation sees CE as a new business model for a restorative and regenerative economy, which is one of the many ways in which it can be conceived of. Another way to think of CE is as a concept that can be thought of in a variety of other ways. The 3R principles serve as the foundation for this study, which takes a look at CE from China's perspective. This change to the Chinese CE is a direct result of the fast growth of the economy over the past 20 years, which has led to a shortage of resources and problems with the environment.

In China, the CE model has been implemented at three distinct levels, including the macro-level (eco-regions), the meso-level (eco-industrial parks), and the micro-level (eco-enterprises) (Yuan, Bi, & Moriguichi 2006; Su et al., 2013; Ghisellini, Cialani, & Ulgiati 2016). In the context of CE, businesses or supply chains should operate in accordance with the concepts of recycling materials, resources, and energy in order to support the functioning of natural systems (Geng et al. 2013, 2016). These SSCM principles, together with the efforts made by SSCF, enable supply chains to reduce the amount of pollution and waste produced. Furthermore, the SSCF approach reduces the amount of energy and other resources consumed while simultaneously improving the supply chain's reputation for being environmentally responsible. For example, being able to make goods that are good for the environment could make it much easier to reuse, recover, and recycle resources.

Previous empirical research suggests that a high level of flexibility will lead to high supply chain agility (Braunscheidel & Suresh, 2009) and better firm performance (Swink, Narasimhan, & Kim, 2005) in high-uncertainty environments. The fact that the research that came before solely looked at economic or operational success is one of the most significant problems with those studies. These studies didn't look at other essential components of sustainability performance, including measurements that focus on the environment, like CE target performance.

These prior studies do not investigate flexibility, nor do they investigate the relationship between flexibility measurements and CE-target achievement. To overcome this, next section will discuss ways to measure how flexible the supply chain is in the long term.

A Way to Measure How Flexible the Supply Chain Is in The Long Term

It is essential to have SSCF in order to react rapidly and affordably to shifts in the market, as well as to make supply chains and organisations more competitive (Bai, Fahimnia, & Sarkis 2017a). Developing flexibility requires a significant investment of time and money from a business. The performance of the circular economy is not always improved in the same way by each and every SSCF endeavour (Dou, Zhu, & Sarkis 2014). Because of this uneven performance and the lack of financial and organisational resources, it is essential to analyse the SSCF adoption made by the focus enterprise (Bai & Sarkis 2017; Bai, Satir, & Sarkis 2019).

In spite of the fact that empirical methodologies are available (Martnez Sánchez and Pérez Pérez 2005; Swafford, Ghosh, and Murthy 2008), there are not a great deal of formal models and tools available for SSCF assessment. Lummus, Vokurka, and Duclos (2005) conducted a Delphi survey over the internet to compile a list of the factors that contribute to the flexibility of a supply chain and determine the relative weight of each factor (Agrawal, Kumar and Mukti, 2021) (Agrawal, Diwakar and Mukti, 2020). Chuu (2011) developed an algorithm for determining the degree to which a supply chain is flexible that is based on the use of fuzzy language. Using a rough set in the neighbourhood, Bai and Sarkis (2013a) elaborated a methodology for evaluating the performance of third-party reverse logistics providers. This model can be found in their publication. Tyagi, Kumar, and Kumar employed the fuzzy DEMATEL technique to construct a flexible supply chain performance monitoring system. They also conducted an analysis of the enablers. Bai and Sarkis (2017) created a hybrid possibility, multiple criteria decision model to allow for flexible evaluation of modern manufacturing technologies.

The methods of analysis that have been discussed up until this point are only capable of measuring the link between SSCF and CE-target performance to a limited degree, if at all. None of these methodologies provides a simple way to conduct a scientifically sound and easily understandable analysis of the relationships that exist between the degree of SSCF measurements and the performance of CE-targets.

Insights on the Circular Economy's Need for Adaptable Supply Chains

In this article, the authors present a multi-stage process for evaluating SSCF using RST and DEMATEL. The RST is employed to ascertain the objective direct relationships between SSCF metrics and CE-target performance. The authors of this study used DEMATEL to create a tree-like structure that shows how the SSCF metrics and CE-target outcomes relate to one another. The SSCF may be used for strategic planning on several levels (Agrawal and Mukti, 2019). In the first place, it may show the connection between SSCF metrics and CE-target performance, which can direct management toward the best SSCF initiatives to pursue. Second, it can help managers decide how to execute SSCF measures by determining the cause-and-effect link among all SSCF measures. Data gathered from production sites in India is utilised to verify the theory in practice. Before diving into the approach, researchers provide some context about the testing environment.

History of the Field Study and The Organizations

An instrument for conducting a survey was developed in order to facilitate the collection of information. The study team initially developed the survey instrument by utilising SSCF components that were taken from previously published papers. The initial survey instrument was distributed for testing and feedback purposes to three high-level managers as well as three specialists in environmentally responsible supply chain management. The ideas for changes were taken into account, and the new questionnaire was simpler and easier for practitioners to understand because of it.

A survey instrument was sent at random to a sample of twenty-three different organizations, which resulted in twenty-three distinct responses. Companies in a wide variety of industries, including electronics (9), pharmaceuticals (3), food (5), and machinery (5), are represented here. Companies in a wide variety of industries, including electronics (9), pharmaceuticals (3), food (5), and machinery (5), are represented here. The majority of those who responded are middle managers with a significant amount of knowledge about environmental management and supply chain management. The demographic profiles of the 23 respondents are presented in Table 1. Some of the individuals that provided responses are at the very top of their respective organisations (CEOs). All of the respondents who were chosen to fill out the questionnaire had at least five years of professional experience among them. The vast majority of them have participated in environmental projects such as obtaining ISO 14001 certification or creating environmental reports or policies.

When completing this questionnaire, respondents were also urged to speak with the buying managers and environmental managers in their businesses, particularly if they weren't sure about their responses. This request was made specifically for those respondents who were unsure about their replies.

Following the gathering of the replies from the respondents, the data were reviewed very carefully to ensure that the respondents had given accurate information. The researchers came to the conclusion that every response was both appropriate and trustworthy. As a consequence of this, researchers have started using the DEMATEL method for further computation. The DEMATEL strategy, which is described in the part below under "methodology implementation,"

METHODOLOGICAL IMPLEMENTATION

The authors have used the DEMATEL Technique because of its versatility in developing the relationship between the factors (Animesh & Mukti, 2019). In this study of sustainable supply chains (e.g., Su et al., 2016), the DEMATEL method has been shown to be a useful way to map out the connections between the many different parts. Even though other methods, such as structural equation models (Fornell & Larcker, 1981), fuzzy cognitive maps (Kosko, 1986), and the analytic hierarchy process (AHP) (Carlucci & Schiuma 2007), are also capable of building causal structures, the DEMATEL method has some advantages that the other methods do not have.

First, DEMATEL is able to find out how to categorise a collection of complicated characteristics into different groups according to the relationships between their causes and effects. Other approaches operate under the presumption that these connections either already exist or are established with a minimum of mental effort. Methods of graphing can be utilised to illustrate the correlations as well as the degrees of effect (Bai, Sarkis, & Dou 2017; Govindan, Khodaverdi, & Vafadarnikjoo 2015; Wu & Tsai 2011; Bai & Sarkis 2013b). Certain methods, such as the AHP, simply consider the direct correlations that exist between qualities. It is possible to map both direct and indirect links between characteristics using this approach. The second benefit of using these DEMATEL findings is that they might help one better comprehend the intricacy of a situation and come to more sound conclusions (Wu and Tsai 2011). In contrast to structural equation modelling, DEMATEL does not require a significant amount of data or knowledge about other factors. Based on these benefits, DEMATEL is used to adapt and evaluate the different SSCF measures as well as the links between SSCF measures and CE-target performance.

In order to assess direct effects that aren't apparent and exact, DEMATEL requires the judgements of subject matter experts. The vast majority of the publications have relied on conventional fuzzy sets to address the issue of ambiguity in regards to expert judgements (e.g., Dou, Zhu, and Sarkis 2014). RST, which is a data mining tool, is utilised in order to work around the dimensions that are absent in DEMATEL. The DEMATEL approach becomes an order of magnitude more sophisticated and requires an order of magnitude more time to complete as the number of factors that need to be assessed increases. This demonstrates how relationships are formed on a personal level. This technique is known as objective DEMATEL, and it is applied in order to conduct an analysis of the correlations between SSCF measurements in a manner that is both more objective and informative. In order to assist in making decisions about SSCM, the impact that a variety of SSCF measures have had on the performance of CE-targets has been presented.

In this particular investigation, the researchers evaluated the connection between SSCF measures and CE-target performance by employing a fresh technique in addition to the framework that was discussed in Section 2.1. This helps to fill in some of the blanks that were left by the study and circumvent some of the constraints that were caused by the techniques. It is helpful for managers to have an understanding of the SSCF metrics to utilise and how those measurements impact CE-target performance. This is especially true given the number and complexity of SSCF measures. Researchers are going to now outline the procedure while keeping in mind the SSCF measuring framework.

The purpose of this article is to advocate the creation of a multiple-attribute decision-making approach utilising proactive and integrated SSCF measures and to illustrate how these SSCF efforts may help with circular economy activities. The study will also explain how SSCF measures can be incorporated. At the level of the supply chain, a collection of multi-items SSCF measures functions as a theoretical foundation for the activities of businesses and the research conducted in academic institutions. This approach is the first of its kind to apply circular economic thinking to investigate how SSCF measures might assist in lowering pollution levels, improving resource use, and constructing a greener public image. The problem lies in effectively managing the connections between activities that, rather than functioning as independent endeavours with their own independent programming efforts and performance results, impact each other in both positive and negative ways. This research study offers a solution to the difficult problem of sustainable supply chain management by providing a mechanism to make a choice (Bai et al., 2019).

To perform the DEMATEL technique, the following steps have been followed:

Step-I Calculate the initial average matrix (A) by scores of respondents.

$$A = \begin{vmatrix} 0 & a_{12} & \cdots & a_{1n} \\ a_{21} & 0 & \cdots & a_{2n} \\ \cdots & \cdots & \cdots & \cdots \\ \cdots & \cdots & \cdots & \cdots \\ a_{n1} & a_{n2} & \cdots & 0 \end{vmatrix} \quad (1)$$

At this point, a_{ij} indicates the importance of ith standard on jth standard. eventually calculating the direct-importance matrix $K_{Id} = \left[K_{Id_{ij}} \right]_{n \times n}$. Equation (2) is exploited to establish the $K_{Id} = \left[K_{Id_{ij}} \right]_{n \times n}$, which is embodied by Equation (3), where the elements in K_{Id} signifies the direct importance matrix of ith and jth standard:

$$K_{Id} = y \times A \quad (2)$$

$$\text{where, } y = \min \left\{ \begin{matrix} \dfrac{1}{Max} \sum_{i=1}^{n} |A_{ij}|, \dfrac{1}{Max} \sum_{j=1}^{n} |A_{ij}| \\ 1 < i < n \qquad\qquad 1 < j < n \end{matrix} \right\}, i = j = 1, 2, 3, \ldots\ldots\ldots n \quad (3)$$

Step-II Calculate the direct influence matrix (K_{Id}) as using equation (2) and (3)

$$K_{Id} = \begin{vmatrix} 0 & K_{Id12} & \cdots & K_{Id1n} \\ K_{Id21} & 0 & \cdots & K_{Id2n} \\ \cdots & \cdots & \cdots & \cdots \\ \cdots & \cdots & \cdots & \cdots \\ K_{Idn1} & K_{Idn2} & \cdots & 0 \end{vmatrix} \quad (4)$$

Calculating the value of K_{Id}, after putting the value of y and A in the equation (2)

Such that

Step-III Subsequently calculating the total impact matrix ($TIM_{n \times n}$) such that.

$$TIM = K_{Id} \times \left(I - K_{Id}\right)^{-1} \tag{5}$$

Where I represent the identity matrix. The elements (TIM_{ij}) in $TIM_{n \times n}$ indicate the probable influence and cause of ith standard on j th standard, such that i , j = 1, 2,......, n. Afterward, given equations calculate the total significance (cause and effect) of i th standard on j th standard:

$$D_i = \sum_{i=1}^{n} TIM_{ij}, \ \left(i = 1, 2, 3 \ldots\ldots.n\right) \tag{6}$$

$$R_i = \sum_{j=1}^{n} TIM_{ij}, \ \left(j = 1, 2, 3 \ldots\ldots.n\right) \tag{7}$$

Step-IV calculating the prominence $\left(C + R\right)$ and relation $\left(C - R\right)$

Consequentially, the row and column vectors, recognized as prominence $\left(C + R\right)$ and relation $\left(C - R\right)$, describe the cause and receiver expansion of criteria (Yang and Tzeng, 2011). The cause and effect relationship diagram structure the influence relationship amid prominence and relation between criteria.

To identify the criteria affecting Sustainable Supply Chain Flexibility (SSCF).

1. To analyse the relationships between the degree of Sustainable Supply Chain Flexibility (SSCF) measures and Circular Economy (CE)-target performance.

Table 2 shows the abbreviation to simplify the calculation and understanding for supply chains for sustainable flexibility measures and CE-target performance measures

Table 1. Demographic profile of the respondents

No. of respondents	Industries	Number of respondents department wise	Department
9	Electronics	5	Purchasing and supply chain managers
		4	Environmental managers
6	Food	2	General managers
		4	Purchasing managers
5	Machinery	3	Quality control managers
		2	Environmental managers
3	Pharmacy	1	General manager
		2	Purchasing managers

RESULT OBTAINED THROUGH DEMATEL TECHNIQUE

The anticipated approach successfully extends the DEMATEL approach by executing relationships between SSCF efforts and CE so that it can efficiently deal with vague and inaccurate judgments. Particularly, the approach can also effectively divide a set of complex factors into a cause group and an effect group and produce a noticeable causal illustration. Through a causal illustration, it's easier to understand how complicated a problem is, which makes it easier to make tough decisions.

To make a decision, the DEMATEL technique involves the following steps: Firstly, SSCF factors have been identified through a rigorous literature review followed by a brainstorming session with the experts. Consequently, the responses from the respondents have been collected in the form of a matrix, which is shown in Table 3

Figure 1. Steps of DEMATEL Techniques

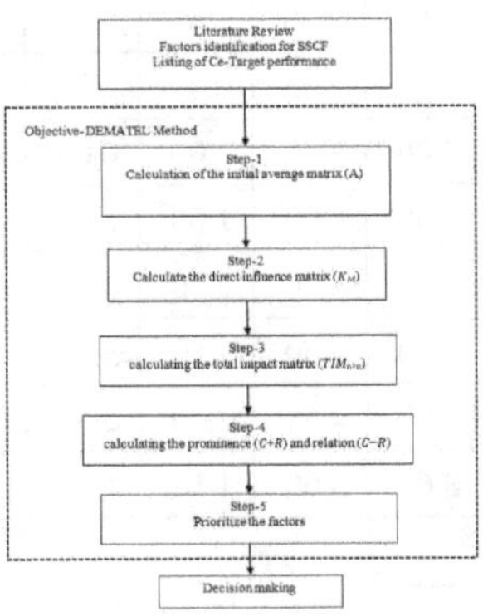

Table 2. Abbreviation used for supply chains for sustainable flexibility measures and CE-target performance measures

S. No.	Variables	Abbreviation Used
1.	Range	S1
2.	Configuration	S2
3.	Volume (expansion flexibility)	S3
4.	Product mix	S4
5.	Schedule	S5
6.	Green Degree	S6
7.	Adaptability	S7
8.	Resource Consumption	S8
9.	Pollution Production	S9
10.	Green Practices	S10
11.	Environmental management system	S11
12.	Reduced Pollution	S12
13.	Decreased Consumption	S13

Definition 1: Calculation of initial average matrix

Table 3. Initial average matrix

0	S1	S2	S3	S4	S5	S6	S7	S8	S9	S10	S11	S12	S13
S1	0.00	5.00	4.00	3.00	3.00	4.00	3.00	3.33	3.00	4.50	2.50	2.00	4.50
S2	3.67	0.00	2.33	2.67	3.67	3.33	1.67	2.33	3.67	3.67	4.67	2.33	1.67
S3	3.67	4.33	0.00	2.00	2.33	3.33	2.67	2.00	2.33	4.00	4.00	2.33	2.33
S4	4.00	2.33	3.33	0.00	3.33	2.33	3.33	2.00	2.00	2.67	2.67	2.00	1.67
S5	4.00	3.00	3.33	3.00	0.00	1.33	3.00	2.33	3.33	2.33	3.00	1.00	1.00
S6	4.33	4.33	4.00	2.33	2.00	0.00	1.67	3.00	3.33	5.00	4.67	2.67	1.67
S7	5.00	2.00	4.67	3.33	1.67	2.33	0.00	3.33	2.33	3.00	3.00	3.33	1.33
S8	3.00	1.33	3.00	1.67	3.00	3.67	2.00	0.00	3.67	3.00	3.33	1.33	1.00
S9	2.50	2.50	3.67	2.50	3.50	4.00	1.00	2.50	0.00	1.50	3.33	2.33	1.33
S10	5.00	4.00	5.00	4.50	2.00	4.00	2.00	3.50	2.67	0.00	4.50	2.33	2.67
S11	4.00	4.50	4.50	4.00	2.50	4.50	2.50	3.00	3.50	4.00	0.00	1.00	2.00
S12	3.50	3.50	4.00	2.50	4.00	4.50	2.50	2.50	3.00	4.00	4.00	0.00	4.50
S13	2.00	4.00	4.00	2.50	3.67	3.00	1.50	1.50	2.50	4.00	3.00	2.33	0.00

Definition 2: Calculate the direct influence matrix (K_{Id})

Table 4. Direct influence matrix (KId)

0	S1	S2	S3	S4	S5	S6	S7	S8	S9	S10	S11	S12	S13
S1	0.000	0.109	0.087	0.065	0.065	0.087	0.065	0.073	0.065	0.098	0.055	0.044	0.098
S2	0.080	0.000	0.051	0.058	0.080	0.073	0.036	0.051	0.080	0.080	0.102	0.051	0.036
S3	0.080	0.095	0.000	0.044	0.051	0.073	0.058	0.044	0.051	0.087	0.087	0.051	0.051
S4	0.087	0.051	0.073	0.000	0.073	0.051	0.073	0.044	0.044	0.058	0.058	0.044	0.036
S5	0.087	0.065	0.073	0.065	0.000	0.029	0.065	0.051	0.073	0.051	0.065	0.022	0.022
S6	0.095	0.095	0.087	0.051	0.044	0.000	0.036	0.065	0.073	0.109	0.102	0.058	0.036
S7	0.109	0.044	0.102	0.073	0.036	0.051	0.000	0.073	0.051	0.065	0.065	0.073	0.029
S8	0.065	0.029	0.065	0.036	0.065	0.080	0.044	0.000	0.080	0.065	0.073	0.029	0.022
S9	0.055	0.055	0.080	0.055	0.076	0.087	0.022	0.055	0.000	0.033	0.073	0.051	0.029
S10	0.109	0.087	0.109	0.098	0.044	0.087	0.044	0.076	0.058	0.000	0.098	0.051	0.058
S11	0.087	0.098	0.098	0.087	0.055	0.098	0.055	0.065	0.076	0.087	0.000	0.022	0.044
S12	0.076	0.076	0.087	0.055	0.087	0.098	0.055	0.055	0.065	0.087	0.087	0.000	0.098
S13	0.044	0.087	0.087	0.055	0.080	0.065	0.033	0.033	0.055	0.087	0.065	0.051	0.000

Definition 3: Total impact matrix ($TIM_{n\times n}$)

as the "initial average matrix." After that, table 4 shows the direct influence matrix,

Table 5. Total impact matrix

0	S1	S2	S3	S4	S5	S6	S7	S8	S9	S10	S11	S12	S13
S1	0.328	0.408	0.413	0.316	0.314	0.379	0.263	0.303	0.324	0.399	0.368	0.230	0.284
S2	0.360	0.268	0.338	0.277	0.294	0.328	0.211	0.254	0.303	0.341	0.366	0.209	0.204
S3	0.360	0.356	0.289	0.264	0.268	0.328	0.230	0.248	0.277	0.350	0.355	0.211	0.218
S4	0.336	0.287	0.326	0.197	0.263	0.279	0.226	0.225	0.245	0.294	0.298	0.186	0.186
S5	0.325	0.289	0.315	0.251	0.188	0.251	0.213	0.224	0.263	0.276	0.294	0.161	0.166
S6	0.400	0.382	0.398	0.292	0.283	0.288	0.228	0.287	0.319	0.395	0.395	0.232	0.222
S7	0.384	0.309	0.381	0.287	0.254	0.309	0.177	0.273	0.275	0.329	0.333	0.230	0.200
S8	0.303	0.256	0.306	0.222	0.245	0.294	0.190	0.174	0.268	0.287	0.299	0.165	0.164
S9	0.297	0.282	0.321	0.240	0.260	0.303	0.174	0.227	0.197	0.263	0.303	0.187	0.173
S10	0.433	0.395	0.436	0.348	0.299	0.385	0.249	0.310	0.322	0.315	0.410	0.238	0.253
S11	0.397	0.386	0.409	0.325	0.294	0.377	0.246	0.288	0.323	0.377	0.303	0.202	0.227
S12	0.407	0.388	0.421	0.312	0.339	0.396	0.258	0.293	0.330	0.398	0.403	0.191	0.290
S13	0.314	0.337	0.355	0.263	0.284	0.308	0.199	0.227	0.269	0.336	0.323	0.203	0.160

Definition 4: Calculating the prominence $(C+R)$ and relation $(C-R)$

Table 6. Prominence, relation and ranking

	C	R	C+R	C-R	Ranking
S1	4.328668	4.644112	8.972781	-0.315444	1
S10	4.393087	4.362179	8.755265	0.030908	2
S11	4.154449	4.450329	8.604777	-0.29588	3
S3	3.755584	4.707775	8.463358	-0.952191	4
S6	4.121962	4.224772	8.346734	-0.10281	5
S2	3.753272	4.343682	8.096954	-0.590409	6
S12	4.425954	2.643387	7.069342	1.782567	7
S4	3.348561	3.596558	6.945119	-0.247997	8
S9	3.228367	3.715904	6.944271	-0.487536	9
S5	3.216116	3.583166	6.799281	-0.36705	10
S7	3.740203	2.864763	6.604966	0.875439	11
S8	3.172919	3.333393	6.506312	-0.160474	12
S13	3.578749	2.747872	6.326621	0.830876	13

which has been calculated using equations (2) and (3). Table 5 discusses the total impact matrix using equation (5). Table 6 discusses the factors' prominence (C+R), relationship (C-R), and priority ranking. With this priority ranking, researchers and practitioners can easily identify the driving factors of SSCF and CE.

After summing up prominence (C+R) and relation (C-R), divided by the thirteen variables to receive the mean, they are respectively 7.571983 and 0, which gives the origin point in figure 2. As the concentrated trend increases, the estimated value of the causal matrix can be used to divide the matrix into four quadrants, as shown in figure 2. Green Practices (S10) of the SSCF factor have the highest prominence and relationship in the I quadrant, according to the analysis of figure 2. Green degree

Figure 2 Prominence and relation diagram

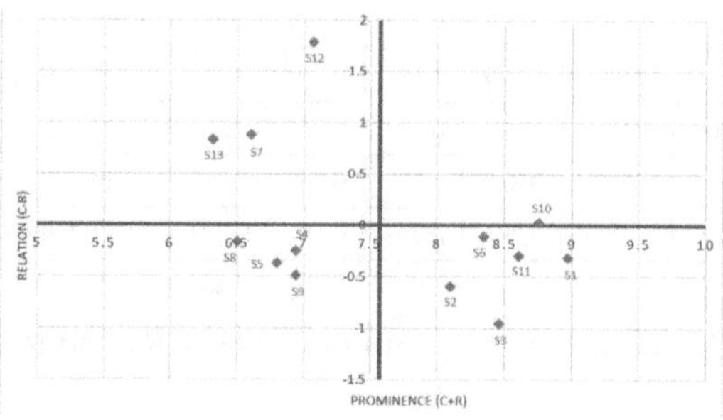

(S6), range (S1), environmental management system (S11), configuration (S2), and volume (S3) have high prominence but low relation in the IV quadrant, indicating that these variables are the core items influenced by other conditions. Other variables in the II quadrant are adaptability (S7), reduced pollution (S12), and decreased consumption (S13), all of which have a low prominence but a high relationship. Due to prominence being lower than the mean (7.571983) and the relation being near the mean (0), it means that these variables impact other variables. The IV quadrant is dominated by the following: resources (S8), scheduling (S5), product mix (S4), and pollution production (S9). As can be seen from the causal diagram, green practises (S10) are the key impact factors that affect sustainable supply chain flexibility and the circular economy.

CONCLUSION

The research concludes based on the results from the causal diagram, which divides the factors into two groups of cause and effect. ***Reduced Pollution*** (S12), ***Decreased Consumption*** (S13), ***Adaptability*** (S7), and ***Green Practices*** (S10) are among the causes that should be addressed, while the effect group, which includes ***Range*** (S1), ***Configuration*** (S2), ***Volume*** (S3), ***Product Mix*** (S4), ***Schedule*** (S5), ***Green Degree*** (S6), ***Resource Consumption*** (S8), ***Pollution Production*** (S9), and ***Environmental Management System*** (S11), should be prioritized. Among these, ***range*** (S1) and ***green practises*** (S10) would be given more attention. The DEMATEL method is comprehensive and applicable to all e-commerce companies facing problems that require group decision-making in a decision making environment.

REFERENCES

Agrawal, A., Diwakar, H. K., & Mukti, S. K. (2020). Employee Productivity and Service Quality Enhancement Using ERP With Knowledge Management: A Study in a Power Distribution Company in the Global Context [IJKM]. *International Journal of Knowledge Management, 16*(3), 89–111. doi:10.4018/IJKM.2020070106

Agrawal, A., Kumar, C., & Mukti, S. K. (2021). Role of Information and Communication Technology (ICT) to Enhance the Success of Knowledge Management (KM): A Study in a Steel Plant. *Journal of the Knowledge Economy, 12*(4), 1760–1786. doi:10.100713132-020-00694-6

Agrawal, A. and Mukti, S.K. (2019) 'Knowledge Management to Minimize the Marketing Challenges using Strategic Management as a Tool'.

Agrawal, A., & Mukti, S. K. (2020). Knowledge Management & It's Origin, Success Factors, Planning, Tools, Applications, Barriers and Enablers: A Review [IJKM]. *International Journal of Knowledge Management, 16*(1), 43–82. doi:10.4018/IJKM.2020010103

Alshurideh, M., Kurdi, B. A., Alzoubi, H. M., Obeidat, B., Hamadneh, S., & Ahmad, A. (2022). 'The influence of supply chain partners' integrations on organizational performance: The moderating role of trust'. *Uncertain Supply Chain Management, 10*(4), 1191–1202. doi:10.5267/j.uscm.2022.8.009

Animesh, A., & Mukti, S. K. (2019). Case study of critical success factors affecting knowledge management in small-and medium-sized enterprises in developing state: Steel sector. In *Advances in industrial and production engineering* (pp. 825–831). Springer.

Bhide, A.V. and Akarte, M.M. (2022) Hybrid manufacturing-remanufacturing system of essential commodity in the context of COVID-19. *Journal of Modelling in Management*.

Cooper, T. (1999) Creating an economic infrastructure for sustainable product design. Journal of sustainable product design, pp. 7–17.

Edwin Cheng, T. C., Kamble, S. S., Belhadi, A., Ndubisi, N. O., Lai, K., & Kharat, M. G. (2022). Linkages between big data analytics, circular economy, sustainable supply chain flexibility, and sustainable performance in manufacturing firms. *International Journal of Production Research, 60*(22), 6908–6922. doi:10.1080/00207543.2021.1906971

Fu, Q., Abdul Rahman, A. A., Jiang, H., Abbas, J., & Comite, U. (2022). Sustainable supply chain and business performance: The impact of strategy, network design, information systems, and organizational structure. *Sustainability, 14*(3), 1080. doi:10.3390u14031080

Geng, Y., Sarkis, J., & Bleischwitz, R. (2019). *How to globalize the circular economy*. Nature Publishing Group. doi:10.1038/d41586-019-00017-z

Geng, Y., Sarkis, J., & Ulgiati, S. (2016). Sustainability, well-being, and the circular economy in China and worldwide. *Science, 6278*(Supplement), 73–76.

Geng, Y., Sarkis, J., Ulgiati, S., & Zhang, P. (2013). Measuring China's circular economy. *Science, 339*(6127), 1526–1527. doi:10.1126cience.1227059 PMID:23539584

Lehtokunnas, T., Mattila, M., Närvänen, E., & Mesiranta, N. (2022). Towards a circular economy in food consumption: Food waste reduction practices as ethical work. *Journal of Consumer Culture*, *22*(1), 227–245. doi:10.1177/1469540520926252

Lukinskiy, V., Lukinskiy, V., Ivanov, D., Sokolov, B., & Bazhina, D. (2023). A probabilistic approach to information management of order fulfilment reliability with the help of perfect-order analytics. *International Journal of Information Management*, *68*, 102567. doi:10.1016/j.ijinfomgt.2022.102567

Makeel, R. (2022). The influence of social supportive culture and performance-based culture on social enterprise performance: the mediation role of Social entrepreneurial orientation. *Journal of Global Entrepreneurship Research*, 1–14.

Mukti, S. K., Ahirwar, A. V., & Agrawal, A. (2021) 'Industry 4.0: Advances and Applications', Industry 4.0: Advances and Applications, pp. 1–35. doi:10.9734/bpi/mono/978-93-5547-287-8

Pushpamali, N. N. C., Agdas, D., Rose, T. M., & Yigitcanlar, T. (2021). Stakeholder perception of reverse logistics practices on supply chain performance. *Business Strategy and the Environment*, *30*(1), 60–70. doi:10.1002/bse.2609

Richey, R. G., Roath, A. S., Adams, F. G., & Wieland, A. (2022). A responsiveness view of logistics and supply chain management. *Journal of Business Logistics*, *43*(1), 62–91. doi:10.1111/jbl.12290

Sarkis, J., & Zhu, Q. (2018). Environmental sustainability and production: Taking the road less travelled. *International Journal of Production Research*, *56*(1–2), 743–759. doi:10.1080/00207543.2017.1365182

Yang, J. L., & Tzeng, G.-H. (2011). An integrated MCDM technique combined with DEMATEL for a novel cluster-weighted with ANP method. *Expert Systems with Applications*, *38*(3), 1417–1424. doi:10.1016/j.eswa.2010.07.048

Zhu, Q., Geng, Y., & Lai, K. (2010). Circular economy practices among Chinese manufacturers varying in environmental-oriented supply chain cooperation and the performance implications. *Journal of Environmental Management*, *91*(6), 1324–1331. doi:10.1016/j.jenvman.2010.02.013 PMID:20223586

Chapter 11
Utilization Process for Electronic Waste in Eco-Friendly Concrete:
Experimental Study

Kedri Janardhana
https://orcid.org/0000-0002-9877-2425
Dayalbagh Educational Institute (Deemed), India

Vikash Singh
Institute of Engineering and Technology, Lucknow, India

Satyendra Narain Singh
Institute of Engineering and Technology, Lucknow, India

T. S. Ramesh Babu
K.G. Reddy College of Engineering and Technology, India

Samreen Bano
Integral University, India

Sampath Boopathi
https://orcid.org/0000-0002-2065-6539
Muthayammal Engineering College, India

ABSTRACT

In this chapter, the authors discuss the utilization of e-waste in the concrete for civil construction activities. Various tests have been used to investigate the effects of e-waste mixed with concrete. The various percentages of e-waste have been mixed with concrete to improve the strength of buildings. An e-waste concrete beam has a maximum tensile strength of 6.23 MPa under sulfuric curing conditions, and the highest flexural strength at 10% e-waste replacement during the hydrochloride curing process. The compressive strength is at its highest value when e-waste replaces 10% of it. After 28 days of curing, the concrete cylinder's maximum split tensile strength was 15%. Thus, the e-waste could be effectively utilized for civil construction purposes to reduce its environmental impacts.

DOI: 10.4018/978-1-6684-7573-7.ch011

INTRODUCTION

E-waste is generated by electronics that are intended for recovery, resale, salvage, recycling, or destruction. The number of natural aggregates used in concrete will decrease if various byproducts are used as substitute materials. Many gadgets that play a key role in our day-to-day lives have been created by humans. This type of technological advancement has considerably increased in the 20th and 21st centuries (Gupta et al., 2008). Today, a major concern is how to handle the enormous volumes of garbage that are produced. Wastes can be created in a variety of ways, including as liquid, solid, organic, toxic, and recyclable materials. This research focuses on potentially hazardous electronic wastes that could seriously impact the environment and public health if they were improperly disposed of outside. Hazardous waste materials are combustible, corrosive, reactive, toxic, volatile, poisonous, and volatile, which causes long-term harm to the environment.

There is still pressure on India to use open landfills and open dumps to dispose of its municipal solid waste. Electrical and electronic equipment (WEEE) is the waste stream with the fastest global growth. Total production of e-waste was over 41.5 million tons in 2011 and 93.5 million in 2016. It includes obsolete domestic equipment such as TVs, refrigerators, aeration and cooling systems, MP3 players, and other electrical and electronic products (PDAs). In developing countries, e-waste management has grown to be a significant issue. Developed countries trade e- waste with developing countries under the guise of recycling, reusing, or gifting. Only the most valuable and separated materials, such as gold plating, aluminum, and copper wiring, can be recovered through controlled reuse and transfer. E-product trash consists of packaging materials and waste-producing products. India has produced roughly 8 lakh metric tonnes of e-waste annually during the past two decades. The Central Pollution Control Board estimates that India creates 0.573 MT, or 1.47 lakh metric tonnes(Needhidasan et al., 2014), of electronic garbage every day. Electronic waste, sometimes known as "E-waste," is produced on a global scale at a rate that is three times faster than other types of waste. It is essential to create effective recycling strategies for its decrease because it is one of the main causes of environmental contamination.

Electronic waste, sometimes known as "E-waste," is produced on a global scale at a rate that is three times faster than other types of waste. It is essential to create effective recycling strategies for its decrease because it is one of the main causes of environmental contamination. Globally, e-waste has grown to be a significant problem. Older PCs and other electronic gadgets produce millions of tons of electronic garbage each year. For the purpose of developing a recycling system that is affordable and environmentally responsible, the waste stream's categorization is essential. Electronic waste contains a wide variety of chemicals and substances that pose a significant risk to human health and the environment (Ankit et al., 2021; Elleuch et al., 2018). E-waste has been connected to brain damage, allergies, and cancer. Electronic waste contains mercury and cadmium, which together account for nearly 70% of the heavy metals in landfills. Arsenic, barium, copper, chrome, cobalt, lead, zinc, nickel, mercury, and silver are the primary contaminants in electrical waste. Significant levels of lead, cadmium, and various poisonous gases are produced when waste is burned or dumped in an open area. Sustainable development integrates the economy, society, and environment to meet current demands without jeopardizing the well-being of future generations. Many industrialized countries significantly damaged the environment in their early stages of growth, which also occurred in emerging countries. The introduction of new waste types, as well as e-waste recycling, has exacerbated the problems associated with solid waste management (Gupta et al., 2008; N. Gupta & Nath, 2020; Needhidasan et al., 2014). The illegal recycling and dumping of e-waste have a substantial impact on both the environment and the health of employees and residents in developing countries. In this study, recycled electronic debris is substituted for coarse aggregate in a range of percentages (0 to 20%) and sizes (10 to 12.5 mm) (crushed).

The goal of the research project is to ascertain whether adding crushed electronic trash will improve the performance of concrete and reinforced beams in terms of strength, durability, and performance. Under different curing circumstances,e-waste in concrete can be optimized by minimizing corrosion of the steel reinforcing. The enhanced toughness of an E-waste concrete specimen was evaluated using measurements of saturated water absorption, sulfuric acid attack, chloride attack, porosity, and a sulphate attack test. The permeability test revealed the concrete specimens' durability and resistance to water and acid infiltration.

RESEARCH OBJECTIVES

- To analyze the problems and the potential for employing polymere-waste in place of a portion of the coarse aggregate in concrete for a high-performance reinforced concrete beam.
- To investigate and assess the performance of non-melted polymer e-waste in freshly poured and hardened concrete that has been cured in various acid solutions.
- To determine the ideal percentage of non-melted e-waste in concrete and determine whether the conditions outlined in the IS standards for concretes under axial and cyclic loading situations are being followed.

MATERIALS PREPARATION

Cement: In this experiment, regular Portland cement of grade 43 was used. According to the IS 4031:1988 code, tests on the cement's various physical qualities were done. Table 1 presents the test findings in tabular format.

Water: Water that is clean enough to drink is necessary for the casting and curing procedures. The pH range of the water used for the entire test was 7.5 to 8.0.

Fly Ash: The fly ash for the class F samples originated from the Mettur Thermal Power Plant in Coimbatore, Tamil Nadu, via ACC, Ltd. (MTTP). The specimens were cast using fly ash with a specific gravity of 2.37. Table 2 shows the results of a fly ash test conducted in accordance with IS methodology.

Table 1. Cement's physical characteristics

Physical properties	Observations
Specific gravity	3.14
Standard Consistency	33%
Initial setting time	30 min
Final setting time	8 hrs 35 min
Compressive strength	43.86 N/mm^2

Table 2. Chemical composition of fly ash (mass %)

SO_3	CaO	MgO	Loss on ignition	SiO_2	Na_2O	$SiO_2+Al_2O_3+Fe_2O_3$
1.8	3.6	1.91	2.0	58	2.0	90.5

Fine Aggregate: The well-graded river sand, which adhered to grading zone II of IS 383-1970 and had a specific gravity of 2.76, was used as fine aggregate.

Coarse Aggregate: Crushed and angular coarse material with a specific gravity of 2.80 was used; it was easily accessible nearby. The evaluation of coarse aggregate's physical qualities was conducted in accordance with IS 383-1970, and the results are shown in Table 3.

E Waste: Scrap from electrical or electronic equipment was collected as e-waste from the sources. Concrete was partially replaced in this study by coarse aggregate made from crushed e-waste.The e-waste is collected at Hosur, Tamil Nadu. Device outer frames, also known as polymer e-wastes, were collected and crushed to a size of between 10 and 12.5 mm. Crushed e-wastes were rinsed with potable water and dried outside in the sun to lessen the effects of the fine powder created after crushing. The physical and chemical features of the identified e-wastes are listed in Table 4.

Table 3. Physical properties of coarse aggregate

Properties	Coarse aggregate
Specific gravity	2.8
Absorption (%)	0.5
Colour	Dark
Crushing value	27.4%
Shape	Angular
Impact value	24.62%

Table 4. Physical Properties of e-waste

Properties	E waste
Specific gravity	1.07
Shape	Crushed and angular
Size	10 – 12.5mm
Absorption (%)	<0.2
Color	Dark
Crushing value	<2%
Impact value	<2%

E-Waste Mixing Proportions

The M25 grade mix percent was estimated using the design approach given in IS 10262-2:1982. With a water-to-cement ratio of 0.45, the expected mix ratio was 1:1.56:2.56. E-waste was used as coarse aggregate in concrete and in varied amounts to cast the cubes. The following amounts were chosen to substitute E trash by volume for the coarse aggregate on top of the controlled concrete with a w/c ratio of 0.45. Depending on the type of concrete, e-waste percentages could be as low as 0% and as high as 20%. The series of mixes were detailed in Table 5.

Table 5. Mix specifications – M25 Grade

Type of concrete	Controlled concrete	M1		M2	M3	M4	M5
% of e-waste added	0%	5%		10%	15%	18%	20%

Casting and Curing

Electronic waste concentrations in concrete samples ranged from 0% to 20%. The slump value of the prepared mix was assessed. Standard tap water was used to cast and cure each specimen. At a temperature of over 27°C (2°C), there were enough cubes, cylinders, prisms, beams, T beams, and RC beams to cast and cure the materials. The cast concrete samples were demolded, immersed in fresh water, treated with chemicals, and left to cure according to the specified time.

EXPERIMENTAL TESTS

This experiment examines how waste polymer E from concrete reacts when coarse aggregate is partially replaced. The deflection and stiffness of M25 grade and high-strength concrete like M40 and M50 were also assessed utilizing electronic waste. 216 cubes, 313 cylinders, and 48 beams, respectively, were cast, cured, and tested for compression, split tensile, and deflection. This study looked at how the amount of coarse aggregate replaced with e-waste affected the strength of concrete (Blinová & Godovčin, 2021; Cherukuri et al., 2018; Julander et al., 2014).

Compressive Strength Test

To test the compressive strength of both controlled and uncontrolled concrete with E trash added, cubes were cast in a 150 ×150 × 150 mm mold. For all tests, 216 concrete cubes were cast and tested for compressive strength. Figure 1 depicts the common concrete cube casting molds.

A compression testing machine with a capacity of 2,000 kN was used to calculate compressive strength. The findings of all tests were averaged among three samples. Table 6 displays the test results for controlled concrete after 14, 28, and 60 days of curing. The testing setups are shown in Figure 2 (Boopathi, Haribalaji, et al., 2022; Boopathi, Jeyakumar, et al., 2022; Haribalaji, Boopathi, Asif, et al., 2021; Sampath et al., 2021).

Figure 1. Cube molds for casting concrete

The results of the compressive strength tests reveal that gradually substituting polymer e-waste for the coarse aggregate in concrete increases the strength of the mixture by up to 10%. The concrete's compressive strength is gradually declining over 10% replacement (Boopathi, Jeyakumar, et al., 2022; Haribalaji, Boopathi, & Asif, 2021). The average compressive strength produced with 10% e-waste substitution in concrete was nearly 5 N/mm^2 higher compared to the controlled concrete. It is abundantly obvious from the data for 14, 28, and 60 days that the strength of all combinations gradually grew. Figure 3 shows the trend of compressive strength in relation to different combinations (Boopathi et al., 2017, 2021, 2023).

Because E-waste has a lower specific gravity than coarse aggregate, the load of the concrete decreases progressively as e-waste replaces coarse aggregate. If e-wastes are added in concentrations greater than 10%, concrete will become more porous and have a lower load. Because too much e-waste impedes the hydration process, concrete loses its ability to be worked. Due to the e-waste concrete's poor workability and high deflection, the concrete becomes porous and weaker. Fly ash may improve the workability of concrete made from E trash because silica can combine with calcium hydroxide to create C-S-H gel in concrete. It seems to enhance the pore structure and heighten the Load of the e-waste-derived concrete matrix.

Figure 2. Compressive strength testing before and after testing

Split Tensile Strength

The cylinder for split tensile strength was manufactured using a mold with internal dimensions of 150 mm in diameter and 300 mm in height for both controlled concrete and e-waste concrete specimens. The split tensile strength of 336 cylindrical concrete specimens was investigated. Concrete casting cylinder molds are depicted in Figure 4.

Table 6. Compressive strength and weight of the concrete cubes with respect to E-waste

E-waste ((%)	Load, kN			Compressive Strength(N/mm^2)			Weight (Kg)	
Days	*14*	*28*	*60*	*14*	*28*	*60*	*Wet*	*Dry*
0	612	679	685	27.20	30.17	30.44	8.72	8.70
5	639	683	692	28.40	30.35	30.76	8.65	8.38
10	683	781	788	30.36	35.02	35.02	8.37	8.28
15	661	778	782	29.38	34.71	34.76	8.27	8.26
18	655	699	706	29.11	31.06	31.38	7.97	7.97
20	585	660	665	26.00	29.33	29.56	7.80	7.78

Figure 3. Variations of compressive strength and weight of specimen by increasing Percentage of E-waste mixed in the concrete

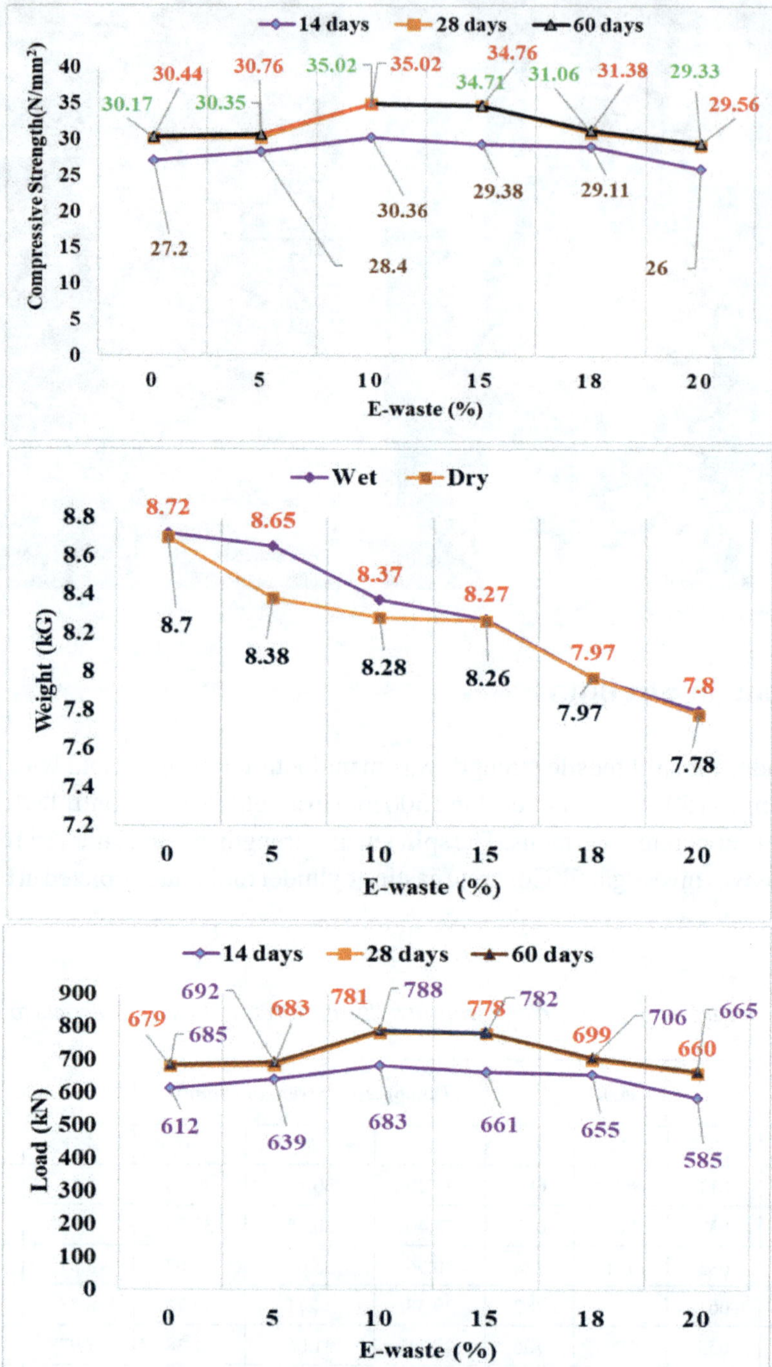

Figure 4. Cylinder molds for casting concrete

The tensile strength of concrete was determined using a split tensile strength in compression testing apparatus (Boopathi, 2022; Boopathi, Haribalaji, et al., 2022; Boopathi, Jeyakumar, et al., 2022). The test results for controlled concrete

Figure 5. Concrete cylinder specimen before and after tensile strength test

Figure 6. Load comparisons for cylinders

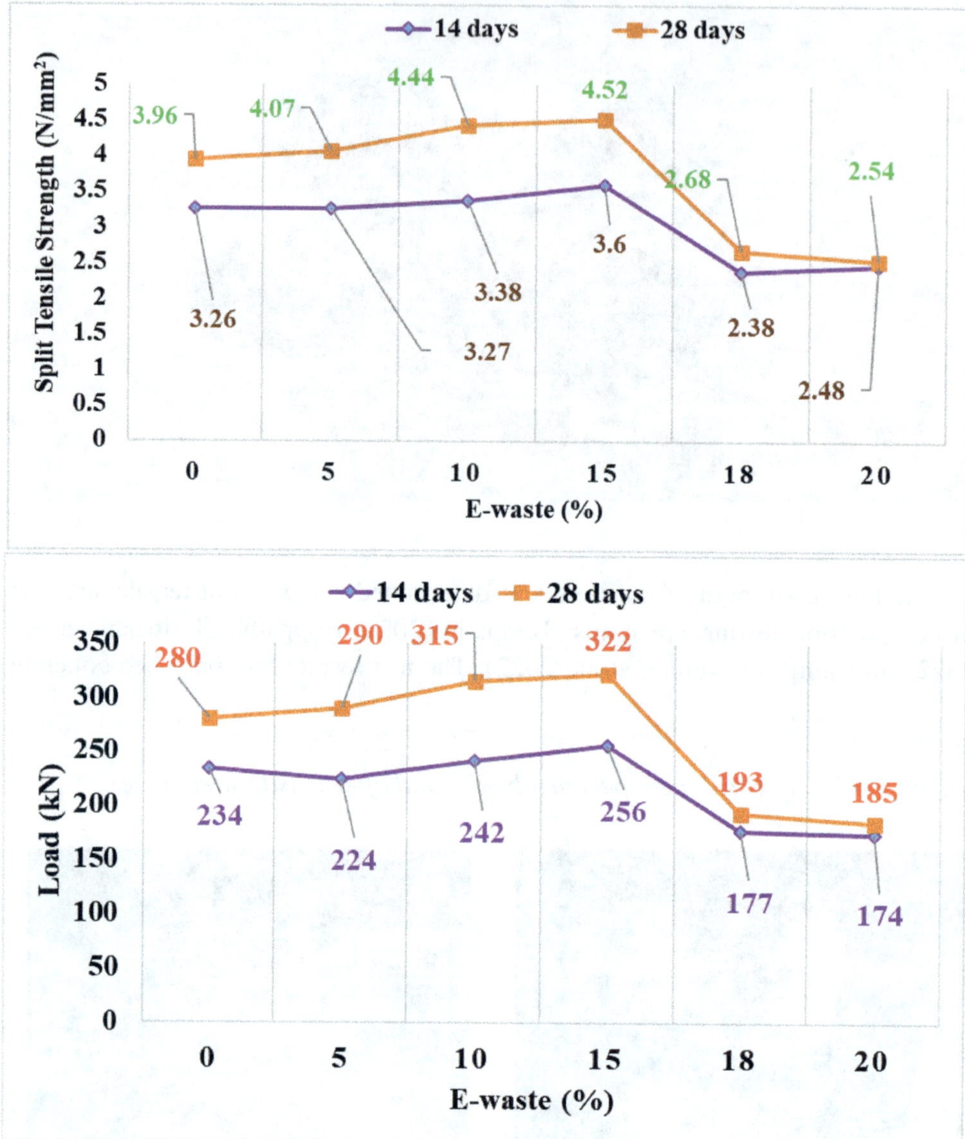

are shown in Table 7 for the 14th and 28th days of curing. The concrete cylinder was 300 mm long and 150 mm in diameter. The cylindrical specimens were cast using various percentages of e-waste, including 0%, 5%, 10%, 15%, 18%, and 20%. After 28 days of curing, the concrete cylinder's maximum split tensile strength, which was attained at 15%, allowed it to support a weight of 323 kN (Boopathi & Myilsamy, 2021; Boopathi & Sivakumar, 2014). The strength has been dropped by

Table 7. Tensile strength result for concrete cylinders

E waste (%)	Failure Load, kN		Split Tensile Strength (N/mm^2)	
	14 days	28 days	14 days	28 days
0	234	280	3.26	3.96
5	224	290	3.27	4.07
10	242	315	3.38	4.44
15	256	322	3.60	4.52
18	177	193	2.38	2.68
20	174	185	2.48	2.54

adding E-waste more than 15% due to reducing the bonding and adhesive property (Boopathi, 2021; Kumar et al., 2023; Sampath et al., 2023). In Figure 5, the testing setups are displayed. Figure 6 shows the split tensile strength for the e-waste cylinder on the 14th and 28th days.

Flexural Strength Test

A beam with the measurements 100 mm in depth, 500 mm in length, and 100 mm in width was made for M25 grade. A total of 48 concrete beams were produced and tested for flexural strength using controlled and e-waste concrete.

The flexural strength of controlled concrete and concrete that had been treated with E-Waste was compared using beam specimens that were tested for M25 grade concrete and had dimensions of 100 mm length, 500 mm depth, and 100 mm breadth. Both ordinary water and water that has been 5% concentrated with HCl, Na_2So_4, and H_2So_4 are being used to cure the beams. Figure 7 shows the beam testing setups. The equation (1) is used to calculate the Flexural Strength of the beam (Boopathi, Jeyakumar, et al., 2022; Boopathi & Myilsamy, 2021; Boopathi & Sivakumar, 2014; Sampath & Myilsamy, 2021).

$$f_b = \frac{pl}{bd^2} \tag{1}$$

Where p - load (force) at the fracture point (kN),

L - length of the support (outer) span (mm), b – width (mm), and d – thickness (mm)

Table 8. Flexural strength for a concrete beam under water curing

% of e-waste	Load in (kN)			Flexural strength (MPa)
	Beam 1	Beam 2	Average	
0%	12	12	12	4.8
5%	13	14	13.5	5.4
10%	16	17	16.5	6.6
15%	15	14	14.5	5.8
18%	14	14	14	5.6
20%	12	13	12.5	5

Table 9. Flexural strength for a concrete beam under H_2SO_4 curing

% of e-waste	Load in (kN)			Flexural strength (MPa)
	Beam 1	Beam 2	Average	
0%	10	11	10.5	4.2
5%	13	12	12.5	5
10%	13	15	14	5.6
15%	12	14	13	5.2
18%	12	13	12.5	5
20%	11	13	12	4.8

Tables 8,9,10, and 11 show the loads and deflection that correspond to the two-point load conditions under which the reinforced e-waste concrete beam was tested. The failure load of the concrete beam was 6.6 MPa in 10% replacement of e-waste. Chloride, sulphuric acid, and Na_2SO_4 were among the chemical solutions used to cure concrete beams manufactured from reinforced E trash. The beam's flexural strength is increased by standard water curing in comparison to other chemical curing conditions (Figure 8).

The greatest flexural strength of an e-waste concrete beam under H_2SO_4 curing conditions was 6.23 N/mm_2 at 10% e-waste. At 10% replacement of e-waste in concrete, the highest flexural strength of the beam during chloride curing was 6.19 N/mm^2(Boopathi et al., 2023; Boopathi, Haribalaji, et al., 2022; Haribalaji et al., 2022; Haribalaji, Boopathi, & Asif, 2021; Myilsamy & Sampath, 2021; Yupapin et al., 2022). The minimum flexural strength of the beam was 3.88 N/mm^2 for H2SO4 curing and 3.81 N/mm^2 for chloride curing, respectively (Figure 8).

Figure 7. Flexural testing – Beam after failure

Flexural strength of the concrete beam reached at 10% e-waste. Due to the high amount of E trash and the poorer link between the waste and concrete than in the control scenario, adding more than 10% e-waste to the beam caused a steady loss of strength (Boopathi et al., 2021, 2023).

Table 10. Flexural strength for a concrete beam under HCl curing

% of e-waste	Load in (kN)			Flexural strength (MPa)
	Beam 1	Beam 2	Average	
0%	9	9	9	3.6
5%	10	9	9.5	3.8
10%	11	12	11.5	4.6
15%	10	11	10.5	4.2
18%	11	10	10.5	4.2
20%	10	8	9	3.6

RESULTS AND DISCUSSIONS

In this study, the many tests used to determine the various physical and chemical characteristics of freshly poured and cured concrete were examined in relation to e-waste management. The following conclusions were drawn from the findings.

- Gradual substitution of coarse aggregate in concrete with polymer e-waste, increasing strength by up to 10%, and then gradually reducing strength.

Figure 8. Flexural strength for beam under water, HCl, H_2So_4, and $NaSo_4$ curing process

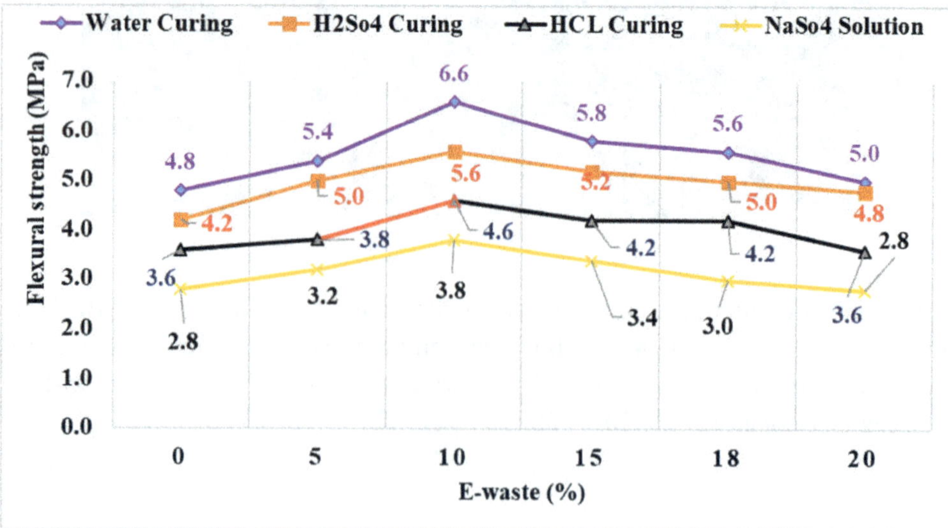

Table 11. Flexural strength for a concrete beam under Na_2SO_4 curing

% of e-waste	Load (kN)			Flexural strength (MPa)
	Beam 1	Beam 2	Average	
0%	7	7	7	2.8
5%	7	9	8	3.2
10%	9	10	9.5	3.8
15%	8	9	8.5	3.4
18%	8	7	7.5	3
20%	6	8	7	2.8

- Due to the lower specific gravity of E trash, concrete Load is steadily declining as e-waste content in concrete rises.
- If e-wastes are added to concrete in amounts greater than 10%, the concrete will become more porous and have a poor load.
- After 28 days of curing, the concrete cylinder's maximum split tensile strength was 15%.
- The compressive strength is at its highest value when e-waste replaces 10% of it.
- A 15% addition of e-waste produced the maximum split tensile strength.
- Under H_2SO_4 curing conditions, the flexural strength of an e-waste concrete beam was 6.23 N/mm^2, which is the maximum value for e-waste.
- The beam's highest flexural strength was 6.19 N/mm^2 at 10% e-waste replacement during chloride curing.
- With the addition of more than 10% of E-waste to the beam, the strength steadily decreased due to the high volume of E-waste and the low bonding between trash and concrete in comparison to the control.

CONCLUSIONS

- Polymer e-waste could be used in place of a portion of the coarse aggregate in concrete for a high-performance reinforced concrete beam. The study aims to investigate and assess the performance of non-melted e-waste in freshly poured and hardened concrete that has been cured in various acid solutions.
- Cement of grade 43 was tested in accordance with the IS 4031:1988 code. Fly ash samples from a thermal power plant in Coimbatore, Tamil Nadu, were cast using a specific gravity of 2.37. The well-graded river sand adhered to grading zone II of IS 383-1970 for fine aggregate.

- Fine Aggregate: Well-graded river sand adhered to grading zone II of IS 383-1970 and had a specific gravity of 2.76, was used as fine aggregate. Fine Aggregate was crushed and angular coarse material with a local gravity of 1.80 was used for evaluation of its physical qualities.
- Electronic waste (e-waste) was used as coarse aggregate in concrete and in varied amounts to cast the cubes. E-Waste is collected in Hosur, Tamil Nadu. Device outer frames were collected and crushed to size of between 10 and 12.5 mm.
- Polymer e-waste can be used to increase the strength of concrete by up to 10%, and then gradually reduce strength over time. The lower specific gravity of E trash also means that concrete load is not as important as it once was, making it more porous and less dense. After 28 days of curing, an e-waste concrete cylinder with a 10% addition of e-waste has the highest split tensile strength and compressive strength in the world, according to the US National Institute of Standards and Technology (NIST) at the University of Pennsylvania.
- An e-waste concrete beam has a maximum tensile strength of 6.23 N/mm^2 under H_2SO_4 curing conditions, and the highest flexural strength at 10% e-waste replacement during chloride curing. The compressive strength is at its highest value when e-waste replaces 10% of it.

REFERENCES

Ankit, S., Saha, L., Kumar, V., Tiwari, J., Sweta, Rawat, S., Singh, J., & Bauddh, K. (2021). Electronic waste and their leachates impact on human health and environment: Global ecological threat and management. *Environmental Technology and Innovation*, *24*, 102049. doi:10.1016/j.eti.2021.102049

Blinová, L., & Godovčin, P. (2021). Importance of Recycling the Waste-Cables Containing Copper and PVC. *Research Papers Faculty of Materials Science and Technology Slovak University of Technology*, *29*(48), 1–21. doi:10.2478/rput-2021-0001

Boopathi, S. (2021). Improving of Green Sand-Mould Quality using Taguchi Technique. *Journal of Engineering Research*. doi:10.36909/jer.14079

Boopathi, S. (2022). Cryogenically treated and untreated stainless steel grade 317 in sustainable wire electrical discharge machining process: A comparative study. *Environmental Science and Pollution Research International*, 1–10. doi:10.100711356-022-22843-x PMID:36057706

Boopathi, S., Balasubramani, V., Kumar, R. S., & Singh, G. R. (2021). The influence of human hair on kenaf and Grewia fiber-based hybrid natural composite material: An experimental study. *Functional Composites and Structures*, *3*(4), 045011. doi:10.1088/2631-6331/ac3afc

Boopathi, S., Haribalaji, V., Mageswari, M., & Asif, M. M. (2022). Influences of Boron Carbide Particles on the Wear Rate and Tensile Strength of Aa2014 Surface Composite Fabricated By Friction-Stir Processing. *Materiali in Tehnologije*, *56*(3), 263–270. doi:10.17222/mit.2022.409

Boopathi, S., Jeyakumar, M., Singh, G. R., King, F. L., Pandian, M., Subbiah, R., & Haribalaji, V. (2022). An experimental study on friction stir processing of aluminium alloy (AA-2024) and boron nitride (BNp) surface composite. *Materials Today: Proceedings, 59*, 1094–1099. doi:10.1016/j.matpr.2022.02.435

Boopathi, S., Kumaresan, A., Manohar, N., & Krishna Moorthi, R. (2017). Review on Effect of Process Parameters-Friction Stir Welding Process. [IRJET]. *International Research Journal of Engineering and Technology*, *4*(7), 272–278.

Boopathi, S., & Myilsamy, S. (2021). Material removal rate and surface roughness study on Near-dry wire electrical discharge Machining process. *Materials Today: Proceedings, 45*, 8149–8156. doi:10.1016/j.matpr.2021.02.267

Boopathi, S., & Sivakumar, K. (2014). Study of water assisted dry wire-cut electrical discharge machining. *Indian Journal of Engineering and Materials Sciences*, *21*(1), 75–82.

Boopathi, S., Venkatesan, G., & Anton Savio Lewise, K. (2023). Mechanical Properties Analysis of Kenaf–Grewia–Hair Fiber-Reinforced Composite. In *Lecture Notes in Mechanical Engineering* (pp. 101–110). Springer. doi:10.1007/978-981-16-9057-0_11

Cherukuri, I., Sultana, N., & Podila, S. P. (2018). Status Of E-Waste In India- A Review. *IOSR Journal of Environmental Science, Toxicology and Food Technology*, *12*(11), 8–16. doi:10.9790/2402-1211010816

Elleuch, B., Bouhamed, F., Elloussaief, M., & Jaghbir, M. (2018). Environmental sustainability and pollution prevention. *Environmental Science and Pollution Research International*, *25*(19), 18223–18225. doi:10.100711356-017-0619-5 PMID:29971753

Gupta, A., Singh, A., Singh, U., Singh, A., & Tomar, C. (2008). A Review Paper on Use of E-Waste in Concrete. *International Research Journal of Engineering and Technology*, 2762. www.irjet.net

Gupta, N., & Nath, M. (2020). Groundwater Contamination by E-Waste and Its Remedial Measure - A Literature Review. *Journal of Physics: Conference Series, 1531*(1), 012023. Advance online publication. doi:10.1088/1742-6596/1531/1/012023

Haribalaji, V., Boopathi, S., & Asif, M. M. (2021). Optimization of friction stir welding process to join dissimilar AA2014 and AA7075 aluminum alloys. *Materials Today: Proceedings, 50*(xxxx), 2227–2234. doi:10.1016/j.matpr.2021.09.499

Haribalaji, V., Boopathi, S., Asif, M. M., Jeyakumar, M., Subbiah, R., & Lewise, K. A. S. (2021). Influences of Friction stir tool parameters for joining two similar AZ61A alloy plates. *Materials Today: Proceedings, 50*(5), 2547–2553. doi:10.1016/j.matpr.2021.12.074

Haribalaji, V., Venkatesan, G., Asif, M. M., Pandian, M., Subbiah, R., & Boopathi, S. (2022). Investigation on corrosion and tensile Characteristics: Friction stir welding of AA7075 and AA2014. *Materials Today: Proceedings, 66*, 743–748. doi:10.1016/j.matpr.2022.04.037

Julander, A., Lundgren, L., Skare, L., Grandér, M., Palm, B., Vahter, M., & Lidén, C. (2014). Formal recycling of e-waste leads to increased exposure to toxic metals: AN occupational exposure study from Sweden. *Environment International, 73*, 243–251. doi:10.1016/j.envint.2014.07.006 PMID:25300751

Kumar, P., Sampath, B., Kumar, S., Babu, B. H., & Ahalya, N. (2023). Hydroponics, Aeroponics, and Aquaponics Technologies in Modern Agricultural Cultivation. In Trends, Paradigms, and Advances in Mechatronics Engineering (pp. 223–241). IGI Global.

Myilsamy, S., & Sampath, B. (2021). Experimental comparison of near-dry and cryogenically cooled near-dry machining in wire-cut electrical discharge machining processes. *Surface Topography: Metrology and Properties, 9*(3), 035015. doi:10.1088/2051-672X/ac15e0

Needhidasan, S., Samuel, M., & Chidambaram, R. (2014). Electronic waste – an emerging threat to the environment of urban India. *Journal of Environmental Health Science & Engineering, 12*(1), 1–9. doi:10.1186/2052-336X-12-36 PMID:24444377

Sampath, B., & Myilsamy, S. (2021). Experimental investigation of a cryogenically cooled oxygen-mist near-dry wire-cut electrical discharge machining process. *Strojniski Vestnik. Jixie Gongcheng Xuebao, 67*(6), 322–330. doi:10.5545v-jme.2021.7161

Sampath, B., Naveenkumar, N., Sampathkumar, P., Silambarasan, P., Venkadesh, A., & Sakthivel, M. (2021). Experimental comparative study of banana fiber composite with glass fiber composite material using Taguchi method. *Materials Today: Proceedings*, *49*(5), 1475–1480. doi:10.1016/j.matpr.2021.07.232

Sampath, B., Sasikumar, C., & Myilsamy, S. (2023). Application of TOPSIS Optimization Technique in the Micro-Machining Process. In Trends, Paradigms, and Advances in Mechatronics Engineering (pp. 162–187). IGI Global.

Yupapin, P., Trabelsi, Y., Nattappan, A., & Boopathi, S. (2022). Performance Improvement of Wire-Cut Electrical Discharge Machining Process Using Cryogenically Treated Super-Conductive State of Monel-K500 Alloy. *Iranian Journal of Science and Technology. Transaction of Mechanical Engineering, 0123456789*. Advance online publication. doi:10.100740997-022-00513-0

Chapter 12
Utilization of E-Waste as Coarse Aggregates in Geopolymer Concrete

Parth Verma
https://orcid.org/0000-0002-8047-1464
Maulana Azad National Institute of Technology, India

Priyanka Dhurvey
Maulana Azad National Institute of Technology, India

Chandra Prakash Gour
Maulana Azad National Institute of Technology, India

ABSTRACT

E-waste is a general term for damaged, surplus, or excess electrical and electronic equipment. Due to quick technological advancement and inexpensive beginning costs, there is a rapidly expanding global surplus of e-waste. If not managed appropriately, the many sorts of compounds and chemicals included in e-waste can seriously harm both human health and the environment. Large volumes of e-waste may be recycled in the concrete sector, solving both the disposal and reuse issues. The most practical application of e-waste is thus thought to be recycling and reuse in the geopolymer concrete (GPC) industry. The experimental study is carried out to enhance the mechanical properties of the GPC (M30 Grade) with varies percentage e-waste replacement for coarse aggregates by casting cubes, cylinders, and beams. Also, an empirical regression-based model is also formulated for the evaluation of the strength of GPC at varying e-waste coarse aggregate (EWCA) percentages.

DOI: 10.4018/978-1-6684-7573-7.ch012

INTRODUCTION

Deepak et al. (2019) has suggested that the geopolymer concrete technology makes it possible to successfully substitute cement as a binding agent with waste products from industry and agriculture that are high in aluminates and silicates. In addition, the cement and aggregate industries may cut their atmospheric CO_2 emissions by over 80% by employing geopolymer technology as reported by Xie et al. (2019). Pouhet et al. (2016) has evaluated that the geopolymer concrete is gaining popularity as a sustainable construction material because it uses fly ash and metakaolin as binding agents in place of cement.

There are many ways to reuse and utilize the enormous amount of E-Waste generated every year. Like, Kiran et al (2020) study creates a framework for utilising waste materials to enhance the soil's properties, so minimising the impact of waste on the environment, and it gives an overview of the research on soil stabilisation. One such waste is electronic waste, and not much attention has been paid to using it to stabilise the soil. This research will provide the framework for using electronic waste to stabilise the soil. Manatkar et al. (2015) examined that then e-waste is used to describe abandoned and disposed of electrical and electronic equipment. Kale et al. (2013) also showed that around 85% of electronic waste is thrown away in incinerators or landfills, which might result in the release or involvement of hazardous materials that affect the environment and other living beings. By employing plastics from e-waste as coarse aggregate in concrete, it may be possible to address the problem of disposing economically and technically of large volumes of wastes. E-waste plastics can be utilised as coarse aggregate, fine aggregate, or fine filler in concrete, depending on their chemical composition and particle size. Anam et al. (2013) investigated several E-waste management techniques as a green computing solution. Nagajothi et al. (2014) studied that an individual's blood, kidneys, and peripheral and central nervous systems may be impacted by the rising lead concentration of electronics. In India, rubbish storage is a significant issue, and just 10 to 12 percent of the country's e-waste is recycled. The most effective method of adding coarse gravel to concrete as shown by Damal et al. (2015). E-waste in concrete is one method for creating a more sustainable environment. The discarded electronic parts are heated in a heating kiln to temperatures below their melting points to make the electronic waste aggregates as proven by Dave et al. (2015). This heated debris is crushed through a certain screen size to create the aggregates for electronic waste. Despite being a lightweight, reasonably priced, and ecologically benign material substitute, e-waste aggregate concrete can only be utilised in non-structural applications due to the loss in strength, which prevents its widespread use. When concrete aggregates include a certain percentage of E-waste, the strength of the concrete is reduced suggested by Sundar et al. (2017). E-waste is produced by the majority of new electronic equipment purchases, including those computers, monitors, TVs, LE/CDs, and TVs. In addition, the overall demand for concrete for construction is predicted to increase daily and reach 59% by the end of 2025 as suggested by Gupta et al. (2015). The creation of devices by everyday people will never cease in the present internet and technology era. As a result, it is important to recycle e-waste leftovers to make sustainable concrete and avoid harming the environment as studied by Bavan et al. (2015). Alanazi et al. (2017) has found that the compressive and split tensile strength tests were performed on 28-day GPC specimens in low, intermediate, and higher grades. The outcomes were evaluated in comparison to standard cement control mixes. Also, Zhang et al. (2020) has studied the regression model, the connection between the split tensile strength and compressive strength of geopolymer concrete is equal to that of conventional Portland cement concrete. The research's findings support the use of the ACI-318-99 model to predict the split tensile strength of geopolymer. Verma et al. (2022) obtained that the e-waste plastic decreases the geopolymer concrete's strength in the mix. However, it is feasible to substitute crushed aggregate by up to 5% without affecting the mechanical properties. The addition of e-waste extends the lifespan of geopolymer concrete by increasing the mix's workability from low to medium and lowering water absorption by roughly 7%. This makes the mixture for the concrete lighter than usual and increases durability. E-waste is used to partially replace fine aggregate in geopolymer concrete. According to their research, e-waste will replace 10%, 20%, and 30% of the sand. Alkaline liquids include sodium hydroxide and sodium silicate (Na_2SiO_3) solutions (12 M). 10% GGBS and 90% fly ash are used as binder ingredients. He investigated the relationship between E-waste and geopolymer concrete's compressive and tensile strength. It was found that the intensity quadrupled when 20% E-Waste was added to ordinary geopolymer concrete of M40 grade as suggested by Palomo et al. (1999). To increase the characteristic strength of geopolymer concrete, researchers evaluated the effects of replacing river sand with 10%, 20%, 30%, 40%, and 50% of E-waste and M-sand as a fine aggregate. The optimal replacement % was then determined. E-waste was used to replace some of the traditional coarse material in studies on the mechanical strength of cement concrete. E-waste was substituted for several types of standard concrete cubes at 10%, 15%, 20%, 25%, and 30%, with a water-cement ratio of 0.5 as suggested by Verma et al. (2022). Ahmad et. al (2022) showed in the study is to add various doses of Nano graphite platelets (NGPs) to concrete generated from e-waste in order to enhance the mechanical properties of recycled plastic coarse aggregate (PCA). It has been shown that producing manufactured PCA with acceptable mechanical characteristics requires replacing up to 25% of the E-waste with 5% NGPs. Additionally, the study's examination of durability characteristics showed an improvement. The resulting FA-GGBS geopolymer concrete's mechanical qualities were evaluated using steel slag coarse particles instead of natural granite aggregates 0%, 25%, 50%, 75%, and 100% of the time. The study has resulted in a slight reduction in mechanical intensity. The permeable void volume and water absorption in the Geopolymer concrete increased but remained below acceptable limits when FA-GGBS was used as the binder and steel slag as the coarse aggregate. Steel slag worked well when utilised as coarse aggregate in paving and structural applications as studied by Gour et al. (2022).

Park et al. (2004) suggested that the vast amount of global e-plastic cannot be recycled using the current recycling techniques. It has been determined. Finding a way to handle the plastic from e-waste is essential to preventing the production of micro-plastics that further harm the environment. Researchers have looked at using plastics made from e-waste as concrete aggregate replacement. Even though it often has worse mechanical properties than conventional concrete due to the nature of e-plastic, the use of e-plastic in concrete offers a potential solution to the vast volume of plastic from e-waste. Concrete that has E-wastes added to it for compressive strength was used for the experiment. The best outcomes are attained when E-wastes are utilised in place of 7.5% of the fine aggregate. However, the compressive strength of concrete begins to decline after this point. The concrete loses compressive strength over time when e-waste replaces fine particles by more than 15%. Gour et al. (2022) studied the different types of waste aggregate are used in place of natural aggregate. For example, bone China waste is used as a fine aggregate in various ratios. The characteristic strength is established based on how much Bone China is replaced. Fine bone China was used in place of fine aggregate to increase the concrete's compressive, split tensile, and flexure strengths. With recycled construction waste (RCA) as the coarse aggregate and bone China fine aggregate (BCA) as the fine aggregate, this study aims to create environmentally friendly concrete that can be utilised for medium-grade strength applications. When comparing the qualities of the concrete, both after it is freshened and when it has hardened, workability, density, compressive, split tensile, and flexural strength are employed. The current study uses experimental and statistical research to examine how RCA and BCA affect concrete performance. Using well-known regression models, three-dimensional response surface plots were made to investigate interactions between the different components and establish the appropriate mixing ratio.

Kong et al. (2010) showed that by reusing e-waste, we may reduce landfill trash and use aggregates and other natural resources less frequently, which could help us address several environmental problems. Efforts have been made worldwide to partially replace coarse aggregate in the construction sector with e-waste components. Substituting crushed E-waste materials for conventional concrete and other building components can reduce concrete manufacturing costs. Recycling is the most efficient approach to reducing the quantity of electronic waste, making ecologically friendly concrete, and protecting the environment from the impacts of pollution.

EXPERIMENTAL NEEDS

The lab test on the geopolymer concrete requires extensive testing and properties determination of the raw materials to be utilized. The ingredients for the geopolymer concrete are very different form the normal cement concrete and hence, each of these materials require proper and thorough testing before incorporating it into the concrete.

Cementitious Agents

Metakaolin, very rich in silicates and aluminates, is the primary material used as the main cementitious material in this experimental study. Also, every material has some limitation and drawbacks associated with them, and hence, metakaolin has very quick setting property, which hinders the smooth casting and placing of the prepared mix. The quick setting limitation of the metakaolin is counteracted by the addition of Fly Ash in the mix, secondary cementitious material. Fly Ash (Class F) is also a very rich source of aluminates and silicates. The materials are tested in the lab for the physical and chemical properties. The physical examination of the raw materials gives the color and the Pycnometer test used to determine the Specific gravity of the materials. Further, X-Ray Fluorescence spectroscopy (XRF) test is used to determine the oxide components of the materials. Calculated properties of these material are shown in "Table-1".

Liquids

The geopolymerization process doesn't begin until the alkaline solution's binder components are activated in the concrete. As an alkaline liquid, use NaOH solution, prepared by dissolved calculated amount of NaOH pellets into the water and Na_2SiO_3 solution.

Sand and Aggregates

The experiment first employed basaltic aggregates with a maximum of 20 mm size, crushed basaltic aggregates with a maximum of 10 mm size, e-waste plastic with a 10 mm maximum size, and locally accessible Narmada River fine aggregates. Table-2 displays these aggregates' physical features.

Table 1. Binder properties

Physical Properties			
S. No.	**Property**	**Metakaolin**	**Fly Ash**
1	Colour	White	Light Grey
2	Specific Gravity	2.23	2.53
Chemical Properties			
S. No.	Oxides	Metakaolin	Fly Ash
1	Silicon Oxide (SiO_2)	51.63	49.86
2	Aluminum Oxide (Al_2O_3)	43.32	29.61
3	Ferrous Oxide (Fe_2O_3)	2.40	10.97
4	Calcium Oxide (CaO)	1.79	3.24
5	Potassium Oxide (K_2O)	0.74	0.96
6	Magnesium Oxide (MgO)	0.71	1.23

Table 2. Aggregate properties

S. No.	Aggregate Type	Fineness Modulus	Water Absorption	Specific Gravity
1	Coarse	6.72	1.05%	2.65
2	Crushed	6.58	0.98%	2.58
3	E-Waste Plastic	6.23	0.12%	1.12
4	Narmada Sand	2.84	2.25%	2.62

TEST PROCEDURE

The authors performed many tests to develop a mixed design with excellent mechanical strength. They finally created the reference mix, which was used to create the test matrix for this study. The dry raw ingredients, which include e-waste plastic, coarse and crushed aggregates, fine aggregates, fly ash, and metakaolin, are combined in an electric concrete mixer for around five minutes. The dry raw materials are then progressively mixed with the alkaline liquid in the running mixer until a homogenous mixture is achieved. Before being moulded into common shapes, including 150 mm cubes, 150 mm diameter and 500 mm height cylindrical shapes, and 100 mm x 100 mm x 500 mm beam moulds, the geopolymer mixture is vibrated on a vibrating table. The samples are kept into the moulds for 24 hours to set and then cured for 7, 14, and 28 days at ambient room temperature after demoulded. All mixes' mechanical properties are evaluated per the normative IS-code rules. The traditional slump test is used to assess the workability of different mixes. Water absorption test is conducted for the samples after 28 days curing. Tests of compressive strength after 7 days and 28 days, as well as tests of splitting tensile and flexural strength after 28 days for various test parameter values, are among the mechanical parameters examined. The quantities of the materials used for the casting are shown in Table-3.

Table 3. Quantities of raw items for 1 m^3 M30 grade

S. No.	Items	Quantity
1	Metakaolin	215 kg
2	Fly Ash	145 kg
3	Sand	810 kg
4	Aggregate (10 mm)	(391 to 460) kg
5	Aggregate (20 mm)	690 kg
6	E-Waste Plastic Aggregate (10 mm)	(0 to 69) kg
7	NaOH Pellets	13 kg
8	Na_2SiO_3 Solution	95 Litres

RESULTS AND DISCUSSION

Water Absorption and Workability

Table 4 shows the workability of the geopolymer concrete mixes. Workability is low in the normal geopolymer concrete but increases proportionately to the EWP percentage increase when EWP is introduced to the mixture. EWP is primarily responsible for making blends more workable since it absorbs hardly any water from the mixture and floats on water due to its low density.

Any material utilised as an aggregate must consider its capacity to absorb water. As seen in Table 2, the EWP absorbs very little water due to the hydrophobic nature of plastic. According to the results in Table 5, which support the logic mentioned above, after 28 days, the mix's ability to absorb water from the cube decreases as the proportion of EWP increases.

Table 4. Workability test for different % EWP

Mix	% EWP as replacement of Crushed Agg.	Slump (mm)	IS Code Range	Remark
S1	0	44	0-50	Low
S2	2.5	52	50-100	Medium
S3	5	68	50-100	Medium
S4	7.5	72	50-100	Medium
S5	10	87	50-100	Medium
S6	12.5	93	50-100	Medium
S7	15	104	100-150	High

Table 5. Water absorption for different % EWP

Mix	% EWP as replacement of Crushed Agg.	W1 (kg)	W2 (kg)	% Water Absorption	% Drop
S1	0	8.324	8.738	4.97	-
S2	2.5	8.212	8.598	4.70	5.43
S3	5	8.032	8.384	4.38	11.87
S4	7.5	7.863	8.183	4.07	18.10
S5	10	7.678	7.983	3.97	20.12
S6	12.5	7.632	7.922	3.78	23.94
S7	15	7.553	7.813	3.45	30.58

Mechanical Strength

Effect of EWP Replacement on Compressive Strength

To partially replace crushed aggregates in the geopolymer concrete mix, varying amounts of EWP are used (maximum size 10 mm). The amount of E-waste plastic decreases the strength of the geopolymer concrete in the mix. However, it is possible to substitute crushed aggregate by up to 5% without significantly reducing the compressive strength. The 5% substitution, there was a 1.2% reduction in compressive strength. Table 6 and Figure 1 shows that the compressive strength of the concrete mix is dramatically reduced if EWP is utilised more than 5%.

Table 6. Test of compressive strength for various % EWP

Mix	% EWP as replacement of Crushed Agg.	3rd day (N/mm²)	7th day (N/mm²)	28th day (N/mm²)	% Drop
S1	0	17.98	28.96	35.75	-
S2	2.5	17.78	28.60	35.55	0.56
S3	5	17.56	28.23	35.32	1.20
S4	7.5	15.47	25.20	31.80	11.04
S5	10	14.39	23.96	29.85	16.50
S6	12.5	12.94	23.60	29.20	18.32
S7	15	10.08	23.24	28.56	20.12

Figure 1. Compressive strength graphs

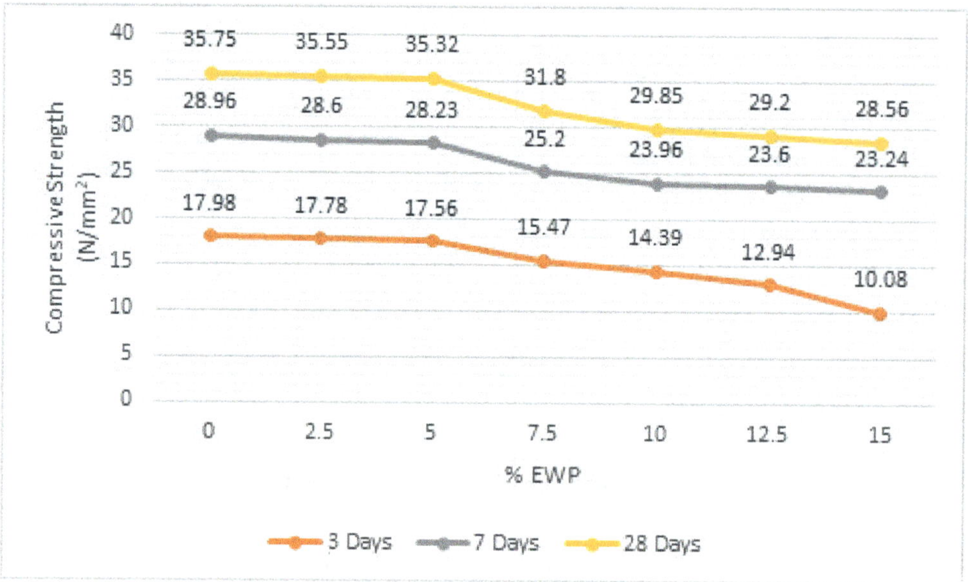

For construction engineers, an accurate prediction of compressive strength is crucial and valuable. A regression-based study is carried out, shown in Figure 2, to anticipate the compressive strength of concrete for various EWP percentages. Regression was used to create a linear curve from Fig. 2 to provide a generalised equation with higher R^2 values for compressive strength evaluation as shown in equation 1.

$F_c = -0.5677x + 36.548, R^2 = 0.9181$ (1)

Figure 2. Relationship between % EWP and compressive strength

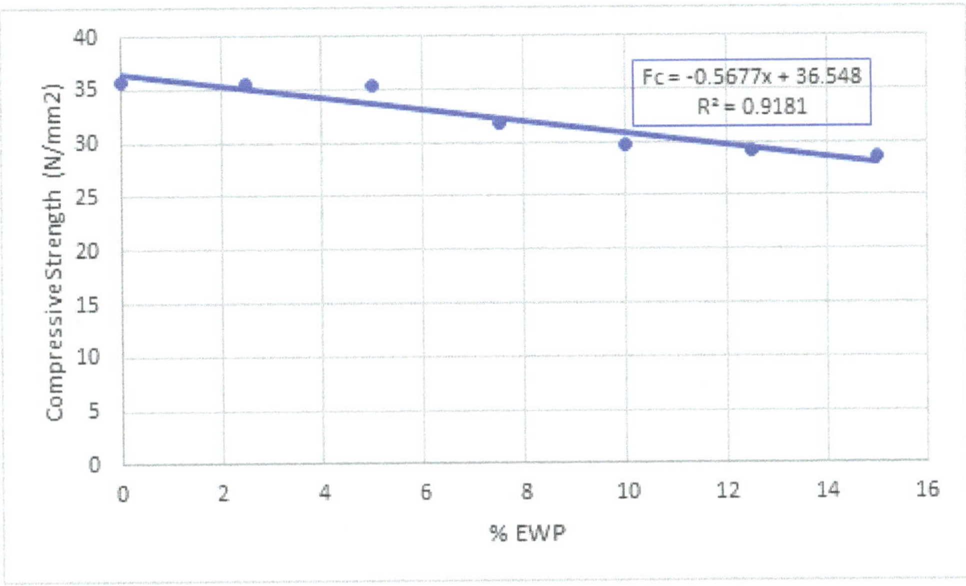

Where, F_c is compressive strength in N/mm^2, and x is the percentage replacement of EWP.

Effect of EWP Replacement on Split Tensile Strength

The geopolymer concrete's mix split tensile strength is decreased by the amount of plastic from e-waste. However, up to 5% can be used in its place without noticeably affecting the split tensile strength. The split tensile strength decreased by 6.1% with

Table 7. Split tensile strength test for different % EWP

Mix	% EWP as replacement of Crushed Agg.	3rd day (N/mm^2)	7th day (N/mm^2)	28th day (N/mm^2)	% Drop
S1	0	1.59	2.52	3.23	-
S2	2.5	1.52	2.46	3.05	5.56
S3	5	1.46	2.32	2.92	6.10
S4	7.5	1.32	2.23	2.75	9.60
S5	10	1.28	2.18	2.44	24.45
S6	12.5	1.18	2.05	2.32	28.17
S7	15	1.10	1.78	2.12	34.36

Figure 3. Relationship between compressive & split tensile strength

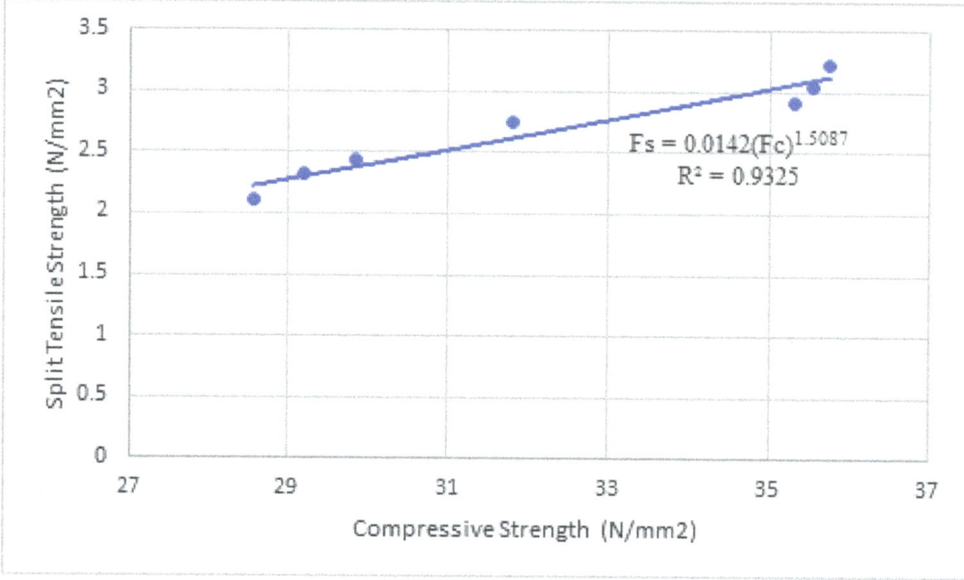

the 5% replacement. Table 7 shows that if EWP is used more than 5%, the concrete mix's split tensile strength is significantly lowered.

Relationship Between Compressive & Split Tensile Strength

The mathematical relationship between the experimental value of splitting tensile strength (F_s) and compressive strength (F_c), shown in Figure 3, of EWP was defined by linear regression analysis with a satisfactory coefficient of determination. The following is a representation of the experiential relationship:

$F_s = 0.0142(F_c)^{1.5087}$ (2)

where F_s is Split tensile strength in N/mm^2, and F_c is compressive strength in N/mm^2. This proposed relationship's Coefficient of determination (R^2) is 0.9325, indicating a strong correlation.

Effect of EWP Replacement on Flexural Strength

The quantity of e-waste plastic reduces the geopolymer concrete's mix flexural strength. The flexural strength won't be much impacted if you use up to 5% in its stead. With a replacement of 5%, the flexural strength dropped by 8.57%. The flexural strength of the concrete mix is dramatically reduced, as Table 8 demonstrates if EWP is employed at a rate of more than 5%.

Table 8. Flexural Strength Test for different % EWP

Mix	% EWP as replacement of Crushed Agg.	3rd day (N/mm²)	7th day (N/mm²)	28th day (N/mm²)	% Drop
S1	0	2.05	3.28	4.08	-
S2	2.5	1.84	2.95	3.96	2.94
S3	5	1.78	2.78	3.73	8.57
S4	7.5	1.63	2.63	3.25	20.34
S5	10	1.48	2.53	3.10	24.01
S6	12.5	1.39	2.35	2.92	28.43
S7	15	1.32	2.22	2.78	31.86

Relationship Between Compressive & Flexural Strength

The relationship between the compressive and flexural strengths of concrete is seen in "Figure-4" Regression analysis revealed the following empirical connection, which may be expressed:

$F_f = 0.0179(F_c)^{1.5092}$ (3)

where F_f is flexural strength in N/mm², and F_c is compressive strength in N/mm².

Figure 4. Relationship between compressive & flexural strength

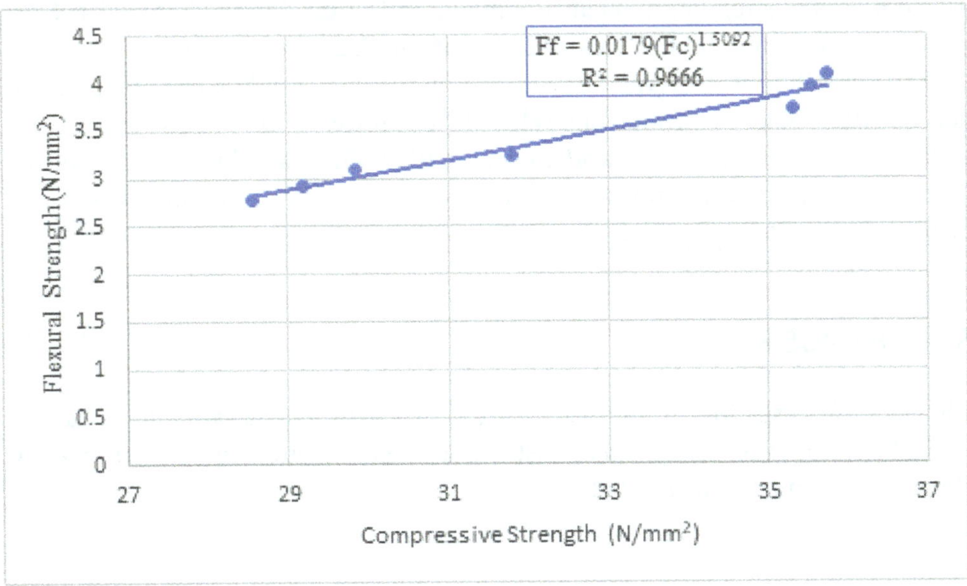

This proposed relationship's Coefficient of determination (R^2) is 0.9666, indicating a strong correlation.

CONCLUSION

This experimental study examined and evaluated the compressive strength, split tensile strength, and flexural strength of geopolymer concrete made from e-waste plastic and a metakaolin base. Also, the durability aspects in the form of workability check and water absorption percentage after 28 days of curing. The study's primary objective was to evaluate the role of geopolymer concrete in utilizing e-waste and convert it from a problem of society to a valuable product. The result of this experimental work proves that the e-waste plastics can be easily used as crushes aggregates in the geopolymer concrete and provide a very durable and mechanical stable mix. According to the study's findings, plastic from e-waste may replace crushed aggregate with up to 5% of the weight of the aggregate. Beyond 5% replacement, the durability parameters keep on increasing as they are directly dependant of e-waste plastic percentage, but the mechanical strengths were decreasing as the basaltic aggregates provide the strength to the mix and below 95% the strength starts reducing sharply.

The metakaolin geopolymer concrete with e-waste plastic has greater mechanical strengths than the standard geopolymer concrete mix. The 3-day strengths were way higher for e-waste geopolymer concrete as compared to normal geopolymer concrete and hence, makes it a very good alternates for the quick repair works and places where early strength gain is required, like road repairs, etc. The overall 28-day compressive strength, split tensile strength and flexural strength for EWP-GPC were higher than normal GPC about 20%, 34% and 32%, respectively.

The durability of the EWP-GPC is also on the higher side when compared to normal GPC. Water absorption of EWP-GPC was lower by about 12% for 5% EWP and 31% for 15% EWP. The workability of EWP-GPC was higher than normal GPC, around 25 mm slump value increased for 5% EWP and 60 mm slump value increased for 15% EWP.

REFERENCES

Ahmad, F., Qureshi, M. I., & Ahmad, Z. (2022). Influence of nano graphite platelets on the behavior of concrete with E-waste plastic coarse aggregates. *Construction & Building Materials*, *316*, 125980.

Alanazi, H., Yang, M., Zhang, D., & Gao, Z. (2017). Early strength and durability of metakaolin-based geopolymer concrete. *Magazine of Concrete Research*, *69*(1), 46–54. doi:10.1680/jmacr.16.00118

Anam, A. & Syed, A. (2013) Green Computing: E-waste management through recycling. *International Journal of Scientific & Engineering Research*, 1103.

Bavan, J. C., & Tandel, Y. (2015). A review on E-waste as construction material. *Indian Journal of Applied Research*, *5*(01), 14–16.

Damal, V. S., Londhe, S. S., & Mane, A. B. (2015). Utilization of electronic waste plastic in concrete. *International Journal of Engineering Research and Applications*, *5*(4), 38–38.

Dave, S. V., Bhogayata, A. C., & Arora, N. K. (2015). Utilization of plastic waste in Geopolymer concrete: State of the art review. *International Journal of Advance Engineering and Research Development*, *2*(12), 6–10.

Deepak, A. L., & Lakshmi, T. V. S. V. (2019). Mechanical properties of geopolymer concrete with flyash and metakaolin. *International Journal of Engineering and Advanced Technology*, *9*(1), 3863–3868. doi:10.35940/ijeat.A1180.109119

Gour, C. P., Dhurvey, P., & Shaik, N. (2022). Design of structural concrete with bone china fine aggregate using statistical approach. *Advances in Materials Science and Engineering*.

Gour, C. P., Dhurvey, P., & Shaik, N. (2022). Optimization and Prediction of Concrete with Recycled Coarse Aggregate and Bone China Fine Aggregate Using Response Surface Methodology. *Journal of Nanomaterials*.

Gupta, R., Puppala, H., & Rajesh, N. (2015). Application of Recycled Coarse Aggregates and E-Waste for Pavements with Low Traffic. *IOSR Journal of Mechanical and Civil Engineering (IOSR-JMCE)*, *12*(2), 64-70.

Kale, S. P., & Pathan, H. I. (2013). Recycling of Demolished Concrete and E-waste. *International Journal of Scientific Research (Ahmedabad, India)*.

Kong, D. L., & Sanjayan, J. G. (2010). Effect of elevated temperatures on geopolymer paste mortar and concrete. *Cement and Concrete Research*, *40*(2), 334–339.

Kumar, J. K., & Kumar, V. P. (2020). Soil stabilization using E-waste: A retrospective analysis. *Materials Today: Proceedings*, *22*, 691–693.

Manatkar, P. A., & Deshmukh, G. P.Pravin A. Manatkar. (2015). Use of Non-Metallic e-waste as a coarse aggregate in a concrete. *International Journal of Research in Engineering and Technology*, *4*(3), 242–246. doi:10.15623/ijret.2015.0403040

Nagajothi, P. G., & Felixkala, T. (2014). Compressive strength of concrete incorporated with E-fiber waste. *International Journal of Emerging Technology and Advanced Engineering*, *4*(4), 23–27.

Palomo, A., Blanco-Varela, M. T., Granizo, M. L., Puertas, F., Vazquez, T., & Grutzeck, M. W. (1999). Chemical stability of cementitious materials based on metakaolin. *Cement and Concrete Research*, *29*(7), 997–1004.

Park, S. B., & Lee, B. C. (2004). Studies on expansion properties in mortar containing waste glass and fibers. *Cement and Concrete Research*, *34*(7), 1145–1152.

Pouhet, R., & Cyr, M. (2016). Formulation and performance of flash metakaolin geopolymer concretes. *Construction & Building Materials*, *120*, 150–160. doi:10.1016/j.conbuildmat.2016.05.061

Prasanna, P. K., & Rao, M. K. (2014). Strength variations in concrete by using E-waste as coarse aggregate. *International Journal of Education and applied research*, *4*(2).

Sundar, M. L., & Raj, S. (2020). *Study on Characteristics of Geopolymer Concrete with E-Waste*. IOS Science.

Verma, P., Dhurvey, P., & Sundramurthy, V. P. (2022). Potential assessment of e-waste plastic in metakaolin based geopolymer using petrography image analysis. *Advances in Materials Science and Engineering*.

Verma, P., Dhurvey, P., & Sundramurthy, V. P. (2022). Structural behaviour of metakaolin geopolymer concrete wall-type abutments with connected wing walls. *Advances in Materials Science and Engineering*.

Xie, J., Chen, W., Wang, J., Fang, C., Zhang, B., & Liu, F. (2019). Coupling effects of recycled aggregate and GGBS/metakaolin on physicochemical properties of geopolymer concrete. *Construction & Building Materials*, *226*, 345–359. doi:10.1016/j.conbuildmat.2019.07.311

Zhang, H. Y., Qiu, G. H., Kodur, V., & Yuan, Z. S. (2020). Spalling behavior of metakaolin-fly ash based geopolymer concrete under elevated temperature exposure. *Cement and Concrete Composites*, *106*, 103483. doi:10.1016/j.cemconcomp.2019.103483

Chapter 13
E-Waste Management Toward Environmental Sustainability in Malaysia

Yusnita Yusof
Universiti Sultan Zainal Abidin, Malaysia

Wan Nor Azilawanie Tun Ismail
Universiti Sultan Zainal Abidin, Malaysia

Noor Aina Amirah Mohd Noor
Universiti Sultan Zainal Abidin, Malaysia

Mazriah Ayu Abu Bakar
Ministry of Environment and Water, Malaysia

ABSTRACT

Industrialized and developing countries today are plagued by the gravity of the problem of e-waste. Managing e-waste is a global environmental problem that has created significant environmental damage in some parts of the world. Although Malaysia has plans to manage e-waste, there are still difficulties, and the need to manage e-waste is now even more pressing. Therefore, to manage e-waste effectively, it is necessary to develop general, structured policy measures to address the issue within the nation and globally. This chapter aims to explore the sustainability issues in Malaysia, challenges towards e-waste management in Malaysia, an overview of e-waste policy implementation, the effects of e-waste on the environment, the activities organized by the government and other organizations, and the mechanism of e-waste management in Malaysia. By making diverse research contributions, it would eventually encourage the growth of sustainable e-waste management in Malaysia.

DOI: 10.4018/978-1-6684-7573-7.ch013

INTRODUCTION

Environmental issues are among the critical problems plaguing the world. Interest in environmental problems is more prevalent in developed countries than in developing countries, particularly regarding community awareness and government policies and regulations (Lita, Surya, Ma'ruf, & Syahrul, 2014).

Environmental management is an essential element in sustainable tourism and is often related to the environmental management system; the activities are undertaken to protect, enhance or reduce the environmental impact. In addition, in a broad sense, environmental management covers issues such as environmental impact (aesthetic, cultural, ecological, and social), sustainability, resource management, and pollution (Rocabella Mykonos Art Hotel & SPA, 2012). Meanwhile, green practices are an essential activator in an environmental management system used to reduce the organization's impact on the environment and make the operations more sustainable (Kim et al., 2017).

Green practices can be defined as the implementation of appropriate programs and activities undertaken and management's influence to reduce the negative environmental impact (Yusof, 2017). Among the criteria for green tourism are encouraging the use of green products and the provision of training programs "to control the introduction of environmental management and waste management techniques," such as recycling. One popular program widely used worldwide to reduce waste is a recycling program. However, the program is sometimes perceived to incur a high cost. Nevertheless, solid waste management is an efficient method to reduce the cost of waste hauling fees (Yusof, 2017; Pirani & Arafat, 2014) and simultaneously contributes to creating a green environment.

Green initiatives or environmental initiatives are a part of the programs adopted in many parts of the tourism sector (Siti-Nabiha et al., 2011). According to Tang (2015), Berezan et al. (2014), and Singh, Cranage, and Lee (2014), green initiatives are related to the recycling program, water-saving devices, and green landscaping. Industrialized and developing countries today are plagued by the gravity of the problem of electrical and electronic equipment waste (E-waste/WEEE). In general terms, WEEE refers to discarded electrical and electronic equipment parts with no potential for further use (Shad et al., 2020; Soo et al., 2013).

In Malaysia, e-waste is categorized as Scheduled Wastes under the Code SW110, First Schedule, Environmental Quality (Scheduled Wastes) Regulations 2005. E-waste is generally divided into two main types, according to its generation sources: e-waste generated from the industrial sector and household e-waste. "E-waste" is a broken, non-working, or old/obsolete electric electronic appliance such as a TV, PC, air conditioner, washing machine, or refrigerator. Code SW110 is defined as waste from electrical and electronic assemblies containing components such as accumulators, mercury switches, glass from cathode-ray tubes, and other activated glass or polychlorinated biphenyl-capacitors. Besides that, it is also contaminated with cadmium, mercury, lead, nickel, chromium, copper, lithium, silver, manganese, or polychlorinated biphenyl (Department of Environment, 2022a; Jayaraman et al., 2019).

According to research by the Department of Environment (2022b), Malaysia will generate 24.5 million units of e-waste in 2025, as illustrated in Figure 1. The high demand for electric and electrical equipment among Malaysian customers is the primary cause of this rise in e-waste. Malaysia is also an attractive country for illegally dumping e-waste (Al-Rahmi et al., 2018).

Figure 1. Forecast of e-waste generator in Malaysia until 2025

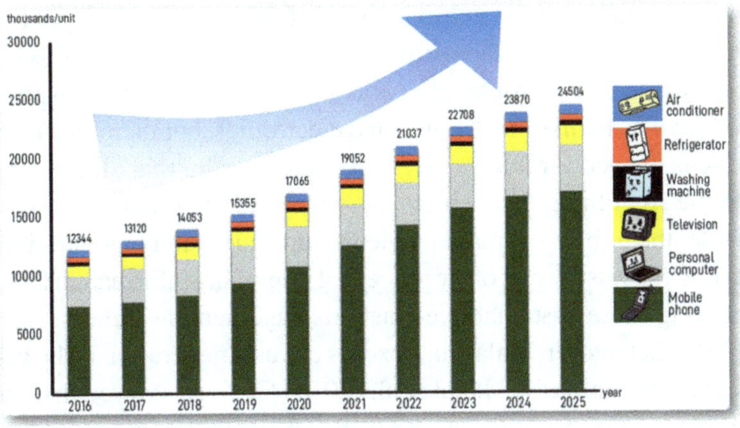

Managing the waste of electrical and electronic equipment (WEEE) is a global environmental problem that has created significant environmental damage in some parts of the world. E-waste must be appropriately disposed of or managed as more electrical and electronic equipment is completed. E-waste disposal requires a complex and different management system from daily solid waste management. E-waste risks human health and the environment due to hazardous materials like mercury and toxic additives. It will harm the brain and coordination system of humans if it is disposed of improperly, such as by sent to the river, landfill, burning, or sending to the informal sector (Shad et al., 2020; Alias et al., 2014; Department of Environment, 2013). Malaysia is paying attention to proper e-waste management. Effective and efficient e-waste management has vital economic and environmental significance for the sustainable development of a country.

Therefore, recycling is one of the best ways to eliminate e-waste. Technology advancements have made it easier to dispose of electronic waste. E-waste disposal is a specialty of many eco-friendly companies in Malaysia, and the procedure has been made easier for society's benefit (Al-Rahmi et al., 2018). Collecting household, institutional, and other e-waste sustainably for recycling is crucial for maintaining its ever-green environment. Stakeholders' involvement is significant in achieving Malaysia's sustainable e-waste recycling system (Shumon & Ahmed, 2013).

In Malaysia, two divisions of the formal and informal sectors are in charge of managing e-waste recycling. Licensed recovery facility that either fully or partially recover e-waste make up the legal sector. The conventional e-waste recovery facility frequently processes and extracts valuable components using less effective methods (Soo et al., 2013). For instance, the consumer or waste generator (citizen) is responsible for disposing of e-waste at the store, registered collectors, or authorized recovery facility to adopt sustainable e-waste management. Also, citizen is urged to send their household e-waste to any licensed recovery facilities or registered collection centers by the Department of Environment (Alias et al., 2014). Thus, e-waste recovery remains the best way to dispose of unwanted electronics, but increased regulation on a global scale is needed for it to succeed. We must ensure that all e-waste recovery facilities do the right thing for their environment and community.

Meanwhile, the public's actual knowledge, attitudes, and practices in managing e-waste are thus a necessary part of the process of improving information dissemination on implementing more sustainable e-waste management awareness. According to Soo et al. (2013), a factor in Malaysia's excess e-waste generation is the general lack of public attention. Similarly, Mahat et al. (2019) findings showed that the public's knowledge and attitude toward e-waste disposal awareness are good but still less practiced in everyday life. Previous research shows that e-waste management among the public is still low compared to other developed countries. Therefore, public awareness of e-waste issues should be increased through awareness campaigns and the deployment of education agents at an early stage, specifically in primary and secondary schools to develop more practical e-waste disposal practices and thereby sustain the environment. This study will contribute to the stakeholders such as national government, local government, industry sector, financial institutions, research institutions, citizens, and NGOs taking part in e-waste management, especially in policy restructuring.

LITERATURE REVIEW

Sustainability Issues In Malaysia

Since the Earth Summit in Rio de Janeiro in 1992, many people have been writing about "sustainability," This trend is still going strong today. Nevertheless, we have lost our way of measuring sustainability throughout this time. Brundtland (1987) proposes the following definition for sustainable development (SD) as "the development that meets the requirements of the present without compromising the ability of future generations to meet their needs." Adopting sustainable development practices is a requirement if we protect our planet from any changes that could negatively affect people and the natural world. The concept of sustainable development encompasses more than just the protection of the natural environment; it also takes into account aspects such as the quality of life, the equitable distribution of resources and benefits, the interactions between the environment and the development process, and planning for the future (Joseph, 2013).

In addition, the social, economic, and environmental elements are the three fundamental components that come together to form the concept of sustainable development. These three aspects are inseparably linked to one another and the quality of life. Everybody should give the highest priority regarding any actions carried out because they are of the utmost significance in forming the way of life of humanity and the highest importance. Because of the rapid changes that we are currently witnessing, which involve a significant trade-off between environmental sustainability and the extraction and utilization of natural resources, sustainable development has become an essential component in the policymaking process of every nation (Balakrishnan et al., 2020). These changes involve a significant trade-off between the sustainability of the environment and the extraction and utilization of natural resources—the results of the tremendous advancements we are presently witnessing in the industry. As a result of the enormous pressures that have been placed on both society and the environment, there is an urgent need for coordinated international action to bring about a more sustainable future (Michael et al., 2020).

As a result of the fact that most countries are currently facing similar socioeconomic and environmental challenges, there is a pressing requirement for the process of formulating global goals to incorporate both a greater degree of public scrutiny and fundamental shifts. In general, the economy is strongly dependent on the environment, to the extent that environmental challenges can lead to a country losing revenue from investment, resulting in a reduced standard of living for the country's population. The real issue with the environment is that advances in science and technology have made it easier for humans to take resources from the natural world, process them, and use them. However, these advancements did not provide comparable insights into how the resources can be returned to the environment, from which they were taken or how they can be used. It is essential to start a new cycle of taking resources from the natural world, processing them, and using them (Zul Ilham et al., 2021). A significant portion of the natural resources is mined in endeavors that are not environmentally friendly and ultimately become waste. If this continues unabated, we might completely deplete our natural resources. Because our environment is deteriorating at an alarming rate, we must adjust how we live and the activities we participate in daily. This fits in perfectly with the plan that was put forth by the United Nations, which is known as the Sustainable Development Goals (SDGs). There are six goals out of 17 goals in SDGs that relate to e-waste, such as Goal 3 (good health and well-being), Goal 6 (clean water and sanitation), Goal 8 (decent water and sanitation), Goal 11 (sustainable cities and communities), Goal 12 (responsible consumption and production) and Goal 14 (life below water).

The United Nations established the Sustainable Development Goals (SDGs) in 2015 to address the changes that are occurring on a global scale while also assuring better and more sustainable activities for the future. Malaysia is one of the countries in Asia that has worked to include Sustainable Development Goals (SDGs) in its national development strategy (Zul Ilham et al., 2021). These aims encompass all three facets of sustainability that were discussed earlier. It is expected that the goals of building environmentally friendly communities and providing inhabitants with high-quality services will be accomplished (Kanapathy et al., 2019). It is anticipated that Malaysia will catch up and enhance its economy due to improved communication and cutting-edge technologies.

There has been a significant amount of work done in Malaysia by the Ministry of Education to integrate discussions about sustainable development (SD) into all of the programs offered at tertiary institutions, either in a subject solely devoted to SD or in an issue that focuses on SD. Some of the higher education institutions in Malaysia provide courses and programs that speak specifically to the significance of sustainable development (SD) for the generations to come. The fact that higher education institutions in Malaysia have made these efforts demonstrates unequivocally that these institutions have attempted to address the problems associated with SD and stress the significance of ESD to the students (Balakrishnan et al., 2020). Before comprehending ESD's ideas, learners must have a foundational understanding of the natural sciences, the humanities, and the social sciences. Environmental, economic, and societal aspects are all included in ESD. Learners can behave in a manner acceptable to the environment once they have obtained the necessary information and abilities. Therefore, it is plausible to assert that issues on sustainable development, which include the social, economic, and environmental elements, have been discussed within the context of ESD in Malaysia (Azhar et al., 2022; Michael et al., 2020).

The findings of the study conducted by Balakrishnan et al. (2020) indicate that undergraduate students in Malaysia have positive perceptions and attitudes toward sustainable development, particularly in the context of environmental issues, except certain aspects of sustainable development that concentrate on economic and social equality. The findings of this analysis make it abundantly evident that undergraduate students in Malaysian higher education institutions received an appropriate education on sustainability issues thanks to the contribution made by the sustainable development education program. However, to realize the goals of Sustainable Development Goal 2030, all aspects of sustainable development must be taught to students studying at the undergraduate level at higher education institutions. As a result, the many stakeholders involved in the design and implementation of ESD should place an equal amount of emphasis on each component and issue of sustainability. Because the relevant information and abilities that span the environmental, social, and economic elements of SD serve as the basis for undergraduate students to acquire the core principles of SD, this is the case.

In the meantime, a significant number of studies have been carried out to investigate how learners' knowledge can influence the attitudes and actions they take concerning the idea of sustainable development. Wee et al. (2017) studied students enrolled in higher education institutions to determine their environmental awareness levels and attitudes toward sustainable development. The findings of this study indicate that respondents' awareness of the concept and issues of sustainable development was well developed; however, respondents had differing views regarding the semantics of sustainable development and what aspects it encompasses. According to their findings, there is only a shaky connection between environmental consciousness and environmentally responsible actions. There is little correlation between their awareness level and their effort to enhance their surroundings. Afroz and Ilham (2020) reported in their earlier research that university students have a high level of awareness regarding Sustainable Development Goals (SDGs). This awareness was demonstrated by a significant number of students who possessed high knowledge and a positive attitude. However, their degree of practice toward the SDGs is relatively poor. It revealed a minor negative association between student knowledge and practice, indicating that university students have a high level of expertise about Sustainable Development Goals (SDGs), but their practice level is slightly lower. This result is because the study revealed a minor negative association between student knowledge and practice. It demonstrates a tight connection between the environment and sustainable development, which is utilized in the broad perspective and the general growth of human people. In addition, every human activity that contributes to a nation's economic development or meets society's requirements will affect the natural environment in some way or another. Environmental sustainability is a prerequisite for social sustainability, even though it is a fundamentally important goal in and of itself. In addition, the transition to environmentally sustainable practices is necessary since the deterioration of global life-support systems, which the environment is comprised of, sets a time constraint (Berglund et al., 2019).

Students are being encouraged to participate in raising awareness of the concept of sustainability, which is becoming an increasingly important issue on a global and international scale. Higher Education Institutions (HEIs) should be modeled after the role of community centers in terms of their efforts to promote sustainability. As a result, higher education institutions in Malaysia should emphasize advancing policy and promoting education for sustainable development in their curricula and activities on campus. These institutions should also work to empower and mobilize young people to accelerate progress toward sustainable development in a manner that is consistent with the Global Action Programme (GAP) and Educational for Sustainable Development (Ariffin & Ng, 2020). In addition, higher education institutions in Malaysia have the potential to play a pivotal role in training, assisting, and preparing younger generations to make contributions to sustainable development that are both necessary and meaningful.

In recent years, the awareness of sustainability and associated issues have been at the forefront of people's interest worldwide. In the 21st century, a growing human population and rising consumption levels per capita will contribute to the depletion of natural resources and increase trash production. As a result, a worldwide focus on sustainability knowledge and practices has encouraged several countries to lessen the severity of irreversible disasters caused by behaviors that are not sustainable and to stop the disasters altogether. Despite this, many emerging nations have not yet developed a sustainability consciousness about conserving and maintaining environmental resources. Because of this, they know sustainability is essential in assuring sustainability initiatives as part of a worldwide effort to benefit future generations. This is especially true in nations like Malaysia, which have rising economies.

Challenges Towards E-Waste Management In Malaysia

The problem of electronic waste is increasingly being recognized on a global scale. Electrical and electronic equipment production increases the quantity of e-waste that must be disposed of or managed appropriately. Suppose electronic waste is disposed of in an environmentally unsound manner, such as by throwing it into a river or landfill, setting it on fire, or sending it to the informal sector. In that case, it can endanger our lives, impact human health, and decrease the overall quality of the environment.

The waste management techniques used in this e-waste recycling are disorganized (Yong et al., 2019). According to Kang et al. (2020), although several industrialized countries use current household e-waste management strategies, such as extended producer responsibility (EPR) schemes, household e-waste legislative frameworks are not yet fully applied in Malaysia. Malaysia has ratified the Basel Convention, and many procedures have been put in place to control the processing and disposal of WEEE, also known as e-waste. However, the rate of e-waste generation in Malaysia is not well understood. An inventory would help obtain a perspective to determine further measures necessary to ensure environmentally sound management of e-waste. This would be accomplished through inventory management software (Shad et al., 2020a).

The problem of e-waste is one of the most rapidly growing problems worldwide. Due to the rapid electronic growth, new technologies substitute older ones resulting in the disposal of vast amounts of e-waste in landfills, ignoring the negative impact this has on the environment. In the recent 20 years, it has been witnessed that newer technology known as the "smart" generations with smart functions is causing previous electronic devices to get old and obsolete. This is true in the case of many electronic devices being replaced with better designs (Soo et al., 2013). According to the Department of Environment (2022a), it is estimated that Malaysia will produce 24.5 million units of electronic waste in the year 2025. The rising demand for the manufacture of electric and electrical equipment from customers in Malaysia is the primary factor contributing to the rise in the amount of e-waste produced. In addition, Malaysia is a desirable destination for the unregulated dumping of e-waste. Because it is located smack dab in the heart of worldwide e-waste trade route 23, Malaysia's geographical position plays a significant part in this situation. Unfortunately, the proliferation of informal recycling practices has increased the likelihood of substantial negative effects on the environment and the public's health (Al-Rahmi et al., 2018).

Furthermore, the low level of public awareness has contributed to the overflow of e-waste produced in Malaysia. According to the study's findings, the percentage of consumers responsibly dispose of their e-waste is still relatively low (only 5% - they either send it to a recycler or return it to the manufacturer). In contrast, more than half of Malaysian consumers disposed of their e-waste by either throwing it away (30%) or putting it away (27%). The lack of an efficient e-waste collection system is the primary contributor to the challenges associated with the management of e-waste from households in Malaysia, even though there have been few studies conducted on the attitudes of Malaysian consumers toward the recycling of electronic waste (Ismail & Hanafiah, 2019). Ismail & Hanafiah (2021) studies showed that direct incineration with energy recovery was the optimum solution for e-waste management. However, according to Jayaraman et al. (2019), the act is punishable by law. Plastic materials manufacture electric electrical devices such as laptops, calculators, and personal computer parts. Injurious dioxins, brominated flame-retardants, and furans are released during the informal and public burning recycling process of plastic e-waste.

Suppose electronic waste is not disposed of in an environmentally responsible manner. In that case, the toxic and hazardous materials contained in its components, such as mercury, lead, cadmium, arsenic, bromine, and beryllium, will seep into the soil and, as a result, water sources, posing a risk to both human and aquatic life. Pan and Li (2016) have mentioned that the disposal of heavy metals in landfills significantly negatively influences the environment. E-waste consists of various components, some of which contain precious metals, including gold, copper, palladium, and silver, each of which has a high recycling value. Because of the adverse effects of improperly managed e-waste disposal activities on human health and the environment, e-waste pollution has emerged as one of the most pressing concerns at the international level. Electronic waste that is not treated correctly poses a risk of cancer and damage to the kidneys and the respiratory system. According to Shad et al. (2020), the absence of a formal framework for disposing of and dismantling WEEE results in an increase in the number of health problems experienced by the community. This carelessness is hurting the functioning of the thyroids and lungs, contributing to the rise in temperature and the risk of premature birth and spontaneous abortion in women.

Therefore, Yong et al. (2019) suggested that the DOE and The National Solid Waste Management Department (JPSPN) can formally collaborate to create mandatory e-waste collection by state and local authorities and to send the sorted e-waste fraction to recovery facilities licensed by the DOE. The funds gathered from EEE importers, manufacturers, and consumers could be used to support e-waste management initiatives. For instance, the current recovery facilities could receive subsidies for transportation and for operating more "ecologically" in terms of recovery. With the mandatory requirement that all e-waste must be delivered to DOE-approved licensed recovery facilities, effective engagement of NGOs may also be vital. Transportation subsidies may also be granted to NGOs that collect e-waste, particularly domestic e-waste in Malaysia.

Overview of E-Waste Policy Implementation

E-waste regulations have been enacted around the world, including in Malaysia. E-waste management in Malaysia has only recently started, in 2005 (Ismail & Hanaðah, 2019. In Malaysia, e-waste is categorized as scheduled waste with code SW110 under the First Schedule, Environmental Quality Regulations (Scheduled Waste) 2005, Environmental Quality Act 1974, enforced by the Department of Environment and the Ministry of Natural Resources, Environment and Climate Change (MNRECC).

A study by Andeobu et al. (2021) indicates that although Malaysia has plans to manage e-waste, there are still difficulties, and the need to manage e-waste is now even more pressing. According to Lehmann (2011), Malaysia has emerged as one of the top importers of e-waste from wealthy nations. The findings showed that despite establishing numerous material recovery facilities (MFR), the country still has serious problems handling the growing volume of e-waste produced (Andeobu et al.,2021).

Therefore, to manage e-waste effectively, it is necessary to develop general, structured policy measures to address the issue within the nation and globally. Incorporating country-specific concerns, the structured policies are expected to provide official infrastructure and processes for the recycling, management, and disposal of e-waste (Andeobu et al., 2021). The local government councils are critical stakeholders in the management and recycling process. Thus policymakers must comprehend the factors that influence e-waste collection and disposal and the costs involved.

In addition, to improve the policy implementation of e-waste disposal, Andeobe et al. (2021) has provided a few suggestions. Due to the diverse socioeconomic nature of growing economies, it is crucial to constantly review and examine current policies to spot flaws and potential areas for development. Australia has also discovered the efficacy of this method. Interdisciplinary research methodologies must also be taken into account when implementing e-waste policies. This would make it easier for policymakers to comprehend and handle the many health and environmental issues related to handling e-waste. Finally, Andeobe et al. (2021) believe that the policy approaches of individual countries must be local and specific, taking into account that nation's inherent socio-cultural, economic, political, and environmental concerns. This is necessary to address the chronic and challenging e-waste issues.

Effects of E-Waste On The Environment

The disposal of e-waste can bring significant effects on the environment. The uncontrolled dumping of e-waste harms the ecosystem because the poisonous metals and components in the equipment contaminate the air, land, and water we drink. As shown in Figure 2, the environmental effect of e-waste disposal can be categorized into four aspects.

Global Warming

Improper e-waste management can contribute to global warming. Global warming refers to the gradual warming of the Earth's surface that has occurred since the pre-industrial era (between 1850 and 1900) and is attributed to human activity, specifically the combustion of fossil fuels, which increases the levels of heat-trapping greenhouse gases in the atmosphere. Greenhouse gases are released when

Figure 2. Environmental Effects of e-Waste

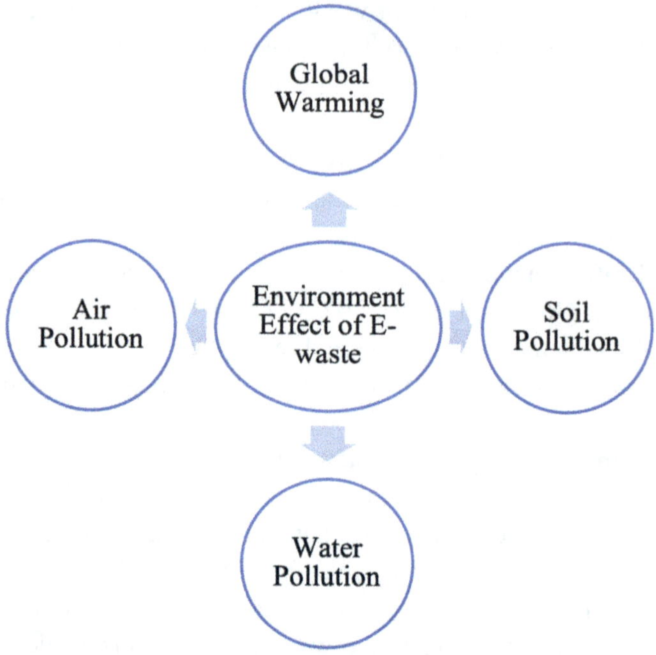

raw materials are extracted and cleaned from e-waste. In this procedure, greenhouse gases, CFCs from refrigerants, and other temperature exchange devices raise the earth's temperature. CFCs weaken the ozone layer, the earth's natural barrier against dangerous UV radiation, allowing them to penetrate. UV exposure can cause many conditions, including cataracts, weakened immune systems, and cancerous skin growth. According to Forti et al. (2020), 98 million metric tonnes of carbon dioxide equivalents were emitted into the environment due to inefficient management of discarded cooling and freezing units.

Soil Pollution

Landfilling, a common practice in many nations, is one of the most popular techniques for e-waste disposal. Improper e-waste disposal reduces soil fertility because it contains few chemicals and toxic materials that are not decomposable for a long time. Misra et. al, (2021). These hazardous substances can also contaminate groundwater. Poisonous gases are released into the environment during the incineration of e-waste (Mihai,2016). On the flat areas during landfilling, pits are created. Waste is buried in dirt excavated from trenches and covered with a substantial layer of soil. The degradation processes in landfills are complex and take a long time.

Townsend et al. (2004) tested 12 electronic devices (CPUs, computer monitors, laptops, TV sets, printers, VCRs, cellular phones, remote controls, etc.). The findings found that lead concentrations in leachates exceeded the regulatory limit of 5 mg/L in many cases, and every device type leached lead above this level in at least one test. The authors concluded that the findings provided sufficient evidence that discarded electronic devices containing a color CRT or printer wiring boards with lead-bearing solder may be hazardous waste for lead.

Besides, the heavy metal such as mercury, lead, and cadmium in e-waste may permeate into the soil, thus causing contamination that leads to soil infertility and disruption to plants and growth. When e-waste is improperly disposed of in regular landfills or illegally dumped, heavy metals and flame retardants can seep directly into the soil, contaminating underlying groundwater or crops planted nearby or in the area in the future. When heavy metals contaminate the soil, crops become vulnerable to absorbing these toxins, which can cause various illnesses and prevent farmland from being as productive as possible. Thus, the disposal of e-waste in landfill should be reduced, and another alternative should be taken.

Water Pollution

As discussed earlier, soil contamination might leak through the earth even further to reach groundwater. When these heavy metals go into groundwater, they may cause toxins infiltration into the river, lakes, ponds, and streams. Consequently, it leads to acidification and toxification of the water. The acidification and toxification created in water are unsafe for humans, animals, and plants. Acidification has the potential to kill both marine and freshwater organisms, disrupt biodiversity, and harm ecosystems. Clear drinking water also becomes problematic to find.

Air Pollution

When e-waste is informally disposed of by dismantling, shredding, or melting the materials, dust particles or toxins, such as dioxins, are released into the environment, causing air pollution and harming respiratory health. E-waste of little value is frequently burned, but burning also serves as a method of extracting valuable metals from electronics, such as copper. Chronic diseases and cancers are more likely to occur when e-waste is burned because it emits fine particles that can travel thousands of miles, posing numerous health risks to humans and animals.

Contamination in the air occurs when e-waste is informally disposed of by dismantling, shredding, or melting the materials, releasing dust particles or toxins, such as dioxins, into the environment that cause air pollution and respiratory damage to health. E-waste of little value is often burned, but burning also serves to get valuable metal from electronics, like copper. Chronic diseases and cancers are at a higher risk when burning e-waste because it releases fine particles, which can travel thousands of miles, creating numerous harmful health risks to humans and animals. Higher-value materials, such as gold and silver, are frequently extracted from highly integrated electronics using acids, desoldering, and other chemicals, which emit fumes in areas where recycling is not adequately regulated. The adverse effects of informal e-waste recycling on the environment are most dangerous for those who handle the waste. Still, pollution can travel thousands of miles away from recycling sites. The air pollution caused by e-waste has a more significant impact on some animal species than others, which may endanger these species and the biodiversity of certain chronically polluted regions. Air pollution can harm water quality, soil, and plant species over time, causing irreversible damage to ecosystems.

E-Waste Activities In Malaysia

Singh and Chawla (2017) stated that no proper methods are being implemented, even in developed countries, to eliminate the problem of e-waste. However, a few general techniques exist to reduce this issue, such as recycling and refurbishing.

Many activities are organized regarding e-waste in Malaysia, especially by the Department of Environment (DOE) Ministry of Natural Resources, Environment and Climate Change (MNRECC), as illustrated in Figure 3. Among them are:

1) The Electrical and Electronic Waste (E-Waste) Collection Competition was organized on 16 June 2022 at the School Level in Negeri Sembilan by The Department of Environmental and Department of Education, Negeri Sembilan, and Cenviro Sdn. Bhd.
2) The Webinar on the Electrical & Electronic Waste (E-Waste) Collection Campaign in Schools was organized by Sekolah Kebangsaan Sultan Yussuf on 15 April 2022 through the Google Meet and Youtube platforms. Encik Mohamad Zamzani Bin Ibrahim, Senior Assistant Director, Department of Environment, Perak, delivered the lecture to more than 200 teachers, students, and parents in Sekolah Kebangsaan Sultan Yussuf, as well as school teachers from all over Perak.
3) Site visits by Jabatan Alam Sekitar (JAS) Negeri Sembilan with the organizer IJM Land Berhad and the agencies, i.e., SWM Environment Sdn Bhd and SW Corp. Malaysia, were held in the housing area around Seremban 2 for the e-waste promotion and collection program. This program is in conjunction with the Negeri Sembilan World Environment Day 2022 on 5 June 2022.
4) The environmental awareness exhibition by Hospital Aurelius Nilai at SJK (C) Kuo Min collaborates with the Department of Environment Negeri Sembilan.

This program aims to raise awareness of the conservation and preservation of the environment. The show was held on 1 June 2022, focusing on the e-waste collection campaign among school children in managing e-waste correctly and according to the appropriate method.

5) The Department of Environment Negeri Terengganu, with the support of KT Wan Recycle Sdn. Bhd held the campaign, namely "Collection of Households Electrical and Electronic Waste (e-waste)," The collection of e-waste that was successfully accumulated in 4 hours, starting at 9:00 am to 1:00 pm, totaled 1 ton. This collaboration between these organizations successfully and the community gives their full cooperation to this program. This result shows that society has become concerned about the risks of e-waste if it is not managed accordingly to the proper and safe method and disposed of in the wrong place, such as a public waste disposal site.
6) The Department of Environment organized E-Waste Campaign called "Jom E-Waste Carnival." This carnival involves several artists, such as Fahrin Ahmad and Bell Ngasri. This program is in conjunction with International E-Waste Day, which is celebrated on October 14 every year (Ibrahim, 2022). The involvement of celebrities over the past year has directly helped deliver information related to e-waste to the public through the social media platform.

Based on the programs or activities are organized to show that the government is taking seriously educating the community in e-waste management. However, the actions still need more, and only some places are in this program. Besides that, these informal sector activities can be linked up with the formal sector. Everybody should take part in reducing these issues starting from home. Thus, stakeholders such as the government sector or NGO can train the community on e-waste handling, especially in rural areas. Tax reductions can be applied to the manufacturer or importer dealing with the program. Other than that, incentive schemes can be given to garbage collectors and waste generators.

Mechanism Of E-Waste Management In Malaysia

The e-waste management mechanism is narrowed to three (3) main flows, as illustrated in Figure 4:

E-Waste Flow

Define the accurate e-waste channel, starting from e-waste creation from the consumer/ waste generator to the registered collection center and ending up at the licensed recovery facility or final disposal.

Figure 3. Programs and activities organized by the department of environment

Deposit and Rebate Flow

Decide on implementing deposit and rebate schemes whereby manufacturers and

Figure 4. Framework of e-waste management mechanism in Malaysia

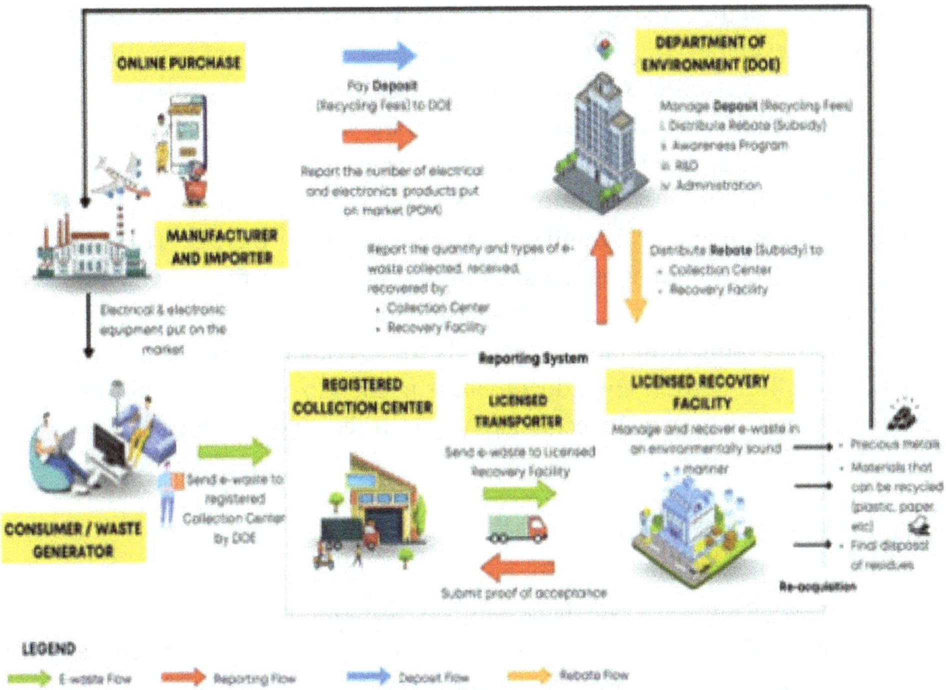

importers will pay deposits (fees) for electrical and electronic equipment placed in the Malaysian market. The payment collected will be paid out as rebates (subsidy) to recovery facilities and collection centers that have managed and handled e-waste environmentally soundly. The accumulated deposits also will be used for administration finance, research and development, and awareness activities.

Reporting Flow

Determine the flow of information through accurate reporting by the relevant stakeholder (manufacturers, importers, collection centers, and licensed recovery facilities). This reporting flow will ensure the flow of e-waste is within the mechanism limit from their creator to the final receiving points and decide the deposits and rebates to be collected and disbursed to the relevant stakeholder.

To strengthen e-waste management in Malaysia, DOE already provides an outline of the involvement of the stakeholders, which is from the consumer, retailer, collector, manufacturer, and recovery facility, as illustrated in Figure 5. DOE (2022) defines consumers as all individuals or entities where household e-wastes are produced and abandoned into the system. These include an individual, households, commercials, and institutions. At the same time, collection centers/collectors refer to some present players with recyclable consumers/dealers and local government recycling centers. Besides that, the collector also includes concessionaire company's collection centers, collection by charity organizations, NGOs, schools or institutions, producers' take-back initiatives, communities and residents associations (RA), etc.

Meanwhile, the retailers refer to the premises where new household electrical and electronic appliances (E-appliances) are sold to the buyers. The household e-wastes shall be collected upon request by the generators. A recovery facility is the primary treatment point for household e-wastes with proper pollution control, safety measures, and recycling in an environmentally sound manner authorized and licensed by DOE. A well-organized legal document is needed to allocate responsibility for each stakeholder with a solid mechanism (Jang, Rhee, & Kim, 2021).

CONCLUSION

In conclusion, managing e-waste successfully is essential and will affect a sustainable environment in Malaysia. There is proof that many activities and programs are created to manage e-waste. However, the events are more concentrated in urban areas than rural areas. Community understanding in urban areas is higher compared to rural communities. Therefore, activities must be increased in rural areas regularly. Raising public awareness is a significant factor in reducing environmental damage and improving recycling procedures. Public awareness is the factor that affects the system for a long-term reduction of waste produced in Malaysia. Therefore, it is advised that government assistance and involvement be required to properly utilize Malaysian academics and specialists in the study and development of effective e-waste management systems in Malaysia rather than solely relying on limited foreign cooperation. By making diverse research contributions, it would eventually encourage the growth of sustainable e-waste management in Malaysia.

Figure 5. Responsibilities of stakeholders in e-waste management

REFERENCES

Afroz, N., & Ilham, Z. (2020). Assessment of university students' knowledge, attitude and practice towards sustainable development goals (SDGs). *The Journal of Indonesia Sustainable Development Planning*, *1*(1), 31–44. doi:10.46456/jisdep.v1i1.12

Al-Rahmi, W., Alias, N., Dahawi, A. A., Yahaya, N., Saleh, M., & Al-Maatouk, Q. (2018). Strategies to manage electronic waste approaches: An overview in Malaysia. [UAE]. *IACSIT International Journal of Engineering and Technology*, *7*(4), 3276–3279. doi:10.14419/ijet.v7i4.15171

Alias, A.-F., Ishak, M. B., Zulkifli, S. N. A. M., & Jalil, R. A. (2014). E-waste management: An emerging global crisis and the Malaysian scenario. *International Journal of Environmental Sciences*, *4*(4), 444. doi:10.6088/ijes.2014040400001

Alias, A.-F., Ishak, M. B., Zulkifli, S. N. A. M., & Jalil, R. A. (2014). E-waste management: An emerging global crisis and the Malaysian scenario. *International Journal of Environmental Sciences*, *4*(4), 444. doi:10.6088/ijes.2014040400001

Andeobu, L., Wibowo, S., & Grandhi, S. (2021). A systematic review of e-waste generation and environmental management of Asia Pacific countries. *International Journal of Environmental Research and Public Health*, *18*(17), 9051. doi:10.3390/ijerph18179051 PMID:34501640

Ariffin, F. N., & Ng, T. F. (2020). Understanding and opinion on sustainable development among youths in higher educational institutions in Penang, Malaysia. *Social Indicators Research*, *147*(2), 421–437. doi:10.100711205-019-02165-1

Azhar, S. N. F. S., Akib, N. A. M., Sibly, S., & Mohd, S. (2022). Students' attitude and perception towards sustainability: The case of Universiti Sains Malaysia. *Sustainability (Switzerland)*, *14*(7), 3925. Advance online publication. doi:10.3390u14073925

Balakrishnan, B., Tochinai, F., & Kanemitsu, H. (2020). Perceptions and attitudes towards sustainable development among Malaysian undergraduates. *International Journal of Higher Education*, *19*(1), 44–51. doi:10.5430/ijhe.v9n1p44

Berglund, T., Gericke, N., Pauw, J. B., Olsson, D., & Chang, T. (2019). A cross - cultural comparative study of sustainability consciousness between students in Taiwan and Sweden. *Environment, De*, 1–27.

Department of Environment. (2013). *The Project for Model Development for E-waste Collection, Segregation and Transportation from Households for Recycling*. Issue March.

Ilham, Z. (2021). Youth awareness level towards Sustainable Development Goals (SDGs) in Greater Kuala Lumpur. *The Journal of Indonesia Sustainable Development Planning*, *2*(3), 217–233. doi:10.46456/jisdep.v2i3.173

Ismail, H., & Hanafiah, M. M. (2019). Discovering opportunities to meet the challenges of an effective waste electrical and electronic equipment recycling system in Malaysia. *Journal of Cleaner Production*, *238*, 117927. doi:10.1016/j.jclepro.2019.117927

Jayaraman, K., Vejayon, S., Raman, S., & Mostafiz, I. (2019). The proposed e-waste management model from the conviction of individual laptop disposal practices-An empirical study in Malaysia. *Journal of Cleaner Production*, *208*, 688–696. doi:10.1016/j.jclepro.2018.10.125

Joseph, C. (2013). Understanding sustainable development concept in Malaysia. *Social Responsibility Journal*, *9*(3), 441–453. doi:10.1108/SRJ-03-2012-0024

Kanapathy, S., Lee, K. E., Sivapalan, S., Mokhtar, M., Syed Zakaria, S. Z., & Mohd Zahidi, A. (2019). Sustainable development concept in the chemistry curriculum: An exploration of foundation students' perspective. *International Journal of Sustainability in Higher Education*, *20*(1), 2–22. doi:10.1108/IJSHE-04-2018-0069

Michael, F. L., Sumilan, H., Bandar, N. F. A., Hamidi, H., Jonathan, V., & Md Nor, N. N. (2020). Sustainable development concept awareness among students in higher education: A preliminary study. *Journal of Sustainability Science and Management*, *15*(7), 113–122. doi:10.46754/jssm.2020.10.011

Shad, K. M., Tan, Y. L., & Karim, M. E. (2020a). Sustainable e-waste management in Malaysia: Lessons from selected countries. *IIUM Law Journal*, *28*(2), 415–447. doi:10.31436/iiumlj.v28i2.517

Shad, K. M., Tan, Y. L., & Karim, M. E. (2020b). Sustainable E-Waste Management in Malaysia: Lessons From Selected Countries. *IIUM Law Journal*, *28*(2), 415–447. doi:10.31436/iiumlj.v28i2.517

Shumon, M. R. H., & Ahmed, S. (2013). Sustainable WEE management in Malaysia: Present scenarios and future perspectives. *IOP Conference Series. Materials Science and Engineering*, *50*, 012066. Advance online publication. doi:10.1088/1757-899X/50/1/012066

Singh, P., & Chawla, M. (2017). E-waste management-Suggested solutions. [IJAERS]. *International Journal of Advanced Engineering Research and Science*, *4*(1), 112–115. doi:10.22161/ijaers.4.1.17

Soo, V. K., Featherston, C., & Doolan, M. (2013). E-waste Assessment in Malaysia. *International Conference on Life Cycle Engineering*, 389–395. 10.1007/978-981-4451-48-2_64

Wee, M. I., Ariffin, F. N., Ng, T. F., & Shabudin, A. F. A. (2017). *Awareness and attitudes towards sustainable development amongst higher education students in Penang*. World Sustainability Series., doi:10.1007/978-3-319-47877-7_4

Compilation of References

Abalansa, S., El Mahrad, B., Icely, J., & Newton, A. (2021). Electronic waste, an Environmental Probl*em Exported to Developing Cou*ntr*ies*: The GOOD, the BAD and the UGLY. Sustainability, 13(9), 5302. doi:10.3390u13095302

Abdelbasir, S. M., Hassan, S. S. M., Kamel, A. H., & El-Nasr, R. S. (2018). Status *of electronic waste recycling tech*ni*ques*: A review. Environmental Science and Pollution Research Internat*ional, 25(17), 16533–16547. doi:10.100711356-018-2136-6 PMID:29737485*

*About Electronics Product Stewardship Canada. (*2022). Electronics Product Stewardship Canada. https://epsc.ca/about-epsc/

Adejumo, O., & Oluduro, O. F. (2021). Ticking Time Bomb: Implications of the *Covid-19 Lockdown on E-Waste Management in Developing Countries. UCL Press., doi:10.14324/111.444/000065.v1*

Afroz, N., & Ilham, Z. (2020). Assessment of university students' knowledge, attitude and practice towards sustainable development goals (SDGs). The Jou*rnal of Indonesia Sustainable Develop*ment Planning, 1(1), 31–44. doi:10.46456/jisdep.v1i1.12

Agarwal, R. (2012). E-Waste Law: New Paradigm or Business as Usual? Economic and Political Weekly, 47(25), 14–16.

Agrawal, A., & *Mukti, S. K. (2020). Knowledge Management & It's Origin, Success Factors, Planning, Tools,* Applications, Barriers and Enablers: A Review. [IJKM]. International Journal of Knowledge Management, 16(1), 43–82. doi:10.4018/IJKM.2020010103

Agrawal, A., Diwakar, H. K., & *Mukti, S. K. (2020). Employee Produ*ct*ivi*ty and Service Quality Enhancement Using ERP With Knowledge Management: A Study in a Power Distribution Company in the Global Context [IJKM]. I*nternational Journal of Knowledge Management, 16(3), 89–111. doi:10.4018/IJKM.202007*0106

Agrawal, A., Kumar, C., & Mukti, S. K. (2021). Role of Information and Communication Technology (ICT) to Enhance the Success of Knowledge Management (KM): A Study in a Steel Plant. Journal of the Knowledge Economy, 12(4), 1760–178*6. doi:10.100713132-020-00694-6*

Agrawal, A. and Mukti, S.K. (2019) 'Knowledge Management to Minimize the Marketing Challenges using Strategic Management as a Tool'.

Ahirwar R, & Tripathi AK (2021) E-waste management: A review of recycling process, environmental and occupational health hazards, and potential solutions. Environ. Nanotechnology, *Monit. Manag. 15. doi:10.1016/j.en*mm.2020.100409

Ahluwalia, P. K., & Nema, A. K. (2006). Optimal planning for computer waste. In: Proceedings of workshop on efficient e-waste management jointly organized by T E R I. The Institute of Waste Management a*nd Contaminated Site Tr*eatment, DUT.

Ahmad, F., Qureshi, M. I., & Ahmad, Z. (2022). Influence of nano graphite platelets on the behavior of concrete with E-waste plastic coarse aggregates. Co*nstruction & Building Materials, 316,* 125980.

Al-Rahmi, W., Alias, N., Dahawi, A. A., Yahaya, N., Saleh, M., & Al-Maatouk, Q. (2018). Strategies to manage electronic waste approaches: An overview in Malaysia. [UAE]. IAC*SIT International Journal of Engineering an*d Technology, 7(4), 3276–3279. doi:10.14419/ijet.v7i4.15171

Alanazi, H., Yang, M., Zh*ang, D., & Gao, Z. (2017). Early strength and durability of metakaolin-based geopolymer concrete. Magazine of Concre*te Research, 69(1), 46–54. doi:10.1680/jmacr.16.00118

Alavi, M., & Leidner, D. E. (2001). Review: Knowledge Management and Knowledge Management Systems: Conceptual Foundations and Research Issues. M*anagement Info*rma*t*ion Systems Quarterly, 25(1), 107. doi:10.2307/3250961

Alias, A. F., Ishak, M. B., Zulkifli, S. N. A. M., & Jalil, R. A. (2014). E-waste management: An emerging global *crisis and the Malaysian scenario. International Journal* of Environmental Sciences, 4(4), 444–457. https://www.ipublishing.co.in/ijesarticles/fourteen/articles/volfour/EIJES410*39.pdf*

Ali H, Khan E, & Ilahi I (2019) Environmental chemistry and ecotoxicology of hazardous heavy metals*: Environmental persistence, toxicity, and bioaccumulation. J. Chem. 2019(Cd): doi:10.1155/2019/6730305*

Almulhim, A. I. (2022). Household's awareness and participation in sustainable electronic waste management practices in Saudi Arabia. Ain Shams Engineering Journal, 13(4), 101729. doi:10.1*016/j.asej.2022.101729*

*Alshurideh, M., Kurdi, B. A., Alz*oub*i*, H. M., Obeidat, B., Hamadneh, S., & Ahmad, A. (2022). 'The influence of supply chain partners' integration*s on organizational performan*ce*:* The moderating role of trust'. Uncertain Supply Chain Management, 10(4), 1191–1202. doi:10.5267/j.uscm.2022.8.009

Anam, A. & Syed, A. (2013) Green Computing: E-waste management through recycling. International Journal of Scientific & Engineering Research, 1103.

Andeobu, L., Wibowo, S., & Grandhi, S. (2021). A Systematic Review of E-Waste Generation and Envi*ronmental Management of Asia Pacific Countrie*s. *I*nternational Journal of Environmental Research and Public Health, 18(17), 9051. doi:10.3390/ijerph18179051 PMID:34501640

Andrade, M., Soares, A. M., Solé, M., Pereira, E., & Freitas, R. (2022). Will climate changes enhance *the impacts of e-waste in aquat*ic systems? Chemosphere, 288, 132264. doi:10.1016/j.chemosphere.2021.132264 PMID:34624793

Animesh, A., & Mukti, S. K. (2019). Case study of critical success factors affecting knowledge management in small-and m*edium-sized enterprises in developing state:* Steel sector. In Advances in industrial and production engineering (pp. 825–831). Springer.

Animesh, A., & Mukti, S. K. (2019). Case Study of Critical Success Factors Affecting Knowledge Management i*n Small-and Medium-Sized Enterprises in D*eveloping State: Steel Sector. In Advances in Industrial and Production Engineering, 825–31. Springer. doi:10.1007/978-981-13-6412-9_76

Ankit, S. L., Saha, L., Kumar, V., Tiwari, J., Sw*eta, Rawat, S., Singh, J., & Baud*dh, *K.* (2021). Electronic waste and their leachates impact on human health and environment: Global ecological threat and management. Enviro*n. Technol. Innov., 24, 102049. d*oi:10.1016/j.eti.2021.102049

Arena, U., Mastellone, M. L., & Perugini, F. (2003). The environmental performance of alternative solid waste management options: A life cycle assessment *study. Chemical Engineering Journal, 96*(3*),* 207–222. doi:10.1016/j.cej.2003.08.019

Ariffin, F. N., & Ng, T. F. (2020). Understanding and opinion on sustainable development among youths in higher educational institutions in Penang, M*alaysia. Soci*al Indicators Research, 147(2), 421–437. doi:10.100711205-019-02165-1

Arshadi, M., Yaghmaei, S., & Mousavi, S. M. (2018). Content evaluation of different waste PCBs to enhance bas*ic metals recycling. Resources, Conservation an*d Recycling, 139, 298–306. doi:10.1016/j.resconrec.2018.08.013

Arya, S., & Kumar, S. (2020). E-waste in India at a glance: Current trends, regulations, challenges and management strategies. Journal of Cleaner Production, 271, 122707. do*i:10.1016/j.jclepro.2020.1227*07

Asamoah, A., Essumang, D. K., Muff, J., Kucheryavskiy, S. V., & Søgaard, E. G. (2018). Assessment of PCBs and exposure risk to infants in breast milk of primiparae and multiparae mothers in an electronic waste hot s*pot and non-hot spot areas in Ghana. The Science of the T*otal Environment, 612, 1473–1479. doi:10.1016/j.scitotenv.2017.08.177 PMID:28903176

Askari, A., & Ghadimzadeh, A. Chandima Gomes, & M.D. Bakri Ishak. (2014). E-Waste Management: Towards an Appropriate Policy. European Journal of Business and Management, 6(1)*, 37–46. http://psasir.upm.edu.my/*id/eprint/37547/

Asokan, V. A., Teah, H. Y., Kawazu, E., & Hotta, Y. (2023). Ambitious EV policy expedites the e-waste and socio-en*vironmental impacts in India. Resources, Conservation and* Recycling, 190, 106829. doi:10.1016/j.resconrec.2022.106829

ASSOCHAM-NEC. (2018). India among the Top ve Countries in E-Waste Generation, 1–4. ASSOCHAM-N*EC Study.*

Atasu, A., & Subramanian, R. (2012). Extended Producer Responsibility for E-Waste: Individual or Collective Producer Responsibility? Production and Operations Management, 21(6), 1042–1059. doi:10.1111/j.1937-5956.2012.01327.x

Atlason, R. S., Giacalone, D., & *Parajuly, K.* (2017). Product design in the circular economy: Users' perception of end-of-life scenarios for electrical and electronic appliances. Journal of Cleaner Production, 168, 1059–1069. doi:10.1016/j.jclepro.2017.09.082

Awasthi, A. K., Zeng, X., & Li, J. (2016). Comparative Examining and Analysis of E-waste Recycling in Typical Developing and Developed Countries. Procedia Environmental Sciences, 35, 676–680. doi:10.1016/j.proenv.2016.07.065

Azhar, S. N. F. S., Akib, N. A. M., Sibly, S., & Mohd, S. (2022). Students' attitude and *perception towards sustainability: The case of Un*iversiti Sains Malaysia. Sustainability (Switzerland), 14(7), 3925. Advance online publication. doi:10.3390u14073925

Balakrishnan, B., Tochinai, F., & Kanemitsu, H. (2020). Perceptions and attitudes towards sustainable development among Mala*ysian undergraduates. In*ternational Journal of Higher Education, 19(1), 44–51. doi:10.5430/ijhe.v9n1p44

Baldé, C. P., D'Angelo, E., Luda, V., Deubzer, O., & Kuehr, R. (2022). Global Transboundary E-waste Flows Monitor - 2022. *United Nations Institute f*or Training and Research. UNITAR.

Baldé, C. P., Forti, V., Gray, V., Kuehr, R., & Stegmann, P. (2017). The Global E-waste Monitor 2017: Quantities, Flows and Resources. United Na*tions Institute for Training and Research.* UNITAR.

Baldé, C. P., Wang, F., Kuehr, R., & Huisman, J. (2015). The global e-waste monitor – 2014. United Nations University, IAS – SCYCLE, Bonn, Germany.

Bandyopadhyay, A. *(2009). Electronics waste manage*ment: Indian practices and guidelines. International Journal of Energy and Environment, 1(5), 793–804. https://www.ijee.ieefoundation.org/

Barton, J. R., Dalley, D., & Patel, V. S. (1996). Life cycle assessment for waste management. J Waste Manage, 161(3), 35–50. doi:10.1016/S0956-053X(96)00057-*8*

Basel Secretariat. (2005). Basel Convention on the Control of Transboundary Movements of Hazardous Wastes and Their Disposal. Basel. http://www.basel.int/text/con-erev.doc

Bassiouny, A. M., Farhan, A. S., Maged, S. A., & Awaad, *M. I. (2021). Comparison of Different Comput*er Vision Approaches for e-waste Components Detection to Automate e-waste Disassembly. 2021 International Mobile Intelligent, and Ubiquitous Computing Conference, (pp. 17–23). IEEE. doi:10.1109/MIUCC*52538.2021.9447637*

*Batoo, K. M., Pan*diara*j*, S., Muthuramamoorthy, M., Raslan, E. H., & Krishnamoorthy, S. (202*1). Behavior-based swarm model using fuzzy controller* for route planning and Ewaste collection. Environmental Science and Pollution Research, 29(14), 19940 19954. doi:10.1007/s11356-021-12824-x

Bavan, J. C., & Tandel, Y. (2015). A review on E-waste as construction material. Indian Journal of Applied Research, 5(01), 14–16.

Behrens, A. (2013). Electronic Waste Management in Canada (Unpublished manuscript). University of Calgary, Calgary, AB. doi:10.11575/PRISM/30091

Berglund, T., Gericke, N., Pauw, J. B., Olsson, D., & Chang, T. (2019). A cross - cultural comparative study of sustainability consciousness between students in Taiwan and Sweden. Environment, De, 1–27.

Berkhout, F., & Howes, R. (1997). The adoption of life cycle approaches by industry: Patterns and impacts. Resources, Conservation and Recycling, 20(2), 71–94. doi:10.1016/S0921-3449(97)01199-3

Bevc, C. A., Marshall, B. K., & Picou, J. S. (2007). Environmental justice and toxic exposure: Toward a spatial model of physical health and psychological well-being. Social Science Research, 36(1), 48–67. doi:10.1016/j.ssresearch.2005.11.001

Bhide, A.V. and Akarte, M.M. (2022) Hybrid manufacturing-remanufacturing system of essential commodity in the context of COVID-19. Journal of Modelling in Management.

Bhutta, M. K. S., Omar, A., & Yang, X. (2011). Electronic Waste: A Growing Concern in Today's Environment. Economic Research International, 2011, 1–8. doi:10.1155/2011/474230

Bimir, M. N. (2020). Revisiting e-waste management practices in selected African countries. Journal of the Air &Amp. Waste Management Association, 70(7), 659–669. doi:10.1080/10962247.2020.1769769 PMID:32401631

Bisschop, L. (2016). How E-Waste Challenges Environmental Governance. In: Tanya Wyatt (eds) Hazardous Waste and Pollution, 27-43 . Springer, Cham. doi:10.1007/978-3-319-18081-6_3

Biswas, A., & Singh, S. G. (2020). E-waste Management in India: Challenges and Agenda. Centre for Science and Environment.

Blinová, L., & Godovčin, P. (2021). Importance of Recycling the Waste-Cables Containing Copper and PVC. Res. Pap. Fac. Mater. Sci. Technol. Slovak Univ. Technol., 29(48), 1–21. doi:10.2478/rput-2021-0001

Boopathi, S. (2021a). An investigation on gas emission concentration and relative emission rate of the near-dry wire-cut electrical discharge machining process. Environmental Science and Pollution Research International, 29(57), 86237–86246. doi:10.100711356-021-17658-1 PMID:34837614

Boopathi, S. (2021b). Improving of Green Sand-Mould Quality using Taguchi Technique. Journal of Engineering Research. doi:10.36909/jer.14079

Boopathi, S. (2022). Cryogenically treated and untreated stainless steel grade 317 in sustainable wire electrical discharge machining process: A comparative study. Environmental Science and Pollution Research International, 1–10. doi:10.100711356-022-22843-x PMID:36057706

Boopathi, S. (2022a). An Extensive Review on Sustainable Developments of Dry and Near-Dry Electrical Discharge *Machining Processes.* Journal of Manufacturing Science and Engineering, 144(5), 1–37. doi:10.1115/1.4052527

Boopathi, S. (2022b). Experimental investigation and multi-objective optimization of cryogenic Friction-stir-welding of AA2014 and AZ*31B alloys usin*g MOORA technique. Materials Today. Communications, 33, 104937. doi:10.1016/j.mtcomm.2022.104937

Boopathi, S., & Myilsamy, S. (2021). Mater*ial removal rate and surface rough*ness study on Near-dry wire electrical discharge Machining process. Materials Today: Proceedings, 45, 8149–815*6. doi:10.1016/j.matpr.2021.02.*267

*B*oopathi, S., & Sivakumar, K. (2013). Experimental investigation and parameter optimization of near-dry wire-cut electrical d*ischarge machining usin*g *multi-objective evolutionary* a*l*gorithm. International Journal of Advanced Manufacturing Technology, 67(9–12), 2639–2655. doi:10.100700170-012-4680-4

Boopathi, S., & Sivakumar, K. (2014). Study of water assisted dry wire-cut electrical discharge machining. Indian Journal of Engineering and Materials Sciences, *21(1), 75–82.*

*Boopathi, S., & Sivakumar, K. (20*16). Optimal parameter prediction of oxygen-mist near-dry Wire-cut EDM. International Journal of Manufacturing Technology and Management, 30(3–*4), 164–178. doi:10.1504/IJMTM.2016.077812*

Boopathi, S., B*al*asubramani, V., Kumar, R. S., & Singh, G. R. (2021). The influence of human hair on kenaf and Grewia fiber-based hybrid natur*al composite material: An expe*rimental study. Functional Composites and Structures, 3(4), 045011. doi:10.1088/2631-6331/ac3afc

Boopathi, S., Haribalaji, V., Mageswari, M., & Asif, M. M. (2022). Influe*nces of Boron Carbide Particles* on the Wear Rate and Tensile Strength of Aa2014 Surface Composite Fabricated By Friction-Stir Processing. Materiali in Tehnologije, 56(3), 263–270. doi:10.17222/mit.2022.409

Boopathi, S., Jeyakumar, M., Singh, G. R., King, F. L., Pandian, M., Subbiah, R., & Haribalaji, V. (2022). An experimental study on friction stir processing of aluminium alloy (AA-2024) and boron nitride (BNp) surface composite. Materials Today: Proceedings, 59, 1094–1099. doi:10.1016/j.matpr.2022.02.435

Boo*pathi, S., Kum*ar, P. K. S., Meena, R. S., Sudhakar, M., & Associates. (2023). Sustainable Developments of Modern Soil-Less Agro-Cultivation Systems: Aquaponic Culture. In Human Agro-Energy Optimization for Bus*iness and Industry (pp. 69–87). IGI Global.*

Boopathi, S., K*um*aresan, A., Manohar, N., & Krishna Moorthi, R. (2017). Review on Effect of Process Parameters-Friction Stir Welding Process. [IRJET]. International Research Journal of Engineering and Technology, 4(7), 272–278.

Boopathi, S., M*yilsamy, S., & Sukkasamy, S. (2021). Experimental Investig*ation and Multi-Objective Optimization of Cryogenically Cooled Near-Dry Wire-Cut EDM Using TOPSIS Technique. Ijamt Preprint. doi:10.21203/rs.3.rs-254117/v1

Boopathi, S., Sivakumar, K., & *Kalidas, R. (2012). Parametric Study of Dry* WE*DM* Using Taguchi Method. International Journal of Engineering Research and Development, 2(4), 1–6.

Boopathi, S., Venkatesan, G., & Anton Savio Lewise, K. (2023). Mechanical Properties Analysis of Kenaf–G*rewia–Hair Fiber-Reinforced Comp*o*si*te. In Lecture Notes in Mechanical Engineering (pp. 101–110). Springer. doi:10.1007/978-981-16-9057-0_11

Borthakur, A., & Govind, M. (2017). Emerging trends in consumers' E-waste disposal behaviour and awareness: A worldwide overvi*ew with special focus on India. Reso*urces, Conservation and Recycling, 117, 102–113. doi:10.1016/j.resconrec.2016.11.011

Borthakur, A., & Sinha, K. (2014). Electronic Waste Management in India: A Stakeholder's Perspective. Electronic Green Journal, 1(36). Advance online publication. doi:10.*5070/G313618041*

Cao, J., L*u,* B., Chen, Y., Zhang, X., Zhai, G., Zhou, G., Jiang, B., & Schnoor, J. L. (2016). Extended producer responsibility system in China improves e-waste recycling: Government policies, enterprise, and public awareness. Renewable & Sustainable Energy Reviews, 62, 882–894. doi*:10.1016/j.rser.2016.04.078*

C*ha*kraborty, P., Selvaraj, S., Nakamura, M., Prithiviraj, B., Cincinelli, A., & Bang, J. J. (2018). PCBs and PCDD/Fs in soil from informal e-waste recycling sites and open dumpsites in India: Levels, *congener profiles and health risk assessment. The Science of* the Total Environment, 621, 930–938. doi:10.1016/j.scitotenv.2017.11.083 PMID:29223911

Chakraborty, S. C., Qamruzzaman, M., Zaman, M. W. U., Alam, M. M., Hossain, M. D., Pramanik, B. K., Nguyen, L. N., Nghi*em, L. D., Ahmed, M. F., Zhou, J. L., Mondal, M. I. H., Ho*ss*ai*n, M. A., Johir, M. A. H., Ahmed, M. B., Sithi, J. A., Zargar, M., & Moni, M. A. (2022). Metals in e-waste: Occurrence, fate, impacts and remediation techn*ologies. Process Safety and Environmental Protection, 16*2, 230–252. doi:10.1016/j.psep.2022.04.011

Chen, H., Wang, L., Hu, B., Xu, J., & Liu, X. (2022). Potential driving for*ces and probabilistic health risks of heavy metal accumulation* in *th*e soils from an e-waste area, southeast China. Chemosphere, 289, 133182. doi:10.1016/j.chemosphere.2021.133182 PMID:34883131

Chen, L. F., & Yee, H. *W. (2011) E-waste management: Are we ready for it? A study on* the awareness of COIT students toward e-waste management. ICIMU 2011: Proceedings of the 5th international Conference on Information Technology & Multimedia, *(pp. 1-5). IEEE. 10.1109/ ICIMU.2011.61*22729

Chen, S. J., Tian, M., Zheng, J., Zhu, Z. C., Luo, Y., Luo, X. J., & Mai, B. X. (2014). Elevated levels of polychlorinated biphenyls in plants, air, and soils at an e-waste site in southern china and enantioselect*ive biotransformation of chiral PCBs* in *pl*ants. Environmental Science & Technology, 48(7), 3847–3855. doi:10.1021/es405632v PMID:24628413

Cherukuri, I., Sultana, N., & Podila, S. P. (2018). S*tatus Of E-Waste In Indi*a- A Review. J. Environ. Sci. Toxicol. Food Technol., 12(11), 8–16. doi:10.9790/2402-1211010816

Cooper, T. (1999) Creating an economic infrastructure for sustainable product design. Journal of sustainable product design, pp. 7–17.

Cooper, T. (2000). WEEE, WEEE, WEEE, WEEE, all the way *home? An evaluation of proposed elect*rical and electronic waste legislation. European Environment, 10(3), 121–130. doi:10.1002/1099-0976(200005/06)10:3<121::AID-EET226>3.0.CO;2-N

Coronavirus, W. H. O. (COVID-19) Dashboard. (n.d.). With Vaccination Data. https://covid19.who.int/?gclid=Cj0KCQiAwMP9BRCzARIsAPWTJ_Hjh ep5qv-AvH0RG2ssFAeoKmTIsrq-_W1wjzQSyh0gBXo1v1aK4r4aAgNwEALw_ wcB

Craighill, A. L., & Powell, J. C. (1996). Lifecycle assessment and economic evaluation of recycling: A case study. Resources, Conservation and Recycling, 17(2), 75–96. doi:10.1016/0921-3449(96)01105-6

Crutzen, P. J., & Stoermer, E. F. (2000). The "Anthropocene". The International Geosphere–Biosphere Program (IGBP): A Study of Global Change of the International Council for Science (ICSU) (Inte*rnational Geosphere-Biosphere Program). New*sletter, 41(May), 17–18.

Cunningham, P. B. K. (2020, November 17). New study highlights the rise in e-waste during global pandemic. Recycling Today. https://www.recyclingtoday.com/article/study-highlights-pandemic-drives-increase-e-waste/

Damal, V. S., Londhe, S. S., & Mane, A. B. (2015). Utilization of electronic waste plastic in concrete. International Journal of Engineering Research and Applications, 5(4), 38–38.

Danish, A., Mosaberpanah, M. A., Ozbakkaloglu, T., Salim, M. U., Khurshid, K., Bayram, M., Amran, M., *Fediuk, R., & Qa*der, D. N. (2023). A compendious review on the influence of e-waste aggregates on the properties of concrete. Case Studies in Construction Materials, 18, e01740. doi:10.1016/j.cscm.2022.e01740

Daum, K., Stoler, J., & Grant, R. (2017). Toward a More Sustainable Trajectory for E-Waste Policy: A Review of a Decade of E-Waste *Research in Accra, Ghana. Interna*tional Journal of Environmental Research and Public Health, 14(2), 135. doi:10.3390/ijerph14020135 PMID:28146075

Dave, S. V., Bhogayata, A. C., & *Arora, N. K. (2015). Utilization of pla*stic waste in Geopolymer concrete: State of the art review. International Journal of Advance Engineering and Research Development, 2(12), 6–10.

Davies, P., Hernandez, M.P. & Wyatt, T. (2019). Economy Versus Environment: How Corporate Actors Harm Both. Critical Criminology 27(1), 85–99. . doi:10.1007/s10612-019-*09433-z*

*Davis, G., & He*rat, S. (2008). Electronic waste: The local government perspective in Queensland, Australia. Resources, Conservation and Recycling, 52(8–9), 1031–1039. doi:10.1016/j.resconrec.2008.04.001

Debnath, B., Dey, R., & Roy, S. (2019) Smart Switching System Using Bluetooth Technology. Proc. – 2019 Amity Int. Conf. Artif. Intell. AICAI 2019 (pp. 760–763). IEEE. 10.1109/AICAI.2019.8701298

Debnath, B., Roychoudhuri, R., & Ghosh, S. K. (2016). E-Waste Management – A Potential *Route to Green Computing. Procedia Environmental Sciences,* 35, 669–675. doi:10.1016/j.proenv.2016.07.063

Deepak, A. L., & Lakshmi, T. V. S. V. (2019). Mechanical properties of geo*polymer concret*e with flyash and metakaolin. International Journal of Engineering and Advanced Technology, 9(1), 3863–3868. doi:10.35940/ijeat.A1180.109119

Denison, R. A. (1996). Environmental life-cycle comparis*ons of recycling, landfilling, and incineration: A review of* recent studies. Annual Review of Energy and the Environment, 21(1), 191–237. doi:10.1146/annurev.energy.21.1.191

Department of Environment. (2013). The Project for Model Development for E-waste Collection, Segregation and Transportation from Ho*useholds for Recycling. Issue March.*

Do*an*, L. T. T., Amer, Y., Lee, S. H., & Phuc, P. N. K. (2019). Strategies for E-waste management: A literature review. International Journal of Energy and Environmental Engineering, 13(3), 157–162.

Doma*konda, V. K., Farooq, S., Chinthamreddy, S., Puviarasi, R., Sudha*kar, M., & Boopathi, S. (2022). Sustainable Developments of Hybrid Floating Solar Power Plants: Photovoltaic System. In Human Agro-Energy Optimization for Business and Industry (pp. 148–16*7). IGI Global.*

*DTSC. (n.d.). Universal Waste. DTSC. https://dtsc.ca.*gov/universalwaste/

Edwin Cheng, T. C., Kamble, S. S., Belhadi, A., Ndubisi, N. O., Lai, K., & Kharat, M. G. (2022). Link*ages between big data an*alytics, circular economy, sustainable supply chain flexibility, and sustainable performance in manufacturing firms. International Journal of Production R*esearch, 60(22), 6908–6922. doi:10.10*80/00207543.2021.1906971

El, A. M. A., Jaber, M. Y., & Bonney, M. (2013). Int. J. Production Economics How Many Times to Remanufacture? Internationa*l Journal of Production Economics, 143(2), 598–604. doi*:10.1016/j.ijpe.2011.11.017

El-NakibI. (2012). Reverse Logistics: A Comparison of Electronic Waste Recycling Between Switzerland and Egypt. Social Science Researc*h Network. https://papers.ssrn.*com/sol3/Delivery.cfm/SSRN_ID2053344_code1603574.pdf?abstractid=2053344&mirid=1&type=2

Elleuch, B., Bouhamed, F., Elloussaief, M., & Jaghbir, M. (2018). Environmental *sustainability and pollution prevention. Environmental Scie*nce and Pollution Research International, 25(19), 18223–18225. doi:10.100711356-017-061*9-5 PMID:29971753*

*Environment Australia. (2001). Developing a product stewardship strategy: for electrical and electr*onic appliances in Australia – Discussion Paper. Commonwealth of Australia.

Environment Australia. (2001). Developing a product steward*ship strategy: for electrical and electronic appliances in Aust*ralia – Discussion Paper. Commonwealth of Australia.

Esenduran, G., Atasu, A., & Van Wassenhove, L. N. (2019). Valuable e-waste: Implications for extended producer responsibility. IISE Transactions, 51(4), 382–396. doi:10.1080/24725854.2018.1515515

European Commission. (2008). Directive 2008/98/EC of *the European Pa*rliament and of the Council of 19 November 2008 on Waste and Repealing Certain Directives. http://eurlex.europa.eu/legalcontent/EN/TXT/PDF/?uri=CELEX:32008L0098&fromEN

Fadaei, A. (in press). E-waste management status worldwide: Major challenges and solutions. International Academy of *Ecology and Environmental Sciences, 12(4)(22208860*). http://www.iaees.org/publications/journals/piaees/onlineversion.asp

Finnveden, G. (1999). Methodological aspects of life cycle assessment of integrated soli*d waste management systems. Resources, Conservatio*n and Recycling, 26(3-4), 173–187. doi:10.1016/S0921-3449(99)00005-1

Fishbein, B. K. (2002). End-of-life management of electronics abroad. Waste in the wireless world*: the challenge of cell phones, INFORM Inc., New York, htt*p://www.informinc.org

Forti, V., Balde, C. P., Kuehr, R., & Bel, G. (2020). The Global E-waste Monitor 2020: Quantities, flows and the circular economy potential.

Forti, V., Ba*ldé, C. P., Kuehr, R., & Bel, G.* (2020). The Global E-waste Monitor 2020: Quantities, flows and the circular economy potential. United Nations Institute for Training and Research (UNITAR). https://ewastemonitor.info/

Forti, V. B. C. (2020). The Global e-waste Monitor *2020: Quantities,flow*s and the circular economy potential. United Nations University (UNU)/United Nations Institute for Training and Research (UNITAR) – co-hosted SCYCLE Programme,International Telecommunication Union (ITU) & International Solid Waste Association (ISWA).

Frazzoli, C. R., Ruggieri, F., Battistini, B., Orisakwe, O. E., Igbo, J. K., & Bocca, B. (2022). E-WASTE threatens health: The *scientific solution adopts the one health strategy. Environmental* Research, 212, 113227. doi:10.1016/j.envres.2022.113227 PMID:35378120

Fu, J., Zhang, H., Zhang, A., & Jiang, G. (2018). E-waste Recycling in China: A Challenging Field. Environmental Science & Amp. Technology, *52(12), 67*27–6728. doi:10.1021/acs.est.8b02329 PMID:29862810

Fu, Q., Abdul Rahman, A. A., Jiang, H., Abbas, J., & Comite, U. (2022). Sustainable supply chain and business performance: The impact of strategy, network design, information systems, and organizational structure. Sustainability, 14(3), 1080. doi:10.3390u14031080

Gaidajis, G., Angelakoglou, K., & Aktsoglou, D. (2010). E-waste: Environmental Problems and Current Management. Journal of Engineering Science and Technology Review, 3(1), 193–199. doi:10.25103/jestr.031.32

Galbreth, M. R., & Blackburn, J. D. (2010). Optimal Acquisition Quantities in Remanufacturing with Condition Uncertainty. *Production and Operations Management, 19(1), 61–69. doi:10.*1111/j.1937-5956.2009.01067.x

Geng, Y., Sarkis, J., & Bleischwitz, R. (2019). How to globalize the circular economy. Nature Publishing Group. doi:10.1038/d41586-019-00017-z

Geng, Y., Sarkis, J., & Ulgiati, S. *(2016). Sustainabilit*y, *w*ell-being, and the circular economy in China and worldwide. Science, 6278(Supplement), 73–76.

Geng, Y., Sarkis, J., Ulgiati, S., & Zhang, P. (2013). Measuri*ng China's circular economy. Science, 339(6*127), 1526–1527. doi:10.1126cience.1227059 PMID:23539584

Goldstein, J. L. (2020). The Spanish 1918 Flu and the COVID-19 Disease: The Art of Remembering and Foreshadowing Pandemics. Cell, 183(2), 285–289. doi:10.1016/j.cell.2020.09.030 PMID:33064981

Gour, C. P., Dhurvey, P.*, & Shaik, N.* (*2*022). Design of structural concrete with bone china fine aggregate using statistical approach. Advances in Materials Science and Engineering.

Gou*r, C. P., Dhurvey, P., & Shaik, N. (2022). Optimizat*ion and Prediction of Concrete with Recycled Coarse Aggregate and Bone China Fine Aggregate Using Response Surface Methodology. Journal of Nanomaterials.

Grant, R., *& Oteng-Ababio, M. (2012). Mapping t*he *I*nvisible and Real "African" Economy: Urban E-Waste Circuitry. Urban Geography, 33(1), 1–21. doi:10*.2747/0272-3638.33.1.1*

Greenpeace. (2009). Where does e-waste end up? https://media.greenpeace.org/archive/Where-Does-E-Waste-End-Up----English-Version-27MZIF2D3F8X.html

Guide, V. Jr, & Daniel, R. (2000). Productio*n Plann*ing *an*d Control for Remanufacturing: Industry Practice and Research Needs. Journal of Operations Management, 18(4), 467*–483. doi:10.*1016/S0272-6963(00)00034-6

Gunasekaran, K., Boopathi, S., & Sureshkumar, M. (2022). Analysis of a Cryogenically Cooled Near-Dry Wedm Process Using Different Dielectrics. Materiali in *Teh*nolo*g*ije, 56(2), 179–186. doi:10.17222/mit.2022.397

Gupta, A., Singh, A., Singh, U., Singh, A., & Tomar, C. (2008). A Review Paper on Use of E-Waste in Concrete. International Research Journal of Eng*ineering and Technology, 2762. www.irjet.net*

Gupta, A., Singh, A., Singh, U., Singh, A., & Tomar, C. (2008) A Review Paper on Use of E-Waste in Concrete. Int. Res. J. Eng. Technol, 2762. www.irjet.net

Gupta, N., & Nath, M. (2020). *Groundwater Contaminati*on by E-Waste and Its Remedial Measure - A Literature Review. Journal of Physics: Conference Series, 1531(1), 01202*3. doi:10.1088/*1742-6596/1531/1/012023

Gupta, R., Puppala, H., & Rajesh, N. (2015). Application of Recycled Coarse Aggregates and E-Waste for Pavements with Low Traffic. IOSR Journal of Mechanical and Civil Engineering (IOSR-JMCE), 12(2), 64-70.

Halder, S., & Roy, S. (2015). Problem, Policies and Management of Electronic Waste in I*ndia: A Review. Coherence (Semin*ar *E*dition, Vol. 2). The Bhawanipur Education Society College.

Hanafi, J., Jobiliong, E., Christiani, A., Soenarta, D. C., Kurniawan, J., & Irawan, J. (2012). Material Recovery and Characterizati*on of PCB from Electroni*c *Wa*ste. Procedia: Social and Behavioral Sciences, 57, 331–338. doi:10.1016/j.sbspro.2012.09.1194

Haribalaji, V., Boopathi, S., & Asif, M. M. (2021). Opt*imization offriction stir weldi*ng process to join dissimilar AA2014 and AA7075 aluminum alloys. Materials Today: Proceedings, 50(xxxx), 2227–2234. doi:10.101*6/j.matpr.2021.09.499*

Haribalaji, V., Boopathi, S., Asif, M. M., Jeyakumar, M., Subbiah, R., & Lewise, K. A. S. (2021). Influences of Friction stir tool parameters for joining two similar AZ61A alloy plates. Mater*ials Today: Proceedings, 50(5), 2547–2553. doi:10.1016/j.matpr.*2021.12.074

Haribalaji, V., Venkatesan, G., Asif, M. M., Pandian, M., Subbiah, R., & Boopathi, S. (2022). Investigation on corro*sion and tensile Characteristics: Fri*ction stir welding of AA7075 and AA2014. Materials Today: Proceedings, 66, 743–748. doi:10.1016/j.matpr.2022.04.037

Harshini Vakkalanka. (2018, June 19). The A-Z of e-waste management. The Hindu. https://www.thehindu.com/sci-tech/energy-and-environment/what-about-e-waste/article24193081.ece

Herat, S., & Agamuthu, P. (2012). E-waste: A problem or an opportunity? Review of issues, challenges, and sol*utions in Asian countries. Waste Managem*ent &Amp; Research. The Journal for a Sustainable Circular Economy, 30(11), 1113–1129. doi:10.1177/0734242X12453378 PMID:22851536

Hopson, E., & Puckett, J. (2016). Scam recycling: E-dumping on Asia by *us recyclers. The E-Trash Transp*arency Project.

Hosoda, J., Ofosu-Anim, J., Sabi, E. B., Akita, L. G., Onwona-Agyeman, S., Yamashita, R., & Takada, H. (2014). Monitoring of organic micropollutants in Ghana by combination of pellet watch with sediment analysis: E-waste *as a source of PCBs. Marine* Pol*l*ution Bulletin, 86(1–2), 575–581. doi:10.1016/j.marpolbul.2014.06.008 PMID:24997873

Hosseini-Motlagh, S. M., Nematollahi, M., & Ebrahimi, S. (2020). Tri-party reverse supply chain coordination with competitive product acquisition process. The *Journal of the Operational* Re*se*arch Society, 73(2), 382–393. doi:10.1080/01605682.2020.1824550

Ilankoon, I., Ghorbani, Y., Chong, M. N., Herath, *G., Moy*o, T., & Petersen, J. (2018). E-waste in the international context – A review of trade flows, regulations, hazards, waste management strategies and technologies for value recovery. Waste Management (New York, N.Y.), 82, 258–275. doi:10.1016/j.wasman.2018.10.018 PMID:30*509588*

*Ilham, Z. (2021). Youth awareness leve*l *to*wards Sustainable Development Goals (SDGs) in Greater Kuala Lumpur. The Journal of Indonesia Su*stainable Development Planning, 2(3), 217–233. do*i:10.46456/jisdep.v2i3.173

Inderfurth, K. (2005). Impact of Uncertainties on Recovery Behavior in a Remanufacturing Environment. International Journal of Physical Distribution & Logistics Management, 35(5), 318–336. doi:10.1108/09600030510607328

Irving, M. (2022, August 21). G*raphene efficiently recov*ers gold from waste electronics. New Atlas. https://newatlas.com/technology/graphene-recovers-gold-waste-electronics/

Isernia, R., Passaro, R., Quinto, I., & Thomas, A. (2019). The Reverse Supply Chain of the E Waste Managem*ent Processes in a Circular Economy Framework:* Evi*d*ence from Italy. Sustainability, 11(8), 2430. doi:10.3390u11082430

Islam, M. T., Dias, P., & Huda, N. (2018). Comparison of E-waste management in Switzerland and in Australia: A qualitative content analysis. International Journal of Environmental and Ecological Engineering, 12(10), 610–*616.*

Islam, M. T., Dias, P., & Hud*a*, N. (2021). Young consumers' e-waste awareness, consumption, disposal, and recycling behavior: A case study of university students in Sydney, Australia. Journal of Clea*ner Production, 282, 124490. doi:10.1016/j.jclepro.2020.1*24490

Ismail, H., & Hanafiah, M. M. (2019). Discovering opportunities to meet the challenges of an effective waste electrical and electronic equipme*nt recycling system in Malaysia. Journal of Cleaner Production, 238,* 117927. doi:10.1016/j.jclepro.2019.117927

Janowski, T., Estevez, E., & Baguma, R. (2018). Platform governance for sustainable development*: Reshapi*ng citizen-administration relationships in the digital age. Government Information Quarterly, 35(4), S1–S16. doi:10.1016/j.giq.2018.09.002

Jayaraman, K., Vejayon, S., Raman, S., & Mosta, I. (2019). The proposed e-waste management model from the convic*tion of indiv*idu*al* laptop disposal practices-An empirical study in Malaysia. The Proposed E-Waste Management Model from the Conviction of Individual Laptop Disposal Practices-An Empirical Stud*y in Malaysia., 208, 688–696. doi:10.1016/j.jclepro.2018.10.125*

Je*ya*raj, P. (2021). Management of E-waste in India – Challenges and recommendations. World J. Adv. Res. Rev., 11(2), 193–218. doi:10.30574/wjarr.2021.11.2.0329

Joseph, C. (2013). Understanding sustainabl*e development concept in Malaysia.* Social Responsibility Journal, 9(3), 441–453. doi:10.1108/SRJ-03-2012-0024

Joseph, K. (2007). Electronic waste management in India–issues and strategies. Eleventh international waste management and landfill symposium, *Sardinia.*

Julander, A., Lundgren, L., Skare, L., Grandér, M., Palm, B., Vahter, M., & Lidén, C. (2014). Formal recycling of e-waste leads to increased exposure to toxic metals: AN occupational exposure study from Sweden. Environment Internati*onal, 73, 243–251. doi:10.1016/j.env*int.2014.07.006 PMID:25300751

Kahhat, R., & Williams, E. (2012). Materials flow analysis of e-waste: Domestic flows and exports of used computers from the United States. Resources, Conservation and Recycling, 67, 67–74. doi:10.1016*/j.resconrec.2012.07.008*

*Kahhat, R., Kim, J., Xu, M., Allenby, B., Williams, E., & Zhang, P. (2008). Exploring e-waste management sys*tems in the United States. Resources, Conservation and Recycling, 52(7), 955–964. doi:10.1016/j.resconrec.2008.03.002

Kale, S. P., *& Pathan, H. I. (2013).* Recycling of Demolished Concrete and E-waste. International Journal of Scientific Research (Ahmedabad, India).

Kalmykova, Y., Sada*gopan, M., & Rosado, L. (2018). Circular economy – From review* of theories and practices to development of implementation tools. Resources, Conservation *and Recycling, 135, 190–201.* do*i*:10.1016/j.resconrec.2017.10.034

Kamberovic, Z., Korac, M., & Ranitovic, M. (2011). Hydrometallurgical Process for Extraction of Metals from Eectronic Waste-Part II: Development of the Processes for the Recovery of Copper from Printed Circuit Boards (PCB*). Metalurgija-MJoM, 17(3), 1*39–149. http://www.metalurgija.org.rs/mjom/vol17/No3/4_Kamberovic_MJoM_1703.pdf

Kanapathy, S., Lee, K. E., Sivapalan, S., Mokhtar, M., Syed Zakaria, S. Z., & Mohd Zahidi, A. (2019). Sustainabl*e development concept in the chemistr*y curriculum: An exploration of foundation students' perspective. International Journal of Sustainability in Higher Education, 20(1), 2–22. doi:10.1108/IJSHE-04-2018-0069

Kannan, E., Trabelsi, Y., *Boopathi, S., & Alagesan, S. (2022). I*nfluences of cryogenically treated work material on near-dry wire-cut electrical discharge machining process. Surface Topography: *Metrology and Properties, 10(1), 015027. Advance online public*ation. doi:10.1088/2051-672X/ac53e1

Kerr, S., & Bryant, L. (2013). Design for Environment Report. Electronic Product Stewardship Canada. https://epsc*.ca/wpcontent/uploads/2022/03/2013_EPSC_De*sign_Env_Report.pdf

Khalid, F., Hashmi, M. Z., Jamil, N., Qadir, A., & Ali, M. I. (2021). Microbial and enzymatic degradation of PCBs from e-waste-contaminated sites: A review. Environmental Science and Pollution Research International, 28(9), 10474–10487. doi:10.100711*356-020-11996-2 PMID:*33411303

Khan, A. H., Tirth, V., Fawzy, M., Mahmoud, A. E. D., Khan, N. A., Ahmed, S., Ali, S. S., Akram, M., Hameed, L., Islam, S., Das, G., Roy, S., & Dehghani, M. H. (2021). COVID-19 transmission, vulnerability, persistence and nanotherapy: A review. Environmental Chemistry Letters, 19(4), 2773–2787. *doi:10.100710311-021-01229-4 PMID:33846683*

Kiddee, P., Naidu, R., & Wong, M. H. (2013). Electronic Waste Management Approaches. An Overview., 33(5), 1237–1250. doi:10.1016/j.wasman.2013.01.006 PMID:23402807

Kirchherr, J., Reike, D., & Hekkert, M. (2017). Conceptualizing the *circular economy: An analysis of 114 definitions.* Resources, Conservation and Recycling, 127, 221–232. doi:10.1016/j.resconrec.2017.09.005

Kirthika, S. K., Singh, S. K., & Chourasi*a, A. (2020). Alternative fine aggreg*ates in production of sustainable concrete- A review. Journal of Cleaner Production, 268, 122089. doi:10.1016/j.jclepro.2020.122089

Kojima, M., Yoshida, A., & Sasaki, S. (2009). Difficulties in applying extended producer responsibility *policies in developing countries: case studies in e-wast*e recycling in China and Thailand. Journal of Material Cycles and Waste Management, 11(3), 263 269. doi:10.1007/s10163-009-0240-x

Kong, D. L., & Sanjayan, J. G. (2010). Effect of elevated temperatures on geopolymer paste mortar and concrete. Cement and Concrete Research, 40(2), 334–339.

Koshta, N., Patra, S., *& Singh, S. P. (2022). Sharing e*conomic responsibility: Assessing end user's willingness to support E-waste reverse logistics for circular economy. Journal of Cleaner Production, *332, 130057. do*i:10.1016/j.jclepro.2021.130057

Krishna, K. (2009). Effective electronic waste management and recycling process involving formal and non-formal sectors. International Journal of Phy*sical Sciences, 4(13), 893–905.*

*Kuma*r, *A.*, & Dixit, G. (2018). An Analysis of Barriers Affecting the Implementation of E-Waste Management Practices in India : A Novel ISM-DEMATEL Approach. Sustainable Production and Consu*mption, 14, 36–52. doi:10.1016/j.s*pc.2018.01.002

Kumar, A., & Holuszko, M. (2016). Electronic Waste and Existing Processing Routes: A Canadian Perspective. Resources, 5(4), 35. doi:10.3390/resources5040035

Kumar, A., Luthra, S., Mangla, S. K., & Kazançoğlu, Y. (2020). COV*ID-19 impact on sustainable production and operatio*ns management. Sustainable Operations and Computers, 1, 1–7. doi:10.1016/j.susoc.2020.06.001

Kumar, D., Agrawal, R., & Sharma, V. (2015). Enablers for Compe*titiveness of Indian Manufac*tur*i*ng Sector : An ISM-Fuzzy MICMAC Analysis. Procedia: Social and Behavioral Sciences, 189, 416–432. doi:10.1016/j.sbspro.2015.03.200

Kumar, J. K., & Kumar, V. P. (2020). Soil stabilizatio*n using E-waste: A retrospect*ive *a*nalysis. Materials Today: Proceedings, 22, 691–693.

Kumar, P., Sampath, B., Kumar, S., Babu, B. H., & Ahalya, N. (2023). Hydroponics, Aeroponics, and Aquaponics Tech*nologies in Modern Agricultural Cultivatio*n. In Trends, Paradigms, and Advances in Mechatronics Engineering (pp. 223–241). IGI Global.

Kumar, U., & Singh, D. (2016). E–Waste Management through Regulations. International Journal of Engineering Inventions, 6–14(3(2)). https://www.researchgate.net/file.PostFileLoader.html?id=579a0881cbd5c2324e13a611&assetKey=AS%3A388818185277444%40146971 2887190

Kumara, V., Mohanaprakash, T. A., Fairooz, S., Jamal, K., Babu, T., & Sampath, B. (2023). Experimental Study on a Reliable Smart Hydroponics System. In Human Agro-Energy Optimization for Business and Industry (pp. 27–45). IGI Global. doi:10.4018/978-1-6684-4118-3.ch002

Kumari, R., & Samadder, S. R. (2022) A critical review of th*e pre-processing and metals recovery meth*ods from e-wastes, Journal of Environmental Management, 320. doi:10.1016/j.jenvman.2022.115887

Laan, V. D., Erwin, M. S., Dekker, R., & Van Wassenh*ove, L. (*1999). Inventory Control in Hybrid Systems with Remanufacturing. Management Science, 45(5), 733–747. doi:10.1287/mnsc.45.5.733

Law, J. (2010). Reality Failures. FESRC Centre *for Research on Socio-Cultural Chan*ge. http://heterogeneities.net/publications/Law2010RealityFailures5.pdf

Lebbie, T. S., Moyebi, O. D., Asante, K. A., Fobil, J., Brune-Drisse, M. N., Suk, W. A., Sly, P. D., Gorman, J., & Carpen*ter, D. O. (2021). E-Waste in Africa: A Serious Threat t*o the Health of Children. International Journal of Environmental Research and Public Health, 18(16), 8488. doi:10.3390/ijerph18168488 PMID:34444234

Lecler, M. T., Zimmermann, F., Silvente, E., Clerc, *F., Chollot, A., & Grosjean, J. (2015).* E*xpo*sure to hazardous substances in Cathode Ray Tube (CRT) recycling sites in France. Waste Management (New York, N.Y.), 39, 226–235. doi:10.1016/j.wasman.2015.02.027 PM*ID:25776743*

*Leclerc, S. H., & Bada*mi, *M.* G. (2020). Extended producer responsibility for E-waste management: Policy drivers and challenges. Journal of Cleaner Production, 25*1, 119657. doi:10.1016/j.jcl*epro.2019.119657

Lee, D., Offenhuber, D., Duarte, F., Biderman, A., & Ratti, C. (2018). Monitor: Tracking global routes of electronic waste. *Waste Management (*New York, N.Y.), 72, 362–370. doi:10.1016/j.wasman.2017.11.014 PMID:29137831

Lehtokunnas, T., Mattila, M., Närvänen, E., & Mesiranta, N. (2022). Towards a circular economy in food consumption: Food waste reduction practices as ethical work. Journal of Consumer Culture, 22(1), 227–245. doi:10.1177/1469540520926252

Li, J., Tian, B., Liu, T., Liu, H., Wen, X., & Honda, S. (2006). Status quo of e-waste *management in mainland China. Journal of Material Cycles and Wast*e *Ma*nagement, 8(1), 13–20. doi:10.100710163-005-0144-3

Lin, S. A., Ali, M. U., Zheng, C., Cai, Z., & Wong, M. H. (2022). Toxic chemicals from uncontrolled e-waste recycling: Exposure, b*ody burden, health impact. Jo*urnal of Hazardous Materials, 426, 127792. doi:10.1016/j.jhazmat.2021.127792 PMID:34802823

Lukinskiy, V., Lukinskiy, V., Ivanov, D., Sokolov, B., & Bazhina, D. (2023). A probabilistic approach to information management of order *fulfilment reliability with the* help of perfect-order analytics. International Journal of Information Management, 68, 102567. doi:10.1016/j.ijinfomgt.2022.102567

Maes, T., & Preston-Whyte, F. (2022). E-waste it wisely: *Lessons from Africa. SN Applied* Sciences, 4(3), 72. Advance online publication. doi:10.100742452-022-04962-9 PMID:35155992

Makeel, R. (2022). The influence of social supportive culture and performance-based culture on social enterprise performance: the media*tion role of Social entrepr*eneurial orientation. Journal of Global Entrepreneurship Research, 1–14.

Manatkar, P. A., & Deshmukh, G. P.Pravin A. Manatkar. (2015). Use of Non-Metallic e-waste a*s a coarse aggregate in a concrete. Internation*al Journal of Research in Engineering and Technology, 4(3), 242–246. doi:10.15623/ijret.2015.0403040

Manomaivibool, P. (2009). Extended producer responsibility in a non-OECD context: The management of *waste electrical and electron*ic equipment in India. Resources, Conservation and Recycling, 53(3), 136–144. doi:10.1016/j.resconrec.2008.10.003

McFadden, C. (2020, September 20). How Much Gold Is in Your Computer and How Efficient It Is to Reclaim It. Interesting Engineering: https://interesting*engineering.com/innovation/how-much-gold-is-i*n-*yo*ur-computer-and-how-efficient-it-is-to-reclaim-it

Meit, Y. (2011, May 12). Ministry of Electronics & Information Technology. *Meity. https://www.* meity.gov.in/writereaddata/files/1035e_eng.pdf

Michael, F. L., Sumilan, H., Bandar, N. F. A., Hamidi, H., Jonathan, V., & Md Nor, N. N. (2020). Sustainable development concept awareness among students in higher education: A preliminary study. Journal of Sustainabili*ty Science and Management, 15(7), 113–122.* doi:10.46754/jssm.2020.10.011

Mihai, F. C., Gnoni, M. G., Meidiana, C., Ezeah, C., & Elia, V. (2019). Waste Electrical and Electronic E*quipment (WEEE): Flows, Quantities, and Management—A Global Sce*nario. Electronic Waste Management and Treatment Technology, 1–34. doi:10.1016/B978-0-12-816190-6.00001-7

Misra, N. R., S. K. (2021). A Review on e-waste: Fostering the Need for Green Electronics. 2021 *International Conference on Computing*, Communication, and Intelligent Systems (ICCCIS), (pp. 1032-1036). IEEE. 10.1109/ICCCIS5*1004.2021.9397191*

Mmereki, D., Li, B., Baldwin, A., & Hong, L. (2016). The Generation, Composition, Collection, Treatment and Disposal System, and Impact of E-Waste. E-Waste in Transition - From Pollution to Resource., doi:10.5772/61332

Moha*buth, N., & Miles, N. (2005). The recovery of re*cyclable materials from Waste Electrical and Electronic Equipment (WEEE) by using vertical vibration separation. Resources, Conservation and Recycling, 45(1), 60–69. doi:10.1016/j.resconrec.2005.02.001

Mohammed, S. I. (2022). E-Waste Management in Different Countr*ies: Strategies, Impacts, and Determinants. In (*Ed.*)*, Advances in Green Electronics Technologies. IntechOpen. doi:10.5772/intechopen.106644

Mohd, S. &. (2018). e-waste management in India: current practices and challenges. ResearchGate A Comprehensive Study on Electrical*, 355.*

*Monika, K. J. (2010, July). e-waste managem*ent: As a challenge to public health in India. Indian Journal of Community Medicine, 35(3), 382–385. doi:10.4103/0970-0218.69251 PMID:210311*01*

Morf, L. S., Tremp, J., Gloor, R., Schuppisser, F., Stengele, M., & Taverna, R. (2007). Metals, non-metals and PCB in electrical and electronic waste – Actual levels in Switzerland. Waste Management *(New York, N.Y.), 27(10), 1306–1316. doi:10.1016/j.wasman.2006.06.014 PMID:17008085*

Moyo, T., Sadan, Z., Lötter, A., & Petersen, J. (2022). Barriers to recycling e-waste within a changing legal environment in South Africa. South African Journal of Science, 118. Advance online publication. doi:10.17159ajs.2022/12564

Moy P., (2005*). A health risk comparison of landf*ill *d*isposal and waste-to-energy treatment of municipal solid wastes in New York city. Research submitted for partial fulfillment of an M.P.H. at the Mailman School of Public Health at Columbia University.

Mukhopadhyay, S. K., & Ma, H. (2009). Joint Procurement and Production Decisions in Remanufacturing under Quality and Demand Uncertai*nty. International Journal of Production Economic*s, 120(1), 5–17. doi:10.1016/j.ijpe.2008.07.032

Mukti, S. K., & Rawani, A. M. (2016). ERP System *Success Models: A Literature Review. J*ournal of Engineering and Applied Sciences (Asian Research Publishing Network), 11(3), 1861–1875.

Mukti, S. K., Ahirwar, A. V., & Agrawal, A. (2021) 'Industry 4.0: Advances and Applications', Industry 4.0: Advances and Applications, pp. 1*–35. doi:10.9734/bpi/mono/978-93-5*547-287-8

Mukti, S. K., Tripathi, P., & Rawani, A. M. (2014). Identification of Factors and Indicators for Success Measurement of ERP System. International Proceedings of Economics Development and Research, *75, 117.*

*Murthy, V., & Ramakris*hna, S. (2022). A Review on Global E-Waste Management: Urban Mining towards a Sustainable Future and Circular Economy. Sustainability, 14(2), 647. doi:10.3390u14020647

Myilsamy, S., & Sampath, B. (202*1). Experimental comparison of near-dry and* cryo*gen*ically cooled near-dry machining in wire-cut electrical discharge machining processes. Surface Topography: Metrology and Properties, 9(3), 035015. doi:10.1088/2051-672X/ac15e0

Nagajothi, P. G., & Felixkala, T. (2014). Compressive strength of concrete incorporated with E-fiber waste. International Journal of Em*erging Technology and Advanced Engineering, 4(4), 23–27.*

*Needhidasan, S., Samue*l, *M*., & Chidambaram, R. (2014). Electronic waste – an emerging threat to the environment of urban India. Journal of Environmental Health Science & Engin*eering, 12(36), 1–9. doi:10.1186/2052-336X-12-36 PMID:24444377*

N*em*a, A. K., & Gupta, S. K. (1999). Optimization of regional hazardous waste management systems: An improved formulation. Waste Management (New York, N.*Y.), 19(7-8),* 44*1*–451. doi:10.1016/S0956-053X(99)00241-X

Ni, Z., Chan, H. K., & Tan, Z. (2021). Systematic Literature Review of Reverse Logistics for E-Waste: Overview, Analysis, and Future Research Agenda. Internationa*l Journal of Logistics Research and Applicati*ons, 0(0), 1–29. doi:10.1080/13675567.2021.1993159

Okunola, A. A., Kehinde, I. O., Oluwaseun, A., & Olufiropo, E. A. (2019). Public and Environmental He*alth Effects of Plastic Wastes Disposal: A Review. J. Toxicol. Risk* Assess., 5(2). doi:10.23937/2572-4061.1510021

Olsen, S. (2006). High Tech: Anything But Green. NEWS.COM. http://news.com.com/High+tech+Anything+*put+green/2008-1041_3 -6090899.html*

*Ongondo, F., Wil*l*iams*, I., & Cherrett, T. (2011). How are WEEE doing? A global review of the management of electrical and electronic wastes. Waste Management, 31(4), 714 730. doi:10.1016/j.wasman.2010.10.023

Organization fo*r Economic Cooperation and Development (OECD). (2001). Exten*ded producer responsibility: A guidance manual for governments.

Osibanjo, O., & Nnorom, I. (2007). The challenge of electronic waste (e-waste) management in developing countries. Waste Management &Amp*; Research. The Journal* for a Sustainable Circular Economy, 25(6), 489–501. doi:10.1177/0734242X07082028 PMID:18*229743*

*O*tieno, I., & Omwenga, E. (2015). E-Waste Management in Kenya: Challenges and Opportunities. Journal of Emerging Trends in Computing and Information Sciences, 6(12), ISSN 2079-8407.

Pak, P. (2008). Haste makes e-waste: *A comparative analy*sis of how the United State should approach the growing e-waste threat. Cardozo J. Int'l & Comp. L., 16, 241.

Palomo, A., Blanco-Varela, M. T., Granizo, M. L., Puertas, F., Vazquez, T., & Grutzeck, M. W. (1999). Chemical stability of cementitious materials based on metakaolin. Cement and Concrete Research, 29(7), 997–1004.

Pandey, R. *U., Surjan, A., & Kapshe, M. (2018). Explori*ng *l*inkages between sustainable consumption and prevailing green practices in reuse and recycling of household waste: Case of Bhopal city in India. Journal *of Cleaner Production, 173, 49–59. doi:10.1016/j.jclepro.2017.03.227*

Pariatamby, A., & Victor, D. (2013). Policy trends of e-waste management in Asia. Journal of Material Cycles and Waste Management, 15(4), 411–419. d*oi:10.100710163-013-0136-7*

Park, S. B., & Lee, B. C. (2004). Studies on expansion properties in mortar containing waste glass and fibers. Cement and Concrete Research, 34(7), 1145–1152.

Patil, R. A., & Ram*akrishna, S. (2020). A compr*ehen*sive analysis of e-waste legislation worldwide. Environmental Science and Pollution Research, 27(13), 14412 14431. 1 doi:10.1007/s11356-020-07992-1

Policy Practices for E-waste Management. (n.d.). ITU. https://www.itu.*int/en/ITUD/Environment/Pages/*T*ool*box/WEEE-A frica-Toolkit.aspx

Pouhet, R., & Cyr, M. (2016). Formulation and performance of flash metakaolin geopolymer concre*tes. Construction & Building Materials, 120, 15*0*–16*0. doi:10.1016/j.conbuildmat.2016.05.061

Powell, J., Craighill, A., & Brisson, I. (1996). The life cycle assessment and valuation of waste management opt*ions: a U.K. study. Proceedi*ng*s* of Air and waste management association's 89th annual meeting and exhibition. IEEE..

Prasad, M. N. V. & Borthakur, *A. (2019). Handbook of Electronic Waste Mana*gement International Best Practices and Case Studies.

Prasanna, P. K., & Rao, M. K. (2014). Strength vari*atio*ns in concrete by using E-waste as coarse aggregate. International Journal of Education and applied research, 4(2).

Press Information Bureau. G. o. (2018, March 23). e-waste Managem*ent Rules amended for effective mana*g*e*ment of e-waste in the country. Union Environment Minister. https://pib.gov.in/newsite/PrintRelease.aspx?relid=177949

Pushpamali, N. N. C., Agdas, D., Rose, T. M., & Yigitcanlar, T. (2021). Stakeholder perception of reverse logistics practices on supply chain performance. Business Stra*tegy and the Environment, 30(1), 60–70. doi:10.1002/bse.26*09

Raharjo, S., & Utomo, A. H. (2021). Comparative Stud*y of Electronic Waste Management in Developed Countries and Indonesia. Andalasian I*nternational Journal of Applied Science. Engineering and Technology, 1(1), 21–32. doi:10.25077/aijaset.v1i1.4

Ramachandran, R., Dharmaraj, K., & Natarajan, P. (2022). Electronic waste: A critical assessment on the unimaginable growing pollutan*t, legislations and environmental imp*act*s*. Environmental Challenges, 7, 100507. doi:10.1016/j.envc.2022.100507

Ramesh Kumar, G., Santhosh, K. S., & Bharani, S. (2020). Influence of E-waste on prope*rties of bituminous mixes. Materials Today: Proceed*ings*, 37(Part 2), 2719–2724.* doi:10.1016/j.matpr.2020.08.539

Raut, R. D., Narkhede, B., & Gardas, B. B. (2017). To Identify the Critical Success Factors of Sustainable Supply Chain Management Practices in the Context of Oil and Gas Industries. Ren*ewable & Sustainable Ene*rgy Reviews, 68, 33–47. doi:10.1016/j.rser.2016.09.067

Ravindra, K., & Mor, S. (2019). E-waste generation and management practices in Chandigarh, India and e*conomic evaluation for susta*inable recycling. Journal of Cleaner Production, 221, 286–294. doi:10.1016/j.jclepro.2019.02.158

Regtech, T. (2016). Environment Protection Act, 1986 and e-waste (Management) Rules, 2016. Team Leaser G Tech. https://www.teamleaseregtech.com*/resources/acts/article/29/e nvironmen*t-*pr*otection-act-1986-and-e-waste-management-rules-
2016/

Rehamn, U., Mack, J., & Alam, P. (2021). From e-waste to robots: a case study on e-waste up cycling in low-to-middle income countries. *2021 IEEE International Huma*nit*ar*ian Technology Conference (IHTC), (pp. 1-4). IEEE. 10.1109/IHTC530*77.2021.9698905*

Ren, Z. F., Huang, B., Liu, M., Bi, X. H., Sheng, G. Y., & Fu, J. M. (2012). The major components and characteristics of particles emitted during recycling of waste printed circuit boards in a typical e-waste workshop of South China. Zhongguo Huanjing Kexue. Zhongguo Huanjing Kexue, 32(8), 1447–1451.

Richey, R. G., Roath, A. S., Adams, *F. G., & Wieland, A. (2022). A responsiveness view of logistics* and supply chain management. Journal of Business Logistics, 43(1), 62–91. doi:10.1111/jbl.12290

RikeTechnologies. (n.d.). Recycled Computer Projects. Pinterest. https://in.pinterest.com/riketech/recycled-computer-projects
/

Robinson, B.H. (2009). e-waste: an assessment of global pro*duction and environmen*tal *impacts. Science of th*e *To*tal Environment, 408(2),183-91. doi:.2009.09.044. doi:10.1016/j.scitotenv

Robotis, A., Boyaci, T., & Verter, V. (2012). Investing in Reusability *of Products of Uncertain Reman*u*fa*cturing Cost: The Role of Inspection Capabilities. Internatio*nal Journal of Production* Economics, 140(1), 385–395. doi:10.1016/j.ijpe.2012.04.017

Rotter VS, Maehlitz P, Korf N, Chancerel P, Huisman J, Habib H, Herreras L, Ljunggren SM, and Hallberg A. (2016*). ProSUM Deliverable 4.1. Waste* Fl*ow* Studies. 1–100.

Roy, H. R.-M., Rahman, T. U., Suhan, M. B. K., Al-Mamun, M. R., Haque, S., & Islam, M. S. (2022). A comprehensive review on hazardous aspects and management strategies of electronic waste: Ba*ngladesh perspectives. Heliyon, 8(7), 09802.* do*i:1*0.1016/j.heliyon.2022.e09802 PMID:35815143

Sahoo, S., Mukherjee, A., & Halder, R. (2021). A unified blockchain-based platform for global e-waste management. International Journal of W*eb Information Sys*tems.

Samikannu, R., Koshariya, A. K., Poornima, E., Ramesh, S., Kumar, A., & Boopathi, S. (2022). Sustainable Development in Modern Aquaponics Cultivation Systems Using IoT Technologies. In Human Agro-Energy Optimization for *Business* and Industry (pp. 105–127). IGI Global.

Sampath, B., & Myilsamy, S. (2021). Experimental investigation of a cryogenically cooled oxygen-mist near-dry wire-cut electrical dis*charge machining process. Strojniski Vestnik.* Jixie Gongcheng Xuebao, 67(6), 322–330. doi:10.5545v-jme.2021.7161

Sampath, B., Naveenkumar, N., Sampathkumar, P., Silambarasan, P., Venkadesh, A., & Sakthivel, M. (2021). Experimental comp*arative study of banana fiber composite with glass fiber* composite material using Taguchi method. Materials Today: Proceedings, 49(5), 1475–1480. doi:10.1016/j.matpr.2021.07.232

Sampath, B., Sasikumar, C., & Myilsamy, S. (2023). Application of TOPSIS Optimization Technique in the Micro-Machining Process. In Trends, Paradigms, and Advances in Mechatronics Engineering (pp. 162–187). IGI Global.

Sampath, B., Yuvaraj, D., & Velmurugan, D. (2022). Parametric ana*lysis of mould san*d *properties for flange* coupling casting. AIP Conference Proceedings, 2460(1), 070002. doi:10.1063/5.0095599

Santos, S. M. (2022). e-waste management in Brazil: Challenges and opportunities of a reverse logistics model. . Environmental Technology & Innovation, 102671.

Saran, R. K., & Ya*dav, S. (2019) Significance* of *E*lectronic Waste Management for Sustainable Industrial Production Rishabh. In Operations Management and Systems Engineering, Lecture Lecture Notes on Multidisciplinary Industr*ial Engineering (pp. 241–249). S*pringer Singapore doi:10.1007/978-981-13-6476-1

Saravanan, M., Vasanth, M., Boopathi, S., Sureshkumar, M., & Haribalaji, V. (2022). Optimization of *Quench Polish Quench (QPQ) Coating Process Usi*ng Taguchi Method. Key Engineering Materials, 935, 83–91. doi:10.4028/p-z569vy

Sarkis, J., & Zhu, Q. (2018). Environmental sustainability and production: Taking the road less travelled. International Journal of Production Research, 56(1–2), 743–759. doi:10.1080/00207543.2017.1365182

Seeberger, J., Grandhi, R., Kim, S. S., Mase, W. A., Reponen, T., Ho, S. M., & Chen, A. (2016). Special Report: E-Waste Management in the United States and Public Health Implications. Jou*rnal of Environmental Hea*lth, 79(3), 8–16. PMID:29120130

Sengupta, D., Ilankoon, I., Dean Kang, K., & Nan Chong, M. (2022). Circular economy and household e-waste mana*gement in India: Integration of formal and i*nformal sectors. Minerals Engineering, 184, 107661. doi:10.1016/j.mineng.2022.107661

Shad, K. M., Tan, Y. L., & Karim, M. E. (2020a). Sustainable e-waste management in Malaysia: Lessons from selected countries. IIUM Law Journal, 28(2), 41*5–447. doi:10.31436/iiumlj.v28i2.517*

Shamee, A. & Shamsuddin, A. (2019). End-of-Life Electrical and Electronic Equipment Remanufacturing Prospects in Malaysia End-of-Life Electrical and Electronic Remanufacturing Prospects in Malay*sia. IOP Science.* . do*i:1*0.1088/1757-899X/530/1/012033

Sharma, M., Joshi, S., & Govindan, K. (2021). Issues and solutions of electronic waste urban mining for circular economy transition: An Ind*ian context. Jour*nal of Environmental Management, 290, 112373. doi:10.1016/j.jenvman.2021.112373 PMID*:33932756*

*Sharma, V., Garg, S. K., & Sharma, P. B. (2016). Identi Fi Cation of Major Drivers and Roadblocks for Remanufacturing in India. Journal of Cleaner Prod*uction, 112, 1882–1892. doi:10.1016/j.jclepro.2014.11.082

Shinkuma, T. (2003). On the Second-best Policy of Household's Waste Recycling. Environmental and Resource Economics, 24(1), 77–95. doi:10.1023/A*:1022842617469*

Shumon, M. R. H., & Ah*med*, S. (2013). Sustainable WEE management in Malaysia: Present scenarios and future perspectives. IOP Conference Series. Materials Science and Engineering, 50, 012066. Advance online public*ation. doi:10.1088/1757-899X/*50*/1/*012066

Singh, M., & Srivastava, R. K. (2018). Resources, Conservation & Recycling Analysis of External Barriers to Remanufactu*ring Using Grey-DEMATEL Approach : An In*dian Perspective. Resources, Conservation and Recycling, 136(February), 79–87. doi:10.1016/j.resconrec.2018.03.021

Singh, P., & Chawla, M. (2017). E-waste manage*ment-Suggested solutions. [IJAERS]. International Journ*al *of* Advanced Engineering Research and Science, 4(1), 112–115. doi:10.22161/ijaers.4.1.17

Singhal, D., Tripathy, S., & Kumar, S. (2018). ScienceDirect ScienceDirect ScienceDirect ScienceDirect School for Costing Models Capacity Optimization between Capacity and Operationa*l Efficiency. Procedia* Ma*n*ufacturing, 20, 452–457. doi:10.1016/j.promfg.2018.02.066

Singhal, D., Tripathy, S., & Kumar, S. (2019). Sustainability through Remanufacturing of E-Waste : Examination of Critical Factors in the In*dian Context. Sustainable Production a*nd Consumption, 20, 128–139. doi:10.1016/j.spc.2019.06.001

Sinha-Khetriwal, D., Kraeuchi, P., & Schwaninger, M. (2005). A comparison of electronic waste recycling in Switzerland and in India. Environmental Impact Assessmen*t Review, 25(5), 492–504. doi:10.1016/*j.*e*iar.2005.04.006

Skelton, S. K. (2020, November 19). Covid-19 shift to remote working adds to Earth's growing e-waste problem. ComputerWeek*ly.com. https://www.computerweekly.com/news/252492372/Covid-19-shi*ft -to-remote-working-adds-to-Earths-growing-e-waste-problem

Solano, E., Dumas, R. D., Harrison, K. W., Ranjithan, S., Barlaz, M. A., & Brill, E. D. (2002). Life cycle-based sol*id waste management. II: Illustrative* appl*i*cations. Journal of Environmental Engineering, 128(10), 993–1005. doi:10.1061/(ASCE)0733-9372(2002)128:10(993)

Soo, V. K., Featherston, C., & Doolan, M. (2013). E-waste Assessment in Malaysia. International Conference on Life Cycle Engineering, 389–395. 10.1007/978-981-4451-48-2_64

Stahel, W. R. (2013). Policy for material efficiency sustainable taxation as a departure fro*m the throwaway society. Philosophical Transaction*s of the Royal Society A: Mathematical, Physical and Engineering Sciences, 371(1986), 20110567. doi:10.1098/rsta.2011.0567

Statista. (2020, July). Retrieved *from Leading countries based on generation of electronic waste worldwide in 2019: https://www.statista.*com/statistics/499952/ewaste-generation -world*wide-by*-major-country/

Statistics Canada. (2015). Manufacturing Sales, by Subsector, by Province and Territory, Statistics Canada http://www.statcan.gc.ca/tables-tableaux/sum-som/l01/cst01/manuf33aeng.htm

StEP solving the E-waste Problem - StEP Initiative. (n.d.). https://www.step-initiative.*org/*

*Sthiannopka*o, S. (2012). Managing E-Waste in Developed and Developing Countries. The Handbook of Environmental Chemistry, 263–278. doi:10.1007/698_2012_195

Sthiannopkao, S., & Wong, M. H. (2013). Handling e-waste in developed and developing countries: Initiatives, practices, and consequences. The Science of the Total Environment, 463–464, 1147–1153. doi:10.1016/j.scitotenv.2012.06.088 PMID:22858354

Streicher-Porte M, Widmer R, Jain A, Bader HP, Scheidegger R, Kytzia S (20*05) Key drivers of the e-waste recyc*ling *syst*em: Assessing and modelling e-waste processing in the informal sector in Delhi. Environ. Impact Assess. Rev., 25(5 SPEC. ISS.), 472–491. doi:10.1016/j.eiar.2005.04.004

Sundar, M. L., & Raj, S. (2020). Study on Characteristics of Geopolymer Concrete with E-Waste. *IOS Science.*

*SWICO. (2011). S*WICO Recycling. https://www.swico.ch/media/filer_public/01/0c/010c5855-4dcc-41d7-9807-3490*1c386158/activity_rapport_2011.pdf*

Tabelin, C. B., Park, I., Phengsaart, T., Jeon, S., Vil*lacorte-Tabelin,* M., Alonzo, D., Yoo, K., Ito, M., & Hiroyoshi, N. (2021). Copper and critical metals production from porphyry ores and e-wastes: A review of resource availability, processing/recycling challenges, socio-environmental aspects, and sustainability issues. Resources, Conservation and Recycling, 170, 105610. doi:10.1016/j.resconrec.2021.105610

Tejedor, S., Cervi, L., Pérez-Escoda, A., & Tusa, F. (2020). Smartphone usage among students d*uring COVID-19 pandemic in Spain, Ita*ly, *a*nd Ecuador. Eighth International Conference on Technological Ecosystems for Enhancing Multiculturality. 10.1145/3434780.3436587

Teunter, R. H., & Simme Douwe, P. F. (2011). Optimal Core Acquisit*ion and Remanufacturing Policies under Uncertain Core Quality Fractions. European Journal* of Operational Research, 210(2), 241–248. doi:10.1016/j.ejor.2010.06.015

Thakur, P., & Kumar, S. (2021). Evaluation of e-waste status, management strategies, and leg*islations. International Journal of Envi*ronm*e*ntal Science and Technology, 19(7), 6957–6966. doi:10.100713762-021-03383-2

Tipre, D. R., Khatri, B. R., Thacker, S. C., & Dave, S. R. (2021). The brighter side of e-waste—A rich secondary source of metal. Environmental Science and Pollution Research International, 28(9), 10503–10518. doi:10.100711356-020-12022-1 PMID:33438127

Trivedi, V., Pandey, K. K., & Trivedi, A. (2022). Analyzing the challenges of e-waste management practices in India during COVID-19. Management of Environmental Quality, 33(6), 1611–1628. Advance online publication. doi:10.1108/MEQ-12-2021-0273

Trojovský, P., Dhasarathan, V., & Boopathi, S. (2023). Experimental investigations on cryogenic friction-stir welding of similar ZE42 magnesium alloys. Alexandria Engineering Journal, 66(1), 1–14. doi:10.1016/j.aej.2022.12.007

Turaga, R. M. R., Bhaskar, K., Sinha, S., Hinchliffe, D., Hemkhaus, M., Arora, R., Chatterjee, S., Khetriwal, D. S., Radulovic, V., Singhal, P., & Sharma, H. (2019). E-Waste Management in India: Issues and Strategies. Vikalpa, 44(3), 127–162. doi:10.1177/0256090919880655

Ullah, S., Qureshi, M. I., Joyklad, P., Suparp, S., Hussain, Q., Chaiyasarn, K., & Yooprasertchai, E. (2022). Effect of partial replacement of E-waste as a fine aggregate on compressive behavior of concrete specimens having different geometry with and without CFRP confinement. Journal of Building Engineering, 50(January), 104151. doi:10.1016/j.jobe.2022.104151

UNEP 2009 annual report. (n.d.). UNEP - UN Environment Programme. https://www.unep.org/resources/annual-report/unep-2009-annual-report

UNEP Annual Evaluation Report 2007. (n.d.). UNEP - UN Environment Programme. https://www.unep.org/resources/synthesis-reports/unep-annual-evaluation-report-2007

VanderPol, M. (2014). Overview of E-waste Management in Canada. *International E-waste Management* Network (IEMN). https://19january2017snapshot.epa.gov/sites/production/files/2014-08/documents/canada_country_presentation.pdf

Vanitha, S., Radhika, K., & Boopathi, S. (2023). Artificial Intelligence Techniques in Water Purification and Utilization. In Human Agro-Energy Optimization for Business and Industry (pp. 202–218). IGI Global. doi:10.4018/978-1-6684-4118-3.ch010

Vennila, T., Karuna, M. S., Srivastava, B. K., Venugopal, J., Surakasi, R., & Sampath, B. (2022). New Strategies in Treatment and Enzymatic Processes: Ethanol Production From Sugarcane Bagasse. In Human Agro-Energy Optimization for Business and Industry (pp. 219–240). IGI Global.

Verma, A. K., & Prakash, S. (2020). e-wastes and their impact on environment and public health. International Journal of Applied Research, 6(9), 164–168. doi:10.22271/allresearch.2020.v6.i9c.7111

Verma, P., Dhurvey, P., & Sundramurthy, V. P. (2022). Potential assessment of e-waste plastic in metakaolin based geopolymer using petrography image analysis. Advances in Materials Science and Engineering.

Verma, P., Dhurvey, P., & Sundramurthy, V. P. (2022). Structural behaviour of metakaolin geopolymer concrete wall-type abutments with *connected wing walls. Advances in Materials* Science and Engineering.

Walls, M. (2011). Deposit-Refund Systems in Practice and Theory. SSRN Electronic Journal. doi:10.2139/ssrn.1980142

Wang, J., Li, W., M*ishima, N., & Adachi, T. (2022). Exploring th*e optimal reverse supply chain for e-waste treatment under Chinese govern*ment subsidy. Wast*e Management (New York, N.Y.), 137, 128–138. doi:10.1016/j.wasman.2021.10.031 PMID:34752946

Wath, S. B., Dutt, P. S., & Chakrabarti, T. (2011). e-waste scenario in India, its man*agement, and implications. Environment*al Monitoring and Assessment, 172(1-4), 249–262. doi:10.100710661-010-1331-9 PMID:20151189

Wath, S. B., Vaidya, A. N., Dutt, P., & Chakrabarti, T. (2010). A roadmap for de*velopment of sustainable E-waste manage*ment system in India. The Science of the Total Environment, 409(1), 19–32. doi:10.1016/j.scitotenv.2010.09.030 PMID:20951410

Wee, M. I., Ariffin, F. N., Ng, T. F., & Shabudin, A. F. A. (2017). Awareness a*nd attitudes towards sustainable dev*elopm*ent amongst higher education students in Penang. World Sustainability Series., doi:10.1007/978-3-319-47877-7_4

Wei, L., & Liu, Y. *(2012). Present Status of e-waste Disposal and Recycling in China. Procedia Environmental Sciences,* 16, 506–514. doi:10.1016/j.proenv.2012.10.070

Widmer, R., Oswald-Krapf, H., Sinha-Khetriwal, D., Schnellmann, M., & Böni, H. (2005). Global perspecti*ves on e-waste. Environmental I*mpac*t Assessment Review, 25(5), 436–458. doi:10.1016/j.eiar.2005.04.001

Williams, E., Kahhat, R., Allenby, B., Kavazanjian, E., Kim, J., & Xu, M. (2008). Environmental*, social, and economic implications of glo*bal reuse and recycling of personal computers. Environmental Science & Technology, 42(17), 6446–6454. doi:10.1021/es702255z PMID:18800513

Wong, M., Wu, S., Deng, W., Yu, X., Luo, Q., Leung, A., Wong, C., Luksemburg, W., & Wong, A. *(2007). Export of toxic chemical*s – A review of the case of uncontrolled electronic-waste recycling. Environmental Pollution, 149(2), 131–140. doi:10.1016/j.envpol.2007.01.044 PMID:17412468

Xiang, W., & Ming, C. (2011). Implementing Extended Producer Responsibility: Vehicle Remanufa*cturing in China. Journ*al *of* Cleaner Production, 19(6–7), 680–686. doi:10.1016/j.jclepro.2010.11.016

Xie, J., Chen, W., Wang, J., Fang, C., Zhang, B., & Liu, F. (2019). Coupling effects of recycled ag*gregate and GGBS/metakaolin o*n *ph*ysicochemical properties of geopolymer concrete. Construction & Building Materials, 226, 345–359. doi:10.1016/j.conbuildmat.2019.07.311

Xing, G. H., Chan, J. K. Y., Leung, A. O. W., Wu, S. C., & Wong, M. H. (2009). Environmental impact and human exposure to PCBs in Guiyu, an electronic waste recycling site in China. Environment International, 35(1), 76–82. doi:10.1016/j.envint.2008.07.025 PMID:18804865

Xu, J., Luo, C., Ng, C. T., Feng, X., & Cheng, T. C. E. (2022). Remanufacturing with Random Yield in the Presence of the Take-Back Regulation. Computers & Industrial Engineering, 168(March), 1–9. doi:10.1016/j.cie.2022.108097

Yang, J. L., & Tzeng, G.-H. (2011). An integrated MCDM technique combined with DEMATEL for a novel *cluster-weighted with ANP method.* Expert Systems with Applications, 38(3), 1417–1424. doi:10.1016/j.eswa.2010.07.048

Yoshida, F., & Yoshida, H. (2014). E-Waste Management in Japan: A Focus on Appliance Recycling. Adv*anced Materials Research, 878, 4*20–423. . doi:10.4028/www.scientific.net/AMR.878.420

Yupapin, P., Trabelsi, Y., Nattappan, A., & Boopathi, S. (2022). Performance Improvement of Wi*re-Cut Electrical Discharge* Machining Process Using Cryogenically Treated Super-Conductive State of Monel-K500 Alloy. Iranian Journal of Science and Technology - Transactions of Mechanical Engineering, 1–17. doi:10.1007/s40997-022-00513-0

Zeng, X., Yang, C., Chiang, J. F., & Li, J. (2017). Innovatin*g e-waste management: From macroscopic to microscopic scales. The Science of the T*otal Environment, 575, 1–5. doi:10.1016/j.scitotenv.2016.09.078 PMID:27723459

Zeng, X. L., Liu, D., & Wu, W. (2022). PM2. 5 exposure and pediatri*c health in e-waste dismantling areas. En*vironmental Toxicology and Pharmacology, 89, 103774. doi:10.1016/j.etap.2021.103774 PMID:34800720

Zhang, H. Y., Qiu, G. H., Kodur, V., & Yuan, Z. S. (2020). Spalling behavior of meta*kaolin-fly ash based geopolymer conc*rete under elevated temperature exposure. Cement and Concrete Composites, 106, 103483. doi:10.1016/j.cemconcomp.2019.103483

Zhang, K., Schnoor, J. L., & Zeng, E. Y. (2012). E-Waste Recycling: Where Does It Go from Here? Envi*ronmental Science &Amp. Techno*logy, 46(20), 10861–10867. doi:10.1021/es303166s PMID:22998401

Zhao, G., Wang, Z., Dong, M. H., Rao, K., Luo, J., Wang, D., Zha, J., Huang, S., Xu, *Y., & Ma, M. (2008). PBBs, PBDEs, and* PCBs levels in hair of residents around e-waste disassembly sites in Zhejiang Province, China, and their potential sources. The Science of the Total Environment, 397(1–3), 46–57. doi:10.1016/j.scitotenv.2008.03.010 PMID:18439655

Zhao, G., Wang, Z., Zhou, H., & Zhao, Q. (2009). Burdens of PBBs, PBDEs, and *PCBs in tissues of the cancer patien*ts *in* the e-waste disassembly sites in Zhejiang, China. The Science of the Total Environment, 407(17), 4831–4837. doi:10.1016/j.scitotenv.2009.05.031 PMID:19539352

Zhong, H., & Schiller, S. (2011). Design of the expense allocation mechani*sm in e-waste recycling deposit sys*tem *un*der EPR framework. ICSSSM11. doi:10.1109/ICSSSM.2011.5959352

Zhu, Q., Geng, Y., & Lai, K. (2010). Circular economy practices among Chinese manufacturers varying in environmental-oriented supply chain c*ooperati*on and the performance implications. Journal of Environmental Management, 91(6), 1324–1331. doi:10.1016/j.jenvman.2010.02.013 PMID:20223586

Ekvall, T. (1999). Key methodological issues for life cycle inventory analysis *of paper recycling. Journal of Clea*ner Production, 74(4), 281–294. doi:10.1016/S0959-6526(99)00149-3

About the Contributors

A. M. Rawani is Professor in Mechanical Engineering and Former Director of National Institute of Technology Raipur (India). He has received his M. Tech. in Industrial Engineering and Management from IIT Kharagpur and Ph.D. in Strategic Management from I.I.T. Delhi. He has about 40 years of teaching/research and administrative experience of educational institutes. He has published more than 120 research papers and produced 10 Ph. D. He has also delivered keynote speeches in academic conferences held in many countries like USA, Switzerland, Singapore, China, Mauritius, Dubai, Malaysia, Hong Kong, Thailand and Indonesia. His areas of interest are Strategic Management, Flexible Systems Management, Fly-ash Management and Quality Management.

Mithilesh Kumar Sahu is a leading researcher in the area of energy conversion and management. He is presently working at OP Jindal University, Chhattisgarh as a senior Assistant Professor. Prior to OP Jindal University, he worked in GVPCE (A) Visakhapatnam for about 3.5 years. He is having more than 7 years of teaching and research experience. Dr. Sahu has completed his M-Tech and Ph.D. from NIT Jamshedpur. He has published more than 40 research papers in reputed journals and attended more than 10 National/International conferences. Dr. Sahu has also participated in more than 25 FDP/SSTP/Workshops. Along with teaching, he is also associated with administrative and research work. He mentored 5 M.Tech and 7 B.Tech projects. He is recognized as an esteemed reviewer for many SCI, SCI-e, and SCOPUS journals. His field of research is energy conversion, energy management, power plant engineering, and exergy analysis of thermal systems.

Siddharth S. Chakrabarti, Director-Centre for Industry-Academia Collaboration, Professor & Head-Mechanical Engineering Department at OP Jindal University is a principal investigator in the Energy and Environmental Analysis Group at OP Jindal University. He is also working as Director for the Centre of Advanced Welding, Vice President-Institutional Innovation Council and his research focus on evaporative heat transfer, flow and heat transfer in porous media, Solar Global

Radiations, Solar thermal applications, nanomaterial, heat transfer with change of phase, profile loss evaluation in turbines and other emerging areas. He has more than 20 years of teaching, research, and industry experience and has also worked as a member of the Board of Management for two consecutive terms at the University. He has visited countries such as the US and China, published around 75 papers in renowned international journals, and has five published and one granted patent. Dr. Chakrabarti has obtained his Ph.D. from the Indian Institute of Technology, Kharagpur, and M Tech from the National Institute of Technology, Warangal.

Ajit Kumar Singh is an Assistant Professor in the Institute of Business Management at GLA University Mathura. He has received his M.Tech in Industrial Engineering and Management from NIT Raipur and PhD in quality management of engineering education from NIT Raipur. He has published many research papers in reputed journals and attended many international conferences. He is a life member of Global Institution of Flexible System Management (GIFT) Delhi, Indian Institution of Industrial Engineering (IIIE) Mumbai, and Institution of Engineer India (IEI). His areas of research are Management of Engineering Education, Quality Management, Optimization, Accreditation and Ranking, etc.

Mazriah Ayu Abu Bakar has extensive knowledge and experience in scheduled wastes management over the past 18 years working in the Department of Environment. Areas of expertise include scheduled wastes identification and classification, waste characteristics, waste treatment and Multilateral Environmental Agreement which are Basel Convention on the Control of Transboundary Movement of Hazardous Wastes and Disposal, Stockholm Convention and Rotterdam Convention. Involved in development of Electronic Scheduled Wastes Information System (ESWIS). Involved in drafting the Scheduled Wastes Regulations, Prescribed Conveyance Regulations, Environmentally Hazardous Substances Regulations and E-Waste Regulations.

Sarma Adithe works in the Department of Electronics and Communication Engineering, B V C Institute of Technology and Science, Bhatlapalem, Andhra Pradesh 533221 India.

Animesh Agrawal is Project Fellow at National Institute of Technology Raipur in Mechanical Engineering Department (Industrial Engineering & Management). He has completed his B.Tech in 2009, M.Tech in 2013 and his Ph.D. in 2022. He is having the 13 years of teaching experience in Engineering college and well known for his rigorous work and high quality of teaching. His area of research is Knowledge Management, Optimization, Multi Criteria Decision Making Techniques, ERP etc. He has published many articles in highly indexed, reputed journals and attended many international and national conference through his carrier. Animesh Agrawal is the active member of Indian Society for Technical Education (ISTE), Institution of Engineers (India) [IEI] and Indian Institution of Industrial Engineering (IIIE) and have life time membership. Recently, Animesh Agrawal is invited as Editorial Manager for Journal of Business Research.

Sanjeeda Atiya is a research scholar, currently pursuing Master of Commerce (MCom) from Christ (Deemed to be University), Bengaluru; with Accounting and Taxation as majors.

Samreen Bano works in the Department of Civil Engineering, Integral University, Lucknow, Uttar Pradesh, India.

Sampath Boopathi completed his undergraduate in Mechanical engineering and postgraduate in the field of Computer-Aided Design. He completed his Ph.D. from Anna University and his field of research includes Manufacturing and optimization. He published 60 more research articles in Internationally Peer-reviewed journals, one Patent grant, and three published patents.He has 16 more years of Academic and research experiences in the various Engineering Colleges.

Ramit Choudhury is currently working as research scholar in the department of Mechanical Engineering at NIT Durgarpur India. His expertise are renewable energy, solar energy. He has published more than 5 journal in the reputed national/ international/conferences journals.

Yvonne Chung is a research scholar pursuing a Master's in Commerce from Christ University (Central Campus), Bengaluru. The subjects that Chung is majoring in are the field of Accounting and Taxation.

Priyanka Dhurvey is a Professor (Assistant) at Maulana Azad National Institute of Technology, Bhopal.

Nadeem Faisal is currently working as a Lecturer at the Central Institute of Petrochemicals Engineering and Technology, India. He has more than 5 years of teaching and research experience. Prior to switching careers to academics, he worked as an Assistant Engineer at TATA Voltas Ltd. He has published more than 20 research papers in Journals of National and International repute many of which have been indexed in Scopus, SCI & WoS. He has also authored 2 books and has contributed 12 Book Chapters in Edited Volumes published by International Publishers like Springer, Wiley, IGI Global & DeGiyuter. He is also a reviewer of 4 International

Journals and a member of 5 Professional Bodies. He has been facilitated with many awards and honors including the Best Teacher Award at ITM University. His areas of interest are Virtual & Digital Manufacturing, Optimization, Biomedical, and Materials Engineering.

Ginu George comes with around 10+ years of academic and industry experience and currently working as an Assistant Professor, Christ (Deemed to be University). Teaching subjects like management, organizational behvaiour, HRM, and research methodology for undergraduate and postgraduate students. Completed Ph.D. in 2017 in the area of Employee Engagement. George's areas of interest are HRM, Marketing, and Organizational Behaviour. Research has always been a passion, and has published papers in Scopus indexed, UGC, and ABDC journals. George has also presented papers in various international and national journals.

Soumya George is an Asst. Professor at St. George's College, Aruvithura. She completed her Ph. D in Computer Science from Cochin University of Science and Technology. Her research areas of interest include data mining, information retrieval, etc. She had many journal publications to her credit.

S. Gokulakannan, Assistant Professor, Muthayammal Engineering College, Rasipuram.

Chandra Gour works in Construction Materials field.

Samson Isaac J. works in the Department of Biomedical Engineering, Karunya Institute of Technology and Sciences, Coimbatore, India.

Sanjay Jain has done B. E. Electrical and M-Tech. in Electrical Engineering. Presently he is Head of Department & Associate Professor, Electrical Engineering, RKDF University, Bhopal, M.P. He has published several papers in international Journal and international / national conferences. One book chapter published with Wiley Publisher, UK in his credit.

Swatantra Jaiswal is a full time Ph.D. researcher in Department of Mechanical Engineering. His topic of the research is E-Waste Management. He has been published more than 3 articles in reputed journal with high indexing and impact factors. His area of interest is not limited to E-Waste Management only, but he also published some paper related to forecasting of short life cycle product. He has participated and presented more than 6 articles in international conference. He is lifetime member of Indian Institution of Industrial Engineering (IIIE).

Kedri Janardhana working as an Assistant Professor (senior grade) in the Department of Electrical Engineering at Dayalbagh Educational Institute (Deemed to be University), has about 15 years of teaching experience. He received his B.Tech degree in Electrical and Electronics Engineering from J.N.T.U College of Engineering, Anantapur and M.Tech degree in Power Electronics and Drives from National Institute of Technology, Kurukshetra and received his Ph.D. in Electrical Engineering from Dayalbagh Educational Institute, Agra. He has published 9 research papers in refereed international journals and 9 research papers in the proceedings of various international conferences. He has published 2 textbooks and 4 lab manuals. He has filed 2 international patents and 15 national patents. His area of research includes Power Electronics, Electric Vehicles, Solar Photovoltaic Power Systems and Microgrid. He is a life member of IEI and SSEM. He is an active member of IEEE.

Shilpa Kalambe has done B. E. Electrical and M-Tech in integrated power systems from RTM Nagpur University Nagpur, Maharashtra, India. She completed Ph.D. from MA NIT Bhopal. Presently she is Head of Department & Associate Professor, Electrical Engineering, Dr. Babasaheb Ambedkar college of Engineering Nagpur. She has published several papers in international Journal and international / national conferences. One book chapter published with Wiley Publisher, UK in her credit. She is a life member of ISTE, IAENG and member of other professional societies.

Bhojraj Kale has done B. E. Mechanical from SGB Amravati University, M-Tech. from RTM Nagpur University Nagpur, Maharashtra, India. He has completed his PhD from NIT Raipur. Presently he is Assistant Professor, Mechanical Engineering, Dr. Babasaheb Ambedkar college of Engineering Nagpur. He has published more than 50 papers in international Journal and 15 paper presented in international / national conferences. He has Author of one books also. One book chapter published with Wiley Publisher, UK in his credit. He is a life member of ISTE, IAENG, IEA and other professional societies.

Amit Kumar is currently working as Assistant professor in Sri Eshwar College of Engineering Coimbatore Tamilnadu India. His expertise are renewable energy, solar energy, materials. He has published more than 25 journal in the reputed national/international/conferences journals.

Dheeraj Kumar is a mechanical engineer. He began working in the Renewable Energy in the early 2012's. He received his bachelor's degree in Mechanical Engineering from the Cambridge Institute of Technology Ranchi, Jharkhand (India) in 2016. He did his Master in Engineering from Birla Institute of Technology Ranchi Jharkhand (India) in 2018. He is also pursuing his research carrier and working as researcher in National Institute of Technology Durgapur India. He has published more than 42 SCI/Scopus/Book Chapters in reputed International and National Journals. His Areas of research interest are Renewable Energy, Solar Energy, Solar Distillation, Composite Fibrous Materials, Solar Air Heater, Nano Composite Materials. He wants to dedicate his carrier in research fields of academic organizations. He is reviewer of few reputed international journals like Hindawi, JMST, Photo energy. He has attained 15 conferences of reputed national and international collabs.

Nagendra Kumar is currently working as research scholar in the department of Mechanical Engineering at NIT Durgarpur India. His expertise are renewable energy, building applications, materials. He has published more than 5 journal in the reputed journals and national/international/conferences journals.

Vikash Kumar is Assistant professor in MANIT, Bhopal. He completed his Ph.D. from NIT Jamshedpur. His area of research is Solar air heater. He completed his MTech from ISM Dhanbad. He published more than 20 research articles in international journals

Harikaran M. currently employed as an assistant professor at Muthayammal Engineering College.

Poonguzhali M., M.E.,PhD, graduated a B.E. degree in Electronics and Communication Engineering from Government College of Technology, Coimbatore, in the year 1997, and Master of Engineering in Applied Electronics in 1999. She obtained her Ph.D. degree in the area of Grid Computing from Anna University, Chennai, in 2014. She has 22 years of teaching experience and presently she is a Professor & Head in the department of Electronics and Communication Engineering, Narasu's Sarathy Institute of Technology, Salem. She has published more than 30 research papers in National and International conferences and Journals. She is a life member in IETE, member of IEEE and IFERP. Her areas of interest include Grid Computing, Mobile ad hoc networks, Digital Signal Processing, Image Processing, Green Computing, Antennas and Cognitive Radios.

Bhagyashri Krishnat Mane has completed his engineering from Dattajirao Kadam Technical Education Society's Textile & Engineering Institute and currently working in research fields in academics.

Anjala Michael is a former BCA Student in the Dept. of Computer Applications, SGC Aruvithura.

Anuprita Mishra is Principal of IES College of Technology, Bhopal (MP), She is having 25 years of teaching experience. She is a member of IEEE and BOS. She has tackled entire NBA work in TIT (2016 and 2018) and OIST (2021). She is also a member of BOS, Raisoni University, Jalgaon. She is a mentor through "Mentor Training Programme," a National Initiative for Technical Teacher Training (NITTT) by AICTE. She contributed as Session Chair in the 'IETE National Conference on "VLSI, Communication and Signal Processing" (NCVCS 2021)' organized by the Electronics and Communication Engineering Department of Maulana Azad National Institute of Technology, Bhopal from 27st to 28th Nov, 2021. She had published 3 papers in SCI Indexed Journals, 3 Scopus Indexed Journal and filed 5 Patents. She also fetched 1 AICTE -MODROB grant of Rs.10,40,000/- in 2013-14.

Noor Aina Amirah Mohd Noor is a Senior Lecturer at Universiti Sultan Zain-alAbidin (UniSZA), Terengganu, Malaysia. She received her Diploma in Business Studies and BBA (Operations Management), respectively, in 2005 and 2007, from Universiti Teknologi MARA, Terengganu, Malaysia. She obtained a Master's in Management (Integrated Coastal Zone Management) and a Ph.D. in Management, respectively, in 2009 and 2014 from Universiti Malaysia Terengganu, Terengganu. She started her employment as an academic in 2009 with PTPL College, Terengganu, before she further studied for her Ph.D. In 2013, she joined UCSI University as a lecturer until August 2015. She has been serving as a Senior Lecturer at UniSZA since September 2015. She is the managing editor of The Journal of Management Theory and Practice. She has received the best research paper award at the 1st Aceh Global Conference (AGC) 2018. She has published her research findings in international journals and conferences related to Operations Management, Occupational Safety and Health, Supply Chain Management, and Human Resource Management as a main author and co-author. She has supervised and co-supervised several postgraduate students. Her principal areas of research interest are Operations Management, Occupational Safety and Health, Supply Chain Management, and Risk Management

Suraj Mukti is an Assistant Professor in the Department of Mechanical Engineering, National Institute of Technology Raipur. He has completed his Ph.D. research work at his own institute i.e. NIT Raipur under the supervision of Dr. A. M. Rawani, Director and Professor NIT Raipur. His fields of interest are Industrial Engineering & Management, Enterprise Resource Planning, Management Information System, Knowledge Management, Change Management, Customer Relationship Management and Production Engineering & Technology. He is lifetime member of Institute of Engineers (India), Indian Society for Technical Education (ISTE), Indian Institution of Industrial Engineering (IIIE), South Asia Institute of Science and Engineering (SAISE) and many more technical and educational bodies. He has teaching and research experience of more than 20 years and also served as an Engineer at Chhattisgarh Electricity Board, in State Government. He has published more than 40 papers in reputed journals and attended & presented research papers in many important International Conferences within the country and abroad as well. He has authored three books viz. Customer Relationship Management: A Management Perspective, Execution of ERP system: A Case Study Approach and Knowledge Management: Implementation Methods and Strategies.

Suraj Mukti is working as an Associate Professor, Department of Mechanical Engineering, National Institute of Technology Raipur. He has completed his Ph.D. research work at his own institute i.e. NIT Raipur under the supervision of Dr. A. M. Rawani, Director and Professor NIT Raipur. His fields of interest are Industrial Engineering & Management, Enterprise Resource Planning, Management Information System, Knowledge Management, Change Management, Customer Relationship Management and Production Engineering & Technology. He is lifetime member of Institute of Engineers (India), Indian Society for Technical Education (ISTE), Indian Institution of Industrial Engineering (IIIE), South Asia Institute of Science and Engineering (SAISE) and many more technical and educational bodies. He has teaching and research experience of more than 20 years and also served as an Engineer at Chhattisgarh Electricity Board, in State Government. He has published, more than 40 papers in reputed journals and attended & presented research papers in many important International Conference within the country and abroad as well. He has authored three books viz. Customer Relationship Management: A Management Perspective, Execution of ERP system: A Case Study Approach and Knowledge Management: Implementation Methods and Strategies.

Satyendra Narain Singh is a part of the Department of Civil Engineering, Institute of Engineering and Technology, Lucknow, Uttar Pradesh 226021, India.

Raghuram Pradhan is an Associate Professor, Department of Mechanical Engineering, PACE Institute of Technology & Sciences, Near Valluramma Temple, Ongole, Andhra Pradesh 523272, India.

T.S. Ramesh Babu is a part of the Department of Civil Engineering, KG Reddy College of Engineering and Technology, Hyderabad, Telangana 501504, India.

S. Selvakumar is a part of the Department of Computer Science and Engineering Aarupadi Veedu Institute of Technology, Vinayaka Mission Research Foundation, Paiyanur, Tamil Nadu 603104 India.

Abhimannyu Sharma is currently working as research scholar in the department of Mechanical Engineering at NIT Durgarpur India. His expertise are renewable energy, solar energy, bio fuels. He has published more than 3 journal in the reputed national/international/conferences journals.

Ravi Kant Singh is currently working as research scholar in the department of Mechanical Engineering at NIT Durgarpur India. His expertise are renewable energy, fluids, materials. He has published more than 3 journal in the reputed journals and national/international/conferences journals.

Vikash Singh is a part of the Department of Civil Engineering, Institute of Engineering and Technology, Lucknow, India.

Dharam Singh, research scholar in mechanical engineering department, MANIT Bhopal. I have completed my master's in engineering in Heat Power and Thermal branch from LNCT, Bhopal.I worked as Assistant Professor in LNCT, Bhopal for 7 yrs.

Wan Nor Azilawanie Tun Ismail is a lecturer at the Faculty Applied of Social Sciences, Universiti Sultan Zainal Abidin (UniSZA). She received a Bachelor of Science (Human Development) from Universiti Putra Malaysia in 2011, and a Master of Science (Philosophy and Civilization Studies) from Universiti Putra Malaysia in 2014. She received her Ph.D. in Sociology of Development from the Universiti Sultan Zainal Abidin (UniSZA) in 2020. The area of her research includes developmental sociology, community development, and environmental sociology.

V. Venkatesh is a part of the Department of Electrical and Electronics Engineering, Rajalakshmi Engineering College, Chennai, India.

Parth Verma is a Ph.D. Reseach Scholar, working in the field of Geopolymer Concrete, at Department of Civil Engineering, Maulana Azad National Institute of Technology, Bhopal

Yusnita Yusof is a senior lecturer in the Faculty of Business and Management at the Universiti Sultan Zainal Abidin, Kuala Terengganu. She received her Ph.D. degree from the Faculty of Applied Social Sciences, Universiti Sultan Zainal Abidin, in the Tourism Management field. She is also actively involved with the research grant and consultation on the tourism field at the national and university levels. She has supervised and co-supervised several postgraduate students. Her current research interest includes Green Tourism, Sustainability Tourism, Hotel Management, Community-based Tourism, and Human Resource Management. She has published her research findings in international journals and conferences related to her research interest.

Index

F

G

H

I

L

M

N

O

P

Q

R

S

T

W

www.ingramcontent.com/pod-product-compliance
Lightning Source LLC
LaVergne TN
LVHW081527300926
849161LV00006B/1334